Receptors, Membrane Transport and Signal Transduction

NATO ASI Series

Advanced Science Institutes Series

A series presenting the results of activities sponsored by the NATO Science Committee, which aims at the dissemination of advanced scientific and technological knowledge, with a view to strengthening links between scientific communities.

The Series is published by an international board of publishers in conjunction with the NATO Scientific Affairs Division

A Life Sciences	Plenum Publishing Corporation
B Physics	London and New York
C Mathematical and Physical Sciences	Kluwer Academic Publishers
D Behavioural and Social Sciences	Dordrecht, Boston and London
E Applied Sciences	
F Computer and Systems Sciences	Springer-Verlag
G Ecological Sciences	Berlin Heidelberg New York
H Cell Biology	London Paris Tokyo

Receptors, Membrane Transport and Signal Transduction

Edited by

A. E. Evangelopoulos
The National Hellenic Research Foundation, Biological Research Center
48 Vassileos Constantinou Avenue, Athens 116 35, Greece

J. P. Changeux
Institut Pasteur
Neurobiologie Moleculaire
28 Rue du Dr. Roux, 75724 Paris Cedex 15, France

L. Packer
University of California, Berkeley
Membrane Bioenergetics Group
2544 Life Sciences Building, Berkeley, CA 94720, USA

T. G. Sotiroudis
The National Rellenic Research Foundation, Biological Research Center
48 Vassileos Constantinou Avenue, Athens 116 35, Greece

K. W. A. Wirtz
Centre for Biomembranes and Lipid Enzymology
State University of Utrecht
Padualaan 8, 3508 TB Utrecht, The Netherlands

Springer-Verlag
Berlin Heidelberg New York London Paris Tokyo
Published in cooperation with NATO Scientific Affairs Division

Proceedings of the NATO Advanced Study Institute on Receptors, Membrane Transport and Signal Transduction held on the Island of Spetsai, Greece, August 16–27, 1988.

ISBN 3-540-50421-4 Springer-Verlag Berlin Heidelberg New York
ISBN 0-387-50421-4 Springer-Verlag New York Berlin Heidelberg

Library of Congress Cataloging-in-Publication Data. NATO Advanced Study Institute on Receptors, Membrane Transport, and Signal Transduction (1988: Nísos Spétsai, Greece) Receptors, membrane transport, and signal transduction/edited by A. E. Evangelopoulos ... [et al.]. p. cm.—(NATO ASI series. Series H, Cell biology; vol. 29) "Proceedings of the NATO Advanced Study Institute on Receptors, Membrane Transport, and Signal Transduction, held on the Island of Spetsai, Greece, August 16–27, 1988"—T. p. verso. "Published in cooperation with NATO Scientific Affairs Division."
ISBN 0-387-50421-4 (U.S.)
1. Cell receptors—Congresses. 2. Biological transport, Active—Congresses. 3. Cellular signal transduction—Congresses. I. Evangelopoulos, A. E. II. North Atlantic Treaty Organization. Scientific Affairs Division. III. Title. IV. Series. QH603.C43N387 1988 574.87'5—dc 20   89-10055

© Springer-Verlag Berlin Heidelberg 1989
Printed in Germany

Printing: Druckhaus Beltz, Hemsbach; Binding: J. Schäffer GmbH & Co. KG, Grünstadt
2131/3140-543210 – Printed on acid-free paper

PREFACE

A NATO Advanced Study Institute on "Receptors, Membrane Transport and Signal Transduction", was held on the Island of Spetsai, Greece, from August 16-27, 1988, in order to consider recent developments in membrane receptor research, membrane transport and signal transduction mechanisms.

These topics were put in the larger context of current knowledge on the structure and function of membranes; connections between different fields of research were established by in-depth discussions of energy transduction and transport mechanisms.

The general principles of regulation by signal transduction and protein phosphorylation/dephosphorylation were presented in the context of specific cellular processes. Discussions included also the role of protein tyrosine kinases which are structurally related to oncogene products and, therefore, implicated in various aspects of cell development and transformation.

This book presents the content of the major lectures and a selection of the most relevant posters presented during the course of the Institute. The book is intended to make the proceedings of the Institute accessible to a larger audience and to offer a comprehensive account of those topics on receptors, membrane transport and signal transduction that were discussed extensively during the course of the Institute.

February 1989   The Editors

CONTENTS

I. G-PROTEINS, ADENYLATE CYCLASE AND PROTEIN
PHOSPHORYLATION

Selective regulation of G proteins by Cell surface receptors.......... 1
    E.M. Ross

Regulation of adenylate cyclase in mammalian cells and
    Saccharomyces cerevisiae 25
    A. Levitzki

Protein kinases, Protein phosphatases and the regulation of
    glycogen metabolism... 38
    T.G. Sotiroudis and A.E. Evangelopoulos

Phosphorylase kinase and protein kinase C:Functional similarities.... 55
    T.G. Sotiroudis, S.M. Kyriakidis, L.G. Baltas, T.B. Ktenas,
    V.G. Zevgolis and A.E. Evangelopoulos

The use of specific antisera to locate functional domains of
    guanine nucleotide binding proteins.......................... 67
    F.R. McKenzie and G. Milligan

Calcium inhibits GTP-binding proteins in squid photoreceptors........ 76
    J.Baverstock, J. Fyles and H. Saibil

Degradation of the invasive adenylate cyclase toxin of bordetella
    pertussis by the eukaryotic target cell-lysate................ 85
    A. Gilboa-Ron and E. Hanski

Identification and characterization of adenylate cyclases in
    various tissues by monoclonal antibodies..................... 95
    S. Mollner, U. Heinz and T. Pfeuffer

The role of G-proteins in exocytosis................................ 102
    J. Stutchfield, B. Geny and S. Cockcroft

Hydrophobic interactions in the calcium-and phospholipid dependent
    activation of protein kinase C............................... 110
    G.T. Snoek

Activation of transducin by aluminum or beryllium fluoride complexes.. 117
    J. Bigay

II. MEMBRANE RECEPTORS AND NEUROTRANSMITTERS

Glutamate receptors and glutamatergic synapses...................... 127
    P. Ascher

Mechanisms of glutamate exocytosis from isolated nerve terminals..... 147
    D. Nicholls, A. Barrie, H. McMahon, G. Tibbs and
    R. Wilkinson

Characteristics of the epidermal growth factor receptor.............. 162
    J. Boonstra, L.H.K. Defize, P.M.P. van Bergen en Henegouwen,
    S.W. De Laat and A.J. Verkleij

Three-dimensional structural models for EGF and insulin receptor
    interactions and signal transduction........................ 186
    T. Blundell, N. McDonald, J. Murray-Rust, A. McLeod, S. Wood

Potentiation of neurotransmitter release coincides with potentiation
    of phosphatidyl inositol turnover - A possible in vitro model
    for long term potentiation (LTP)............................ 196
    D. Atlas, S. Diamant and L. Schwartz

Purification and localization of kainate binding protein in
    pigeon cerebellum.. 214
    A.U. Klein and P. Streit

The Norepinephrine analog meta-iodo-benzylguanidine (MIBG) as a
    substrate for mono(ADP-ribosylation)....................... 223
    C. Loesberg, H.V. Rooij and L.A. Smets

The synaptic vesicle vesamicol (AH5183) receptor contains a low
    affinity acetylcholine binding site........................ 233
    B.A. Bahr and S.M. Parsons

Purification of the D-2 dopamine receptor and characterization of
    its signal transduction mechanism.......................... 242
    Z. Elazar, G. Siegel, H. Kanety and S. Fuchs

Downregulation of M1 and M2 muscarinic receptor subtypes in Y1
    mouse adrenocarcinoma cells................................ 251
    N.M. Scherer, R.A. Shapiro, B.A. Habecker and N.M. Nathanson

Uptake of GABA and L-glutamate into synaptic vesicles............... 263
    E.M. Fykse, H. Christensen and F. Fonnum

Deactivation of laminin-specific cell-surface receptors accompanies
    immobilization of myoblasts during differentiation......... 272
    S.L. Goodman , V. Nurcombe and K. von der Mark

III. MEMBRANE TRANSPORT AND BIOENERGETICS

Signal Transduction in Halobacteria.....................282
    D. Oesterhelt and W. Marwan

Control of bacterial growth by membrane processes.....................302
    K.van Dam, P.W. Postma, H.V. Westerhoff, M.M. Mulder and
    M. Rutgers

Carbonylcyanide-3-chlorophenylhydrazone, a prototype agent for the
    selective killing of cells in acidic regions of solid
    tumours.....................320
    K. Newell and I. Tannock

Ca^{2+} and pH interactions in thrombin stimulated human platelets.......329
    M.T. Alonso, J.M. Collazos and A. Sanchez

Structure-function relationships of the pCloDF13 encoded BRP...........338
    J. Luirink

Binding of a Bacillus Thuringiensis delta endotoxin to the midgut
    of the tobacco hornworm (Manduca sexta).......................344
    K. Hendrickx, H.van Mellaert, J.van Rie and A. De Loof

Functional reconstitution of photosynthetic reaction centre complexes
    from Rhodopseudomonas Palustris.....................352
    D. Molenaar, W. Crielaard, W.N. Konings and K.J. Hellingwerf

Na^+/H^+ exchange in cardiac cells: Implications for electrical and
    mechanical events during intracellular pH changes...............362
    F.V. Bielen, S. Bosteels and F. Verdonck

Receptor-mediated inhibition of reproductive activity in a schistosome-
    infected freshwater snail.....................372
    P.L. Hordijk, R.H.M. Ebberink, M. De Jong-Brink and J. Joosse

NMR study of gramicidin cation trasnport across and integration into
    a lipid membrane.....................382
    P.L. Easton, J.F. Hinton and D.K. Newkirk

SELECTIVE REGULATION OF G PROTEINS BY CELL SURFACE RECEPTORS

Elliott M. Ross
Department of Pharmacology
University of Texas Southwestern Medical Center
5323 Harry Hines Boulevard
Dallas, Texas 75235-9041, U.S.A.

NATO ASI Series, Vol. H29
Receptors, Membrane Transport and Signal Transduction
Edited by A. E. Evangelopoulos et al.
© Springer-Verlag Berlin Heidelberg 1989

A typical cell must respond appropriately to multiple hormonal signals. These signals, which may be mutually potentiative or antagonistic, must be integrated with each other and with the cell's current status to yield appropriate intracellular metabolic signals. Although receptors are responsible for detecting extracellular signals, integration and initial amplification of the signal frequently utilizes a group of GTP-binding transducer proteins known as G proteins.

The large majority of mammalian cell surface receptors use G proteins to convey their messages to intracellular effector proteins, which then generate cytoplasmic second messengers. In such three-protein relays, messages are sorted both convergently and divergently. Several receptors on a cell may trigger activation of a single effector, such as adenylate cyclase; a single receptor may also trigger activation of several effectors, such as a phospholipase and an ion channel. The pattern of a cell's responses to incoming information therefore reflects its complement of receptors, G proteins and effectors, as well as their relative selectivity for each other.

The past few years have seen both the clarification of how G protein-mediated signaling systems work and of their remarkable complexity at the cellular level. This chapter will present an essentially biochemical view of how these systems allow a cell to respond to its environment. Extensive reviews are available as well (Gilman, 1987; Stryer and Bourne, 1986).

THE REGULATORY GTPASE CYCLE

About ten G proteins are now known, probably over one hundred receptors talk to them, and they in turn probably talk to more than ten different effectors. However, a single biochemical mechanism is used for hormonal regulation of G protein function. This mechanism was elucidated using two prototypical experimental systems, the hormone-sensitive adenylate cyclase and the light-sensitive cyclic GMP phosphodiesterase in the retina (Ross and Gilman, 1980; Stryer, 1985, for reviews).

To cause activation of an effector protein, a G protein must itself be activated by binding one molecule of GTP. However, G proteins display intrinsic GTP-hydrolyzing activity, such that bound GTP is rapidly ($t_{1/2}$~15sec) converted to bound GDP, which does not activate. Subsequent reactivation involves the release of the tightly bound GDP hydrolysis product and the binding of a second GTP molecule, both far slower reactions. Thus, a steady-state GTP hydrolytic cycle defines the relative extent of G protein activation as the relative steady-state concentration of the enzyme-substrate (G protein-GTP) intermediate. Under resting conditions, GTP-activated G protein represents less than 1% of the total. However, an agonist-liganded receptor can bind to the G protein and catalyze both GDP release and GTP binding to increase the steady-state amount of G protein-GTP to greater than 60% of the total (Brandt and Ross, 1986). Receptor-catalyzed G protein activation is then

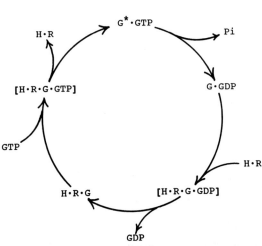

Fig. 1. The regulatory GTPase cycle. The GTP-bound, active form of a G protein, G·GTP, is deactivated by hydrolyzing GTP to GDP. Agonist-liganded receptor, H·R, binds to this species, transiently forming an unstable [H · R · G · GDP] complex that rapidly decomposes to the agonist-stabilized receptor-G protein ternary complex, H·R·G. It is this species that, for most receptors and G proteins, has the highest affinity for agonist (relative to isolated receptor). This species also displays low affinity for guanine nucleotides, but it can bind nucleotides rapidly because the binding site is exposed. GTP binds to H·R·G to form a second transient complex and, when GTP is in excess, drives the reaction to recreate the active G·GTP species, displacing a low-affinity agonist-receptor complex.

expressed as an activation of the downstream intracellular effector proteins (see Fig. 1).

In contrast to GTP binding and GDP release, which can be accelerated up to 100-fold by receptor, the rate of hydrolysis of bound GTP appears to be relatively constant in the receptor-coupled G proteins. The rate of this reaction can be modulated experimentally or its deactivating effect can be obviated altogether. When poorly-hydrolyzed analogs of GTP, such as Gpp(NH)p or GTPγS[1], bind to G proteins, they activated essentially irreversibly, causing persistent and extensive activation of effector proteins. As predicted by the scheme in Fig. 1, G protein activity in the presence of such analogs is not markedly altered by receptor, although the receptor can increase the rate of activation by increasing the rate of the binding of analog. For G_s, the G protein that stimulates adenylate cyclase, cholera toxin exerts an effect similar to that of non-hydrolyzable GTP analogs. By catalyzing the ADP-ribosylation of G_s (on an arginine residue in the α subunit), cholera toxin inhibits the ability of G_s to hydrolyze bound GTP and thereby causes persistent activation of adenylate cyclase in the presence of cytoplasmic concentrations of GTP itself. (Such activation of intestinal adenylate cyclase initiates the classic symptom of cholera.) Al^{+3} and F^- can stimulate G proteins by thwarting the deactivating hydrolysis reaction. Al^{+3} and F^- bind GDP-liganded G protein at the site normally occupied by the γ-phosphoryl group of GTP to form a $GDP-Al^{+3}-F_3^-$ complex that mimics bound GTP (Sternweis and Gilman, 1982; Bigay et al., 1985, 1987). Activation of G proteins by F^- plus Al^{+3} is a convenient and general means of stimulated G protein-mediated pathways.

In the physiological state, with GTP as the relevant nucleotide, the agonist-liganded receptor acts as a formal catalyst of GDP/GTP exchange. A single receptor can interact sequentially with multiple G protein molecules in a process involving lateral diffusion in the plasma membrane. Under optimal conditions, receptor-stimulated GTP exchange is sufficiently rapid that one receptor can maintain the activation of multiple G proteins. Such amplification of the signal of a single receptor through many G proteins has been measured as up to 20-fold for the β-adrenergic receptor (Asano et al., 1984b; Brandt et al., 1986) and about 1000-fold for the rhodopsin-transducin system of the retina (Stryer, 1985).

Receptors catalyze the exchange of GDP for GTP by transiently converting the nucleotide binding site from a high affinity ("closed") conformation to a lower affinity ("open") state. In such a system, basic thermodynamics holds that if one ligand (receptor-agonist complex) reduces the affinity of a protein for another ligand (nucleotide), then the binding of the second ligand will reciprocally and equally decrease the affinity of binding of the first ligand. This relationship is displayed in G protein systems as the ability of guanine nucleotides to decrease the affinity of a receptor for its agonist. A decreased affinity for agonist in the presence of GTP is a good general indication that the agonist's receptor is acting through a G protein. In many cases, the species with the highest affinity for agonist is a very stable receptor-G protein complex. The stabilization of a G protein-receptor complex by agonist is also characteristic of G protein coupled systems. The kinetic and thermodynamic aspects of these interactions have been discussed in depth (Ross and Gilman, 1980; Smigel et al., 1984a,b; Stadel et al., 1982).

In contrast to our ability to describe the interactions of G proteins and receptor in great detail, relatively little is known about how GTP-activated G proteins stimulate their effectors. In fact, the two best studied examples yield conflicting pictures. In the case of adenylate cyclase and G_s, the active species is probably a complex of the cyclase and GTP-activated G_s (Pfeuffer, 1979; Pfeuffer et al., 1983; Neer et al., 1980). Non-hydrolyzable GTP analogs promote the binding of G_s to adenylate cyclase, sometimes forming a complex that is stable to chromatography. In contrast to the cyclase system, the retinal G protein transducin (or G_t) appears to activate a cyclic GMP phosphodiesterase (PDE) by binding and thereby neutralizing one or both of the PDE's inhibitory γ subunits (see Stryer, 1985; Deterre et al., 1986). Although some data argue for direct interaction of the subunits of the PDE with G_t (Hingorani et al., 1988, for example), the PDE can also be activated in the absence of transducin simply by proteolysis of its γ subunits.

At this point, it is not obvious that either mechanism of effector activation by G proteins is more general or that there are not several other ways in which activated G proteins can stimulate effectors. Analogy with the PDE system suggests that G_s might bind to an inhibitory domain of the cyclase. Such a situation has already been noted in the case of the cyclic nucleotide-activated protein kinases. The cyclic AMP-dependent protein kinase has an inhibitory, regulatory subunit that dissociates upon binding cyclic AMP. However, the cyclic GMP-dependent protein kinase is a single polypeptide whose regulatory, cyclic GMP-binding region is homologous to the regulatory subunit of the cyclic AMP-dependent enzyme (Edelman et al., 1987). Whether the G protein-regulated effectors display such patterns should be known within the next few years.

COMPONENTS OF G PROTEIN SIGNALING NETWORKS

The last five years have seen a dramatic increase in what we know about the structural components of G protein-mediated signaling systems. Many G proteins have been purified and their sequences determined from cloned cDNA's.. Receptors have also been purified and cloned. Again, effectors lag behind, reflecting both their diversity and, for most, the recency of their discovery.

G Proteins and their Subunits

G proteins are composed of three subunits, denoted α (largest), β and γ (smallest). The α subunit can be readily separated from the $\beta\gamma$ subunits, which form a stable complex that has not been resolved under non-denaturing conditions. The α subunit binds and hydrolyzes GTP. Furthermore, isolated α subunit can be activated by GTP such that it can activate its appropriate effector in the absence of $\beta\gamma$. Selectivity for receptor also seems to reside in the α subunit; α_s can activate adenylate cyclase but not the retinal PDE and the converse is true for α_t. The $\beta\gamma$ subunits may also regulate certain effectors (see below). However, their more common role is to regulate the binding of nucleotides to α, to help anchor α to the plasma membrane and to mediate regulation of the α subunit by receptor (Gilman, 1986, for review).

G protein α subunits are a family of homologous proteins, ranging in size from about 39,000 Da to 45,000 Da. Homology is most highly conserved in three regions of sequence, and these sequences are also conserved in other GTP-binding proteins. X-ray crystallography of the bacterial elongation factor Tu and the oncogene product p21[ras] indicates that these three regions surround the GTP binding site (Jurnak, 1985; de Vos et al., 1988). The α subunits, while homologous, are distinct for each G protein. Thus, α_s determines G_s, $\alpha_{i,2}$ determines $G_{i,2}$, etc. Sequence homology among the α subunits and the similarity of their biochemical properties suggests that they have a similar tertiary structure. It is reasonable to assume that each α subunit will have a definable receptor-binding domain that is reasonably conserved with respect to structure and function, but which is adequately distinct such that selectivity of the G protein for receptor is maintained. A similar argument can be made for an effector-binding domain. Numerous studies, both biochemical and genetic, have focused on the carboxy terminal region of α subunits as contributing to the receptor-binding site (Sullivan et al., 1987; Masters et al., 1988; Hamm et al., 1988; Stryer and Bourne, 1986; West et al., 1985). The location or structure of an effector-binding domain remains speculative.

In contrast to the obvious functional individuality of the α subunits, the $\beta\gamma$ subunits are frequently considered as a common pool that is shared among the α's. This view is probably correct, at least in part, but it may also reflect our difficulty in separating distinct species of $\beta\gamma$ (see Hildebrandt et al., 1985, for one approach). When the $\beta\gamma$ subunits are prepared from G protein trimers or are isolated from chromatographic fractions of plasma membrane extracts that do not include α subunits, these fractions appear able to regulate any of the α subunits. (This is not true of retinal $\beta\gamma$ (Cerione et al., 1987).) $G_{\beta\gamma}$ fractions contain two different β subunits, the product of separate but nearly identical genes (Fong et al., 1987). Fortunately, the two β subunits can be (barely) separated by SDS-polyacrylamide gel electrophoresis as apparent 35,000 Da and 36,000 Da bands. The number of γ subunits and the differences among them is unknown. Many highly purified preparations of G proteins display up to three bands with electrophoretic mobilities in the 6000-10,000 Da range. These small proteins cofractionate with β subunits under a wide variety of conditions. However, conclusive sequence data is available only for the γ subunit of transducin, a single γ subunit from brain, and a γ subunit in yeast (Hurley, 1984; Whiteway, 1988). The number and diversity of γ subunits in non-retinal tissues is unknown. At very least, a mixture of two β and three γ subunits would yield a diverse group of regulatory $\beta\gamma$ dimers.

The $\beta\gamma$ subunits are required for the regulation of α subunits by hormone receptors (Fung, 1983). This may reflect their importance in anchoring the relatively hydrophilic α subunit to the plasma membrane (Sternweis, 1986). In addition, the $\beta\gamma$ subunits have at least two biological regulatory roles. First, $\beta\gamma$ subunits inhibit the activation of α by inhibiting both GTP binding and GDP release. Conversely, activation of the α subunit by GTP promotes release of $\beta\gamma$ and, in the case of non-hydrolyzable GTP analogs, can force complete dissociation of $\beta\gamma$ from α. Northup and coworkers (1983 a,b; Smigel et al., 1984a) pointed out that activation-driven release of $\beta\gamma$ provides a mechanism for agonist-mediated inhibition of a G protein signaling pathway, an effect exemplified by the receptor-promoted inhibition of adenylate cyclase. G_i was, in fact, first identified as the inhibitory mediator of this activity. Rather than α_i's acting as an inhibitor of cyclase, however, this inhibition primarily reflects release of $\beta\gamma$ from α_i when it is activated. The free $\beta\gamma$ will then inhibit the activation of G_s, preventing its stimulation of adenylate cyclase (Katada et al., 1984a,b; Cerione et al., 1986b). This mechanism is consistent with the high affinity with which $\beta\gamma$ binds α_s relative to α_i and with the large molar excess of G_i over G_s in plasma membranes. It is not the only means of hormonal inhibition of a G protein-mediated signal, however (Jakobs et al., 1983, 1985).

The $\beta\gamma$ subunits may also convey stimulatory signals to effectors. Neer, Clapham, and colleagues have provided extensive evidence that $\beta\gamma$ subunits can activate the M-type potassium channel in muscle (Logothetis et al., 1987, 1988). In yeast, $\beta\gamma$ mediates signals from the receptors for mating pheromones to cause the cell cycle arrest that is characteristic of the mating response (Whiteway, 1988). Other data suggest that $\beta\gamma$ activates a phospholipase A_2 (Jelsema and Axelrod, 1987). It is not clear that $\beta\gamma$ binds directly to the effector protein in any of these instances, however. In the case of the K^+ channel, more potent activation by α_i has been observed (Codina et al., 1987; Yatani et al., 1987). One might also postulate the existence of an inhibitory α whose activity is diminished by $\beta\gamma$. Answers here should be available in short order.

Effectors

Cellular effector proteins that are regulated by G protein signals are diverse, and no unified picture of their structure or mode of regulation is yet apparent. Their common properties are their association with a membrane (not surprising for a receptor-effector system) and a tendency toward cleaving phosphate bonds (cyclic GMP phosphodiesterase, adenylate cyclase, phospholipase C). Only two effectors have been purified. The

sequence of adenylate cyclase from yeast is known, but the
yeast enzyme is twice as large as a mammalian cyclase (Kataoka
et al., 1985) and homology between the two is presumably not
great. As discussed above for the G protein α subunits, the
effectors may be expected to share at least some structural
similarity in a G protein-binding domain. However, this expec-
tation is reasonable only if there is a single mechanism of
regulation of effectors. If not, the effector proteins may be
only a conceptual family, not a structural one.

G Protein Coupled Receptors

Receptors that regulate G proteins are a large family of
highly homologous, hydrophobic, integral plasma membrane proteins. Our understanding of the structure of these proteins is
based largely on biochemical and genetic studies of rhodopsin
and similar visual pigments (Findlay and Pappin, 1986, for
review). Rhodopsin consists of a bundle of seven hydrophobic,
membrane-spanning, largely α-helical segments that are connected by relatively short, hydrophilic sequences on either
side of the photoreceptor membrane. Its short, amino terminal
domain is N-glycosylated and lies on the intradiscal side of
the membrane (topologically equivalent to the extracellular
face of the plasma membrane). The carboxyl terminal region,
also quite hydrophilic, lies on the cytoplasmic face of the
membrane. The retinal chromophore lies within this bundle of
helices, roughly half-way through the membrane, where it is
covalently bound to a lysyl residue in the seventh membrane
span. The few charged amino acid residues that are included in
the membrane-spanning regions are thought to form ion pairs
with each other and to contribute to the spectral selectivity
of the photoreceptive response (Nathans et al., 1986).

When the cDNA's for two β-adrenergic receptors were cloned
in 1986 (Yarden et al., 1986; Dixon et al., 1986), the similarity of their sequences to those of the rhodopsins indicated
that these receptors and rhodopsin share both strong sequence

homology and overall similarity of structure. Knowledge of the amino acid sequences of several muscarinic cholinergic receptors, α_1- and α_2-adrenergic receptors, two serotonin receptors and a receptor for substance K has now enhanced our ability to generalize about common or distinctive structural elements (see Dohlman et al., 1987; Parker and Ross, 1989). Homology among the receptors is not uniform; it is preferentially displayed in the membrane-spanning helices and, to a lesser extent, in the shorter cytoplasmic loops. There is little sequence similarity in the extracellular amino-terminal domain except for the presence of one or more consensus sequences for asparagine-linked glycosylation. The large cytoplasmic loops that connect spans five and six and the carboxy terminal cytoplasmic domains also display little homology when receptor sequences are compared. Even the lengths of these non-homologous regions vary dramatically. The overall impression of the G protein-coupled receptors is therefore one of a highly conserved hydrophobic core of membrane spans joined by short hydrophilic loops, a extracellular amino terminus that might be involved with routing the receptor to the cell surface, and two large cytoplasmic regions that apparently have no uniform function (Fig. 2).

To test these speculations, our laboratory has initiated a number of detailed probes into the functions of the β-adrenergic receptor's definable structural domains. The homology among G protein-coupled receptors in their hydrophobic core regions suggested that this core is important for mediating signal transduction. This idea is confirmed by the functional resistance of the receptor to treatment with proteases (Rubenstein et al., 1987). When purified, detergent-solubilized β-adrenergic receptor is treated with any of several proteases, a limit digest results that contains only two significant peptides. The largest includes the first five membrane-spanning regions and four short interconnecting loops; the smaller contains membrane spans six and seven, their extracellular loop, and short cytoplasmic stalks. These two peptides remain noncovalently bound to each other in detergent solution under non-denaturing conditions.

EXTRACTELLULAR SPACE

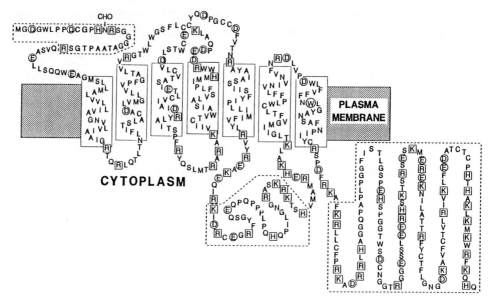

Fig. 2. The amino acid sequence of the ß-adrenergic receptor is shown in a pattern that displays its membrane-spanning helical regions and the associated cytoplasmic loops in relation to the plasma membrane bilayer (Yarden et al., 1986.) The site of glycosylation (CHO) near the amino terminus and one site of ß-adrenergic affinity labeling, Trp^{330} (Wong et al., 1988), are both shown. Basic residues are shown in squares and acidic residues are shown in light circles. Those portions of the receptor that can be proteolytically or genetically deleted without loss of function are outlined by the dashed lines. For each non-essential region, the carboxyl terminus is noted exactly from sequencing experiments, but the amino terminus is estimated based on the retention or loss of chemical labels or defined epitopes (Rubenstein et al., 1987).

Surprisingly, this limit digest retains the ability to bind ß-adrenergic ligands and to regulate G_s in response to agonist. When the purified complex of the two peptides is purified to remove smaller proteolytic fragments and reconstituted with purified G_s into unilamellar phospholipid vesicles, it catalyzes nucleotide exchange by G_s as well as does native receptor. This agonist-stimulated regulatory activity shows that large segments of the cytoplasmic domains of the receptor are not required for interaction with G proteins.

These results have been confirmed by genetic deletion analysis by Dixon and coworkers (1987a,b). Initially, those authors reported that deletion of the DNA that encodes the large cytoplasmic loop of the β-adrenergic receptor resulted in a receptor that retained ligand binding activity but did not regulate G_s. However, when smaller portions of the large loop were individually removed, no single sequence in the center of the loop was required for regulation of G_s. Hypothetically, removal of the entire loop caused a strain on the overall structure of the receptor such that it could no longer function properly. The most startling extension of these studies has been that of Kobilka and coworkers (1988). They prepared and expressed two separate cDNA's, one that encodes the amino terminal portion of the receptor through span five and another that encodes the carboxyl terminal portion beginning with span six (omitting most of the large loop). When both fragments of the receptor were translated in frog oocytes, they were transported to the plasma membrane and associated to form active receptor. Thus, not only does the bulk of the large cytoplasmic loop not contribute to regulation of G proteins, it is not absolutely required for the folding and routing of the receptor protein itself.

Mapping of the binding site for β-adrenergic ligands within the hydrophobic core of the receptor also points out its structural similarity to rhodopsin. Wong and coworkers (1988) found that when the β-adrenergic receptor was photoaffinity labeled with either of two antagonist ligands, label was covalently incorporated at two distinct positions in the receptor's primary amino acid sequence. Because only one molecule of ligand binds per receptor, this pattern of labeling suggests that the two labeled positions are juxtaposed to form the three dimensional ligand-binding site in the native receptor. As it turns out, one of these sites is tryptophan[330], which lies in the seventh membrane span at a site near to that of the lysine residue in rhodopsin to which retinal is covalently bound. The second labeled site has not yet been mapped precisely, but it lies somewhere between spans two and four, suggesting that these spans form the opposite side of the catecholamine binding site. These data suggest that the seven membrane-spanning helices of the β-adrenergic receptor are bundled to form the catecholamine binding site in a structure that is generally similar to that of rhodopsin, with the positively charged catecholamine replacing the retinal-lysine adduct in the binding site. A key question is now how a β-adrenergic agonist or trans-retinal-lysine can alter the structure of their apoproteins to convey a signal to the G protein bound on the cytoplasmic surface.

THE G PROTEIN-REGULATORY SITE OF RECEPTORS

Although the selectivity of receptor-G protein interactions is striking, the homology of the interactive domains suggests that selectivity should not be absolute. It is not. Asano and coworkers (1984a) showed that the β-adrenergic receptor, which usually regulates G_s, could promote activation of G_i when both were reconstituted into phospholipid vesicles. Abramson and Molinoff (1985; Abramson et al., 1987) confirmed these findings in native biological membranes, using a G_s-deficient mutant of S49 lymphoma cells. In these membranes, G_i was able to regulate the affinity with which the β-adrenergic receptor binds agonist. Furthermore, and the addition of agonist could stabilize a solubilized receptor-G_i complex. More recent reconstitution studies indicate that muscarinic cholinergic receptor or α-adrenergic receptor can regulate either G_o or G_i (Florio and Sternweis, 1985; Haga et al., 1986; Cerione et al., 1986a) and that rhodopsin can regulate either G_o and a G_i in addition to G_t (Cerione et al., 1985a). The work of Capon and colleagues now suggests that communication from the four isoforms of the muscarinic receptor through G proteins may be a highly branched pathway (Ashkenazi et al., 1987; Peralta et al., 1988).

Recently, we have approached this problem in a somewhat better defined system using purified, recombinant α subunits of

different G proteins that have been expressed in E. coli (R.C.
Rubenstein, M. Linder and E.M. Ross, in preparation). These
purified α subunits can be combined with purified $\beta\gamma$ subunits,
purified β-adrenergic receptor, and phospholipid to yield active, well-coupled receptor-G protein systems. Using each of
the three α_i's, α_o, and either the short or long form of α_s, we
have shown a distinct pattern of selectivity of the receptor
for different G proteins. G_s is preferred over the G_i's
(1>3>2) and the receptor is essentially unable to regulate G_o.
Further studies are needed to quantitate this order of selectivity, and it is likely that competition studies using two G
proteins in the same vesicle will be required to probe the
physiologic selectivity of receptor for G protein. Nevertheless, these data demonstrate both that the regulation of G_α's
by the receptor displays an appropriate rank order of selectivity in reconstituted systems and that such systems can be used
to probe the structural basis of this selectivity.

Analysis of how a receptor selects among G proteins has
focused on the cationic cytoplasmic face of the receptor as the
site of interactions. Because the bulk of the third cytoplasmic loop and the carboxyl terminal domain of the receptor can
be removed without loss of regulatory function, the more homologous regions near the bilayer surface must form the site that
selectively binds and regulates G proteins. Several laboratories are using reverse genetics to define functionally important determinants in this region.

Recently, our laboratory has begun to exploit an alternative experimental probe for the receptor's G protein-regulatory
domain, the wasp venom peptides called mastoparans. Mastoparans are tetra-cationic, amphiphilic, tetradecapeptides that
were first described by Nakajima and coworkers (Hirai et al.,
1979; Kuroda et al., 1980) as potent stimulators of histamine
secretion from mast cells and, subsequently, secretion of diverse agents from a wide variety of secretory cells. The likely role of the G protein in controlling secretion suggested
that mastoparan might act directly on a G protein-mediated

pathway and led us (Higashijima et al., 1988) to study the effect of mastoparan on individual purified G proteins. We found that mastoparan increases the rate with which G proteins release GDP and bind GTP, thereby stimulating the overall rate of steady-state of GTP hydrolysis and the rate at which G proteins are activated. This essentially receptor-mimetic activity displays several characteristic properties of receptor-mediated control of G proteins. Most significant of these receptor-like behaviors is that most of the action of mastoparan on G_i or G_o is blocked by pertussis toxin-catalyzed ADP-ribosylation of the α subunit. Such ADP-ribosylation generally does not alter the function of an isolated G protein but generally blocks the ability of the G protein to be regulated by receptor (van Dop, 1984; Okajima et al., 1985). Mastoparan also stimulates nucleotide exchange at low concentrations (below 100 nM) of Mg^{2+} without altering the intrinsic k_{cat} for hydrolysis of bound GTP. These data suggest that the mastoparans may be structurally as well as functionally similar to the G protein-binding domain of cell surface receptors.

Although there is no significant similarity between the sequence of mastoparans and those of the cytoplasmic face of the G protein-coupled receptors, it is likely that there is overall similarity of tertiary structure. The three dimensional structure of mastoparan-X bound to a phospholipid bilayer was shown to be a short α helix oriented such that its four positive charges are directed toward the aqueous solvent (Wakamatsu et al., 1983; Higashijima et al., 1983). This ordered array of positive charges may be similar to that of the cytoplasmic face of a receptor. Each of the regions that are candidates for G protein binding sites -- the first and second intracellular loops, both origins of the large, third intracellular loop, and the beginning of the cytoplasmic carboxy terminal region -- are all strongly positively charged. We should now be able to use synthetic mastoparan analogs to determine what physical properties allow the activation of G proteins by positive charge clusters and what other properties allow selectivity.

Not surprisingly, selectivity of mastoparan among G proteins is not as great as that displayed by most cell surface receptors, but it is still impressive. For example, the potency and efficacy of mastoparan is greatest for G_o, somewhat less for G_i, and 10-fold less for G_s and transducin. Mastoparan-X has a similar selectivity, although it is much more active on transducin than on G_s. At this time, our group has discovered cationic peptides that selectively activate G_s as well as other peptides and cationic compounds that selectively antagonize the regulatory effects of "agonist" peptides. We should now be able to use synthetic mastoparan analogs and mutants of the β-adrenergic receptor to focus on what key features of these cationic arrays are important for G protein binding and activation. We propose that the binding of an agonist to a receptor initiates a reorientation of the membrane-spanning helices either to cause the exposure of positive charges on the cytoplasmic face or to alter the orientation of one charge cluster with respect to another. By probing the selectivity of chimeric receptors, mutated receptors, and novel synthetic mastoparans we should be able to approach this problem at a detailed sub-molecular level.

Acknowledgement

Studies from the author's laboratory were supported by NIH grant GM30355.

REFERENCES

Abramson SN, Molinoff PB (1985) Properties of β-adrenergic receptors of cultured mammalian cells. Interaction of receptors with a guanine nucleotide-binding protein in membranes prepared from L6 myoblasts and from wild type and cyc⁻ S49 lymphoma cells. J Biol Chem 260:14580-14588

Abramson SN, Shorr RGL, Molinoff PB (1987) Interactions of β-adrenergic receptors with a membrane protein other than the stimulatory guanine nucleotide-binding protein. Biochem Pharmacol 36:2263-2269

Asano T, Katada T, Gilman AG, Ross EM (1984) Activation of the inhibitory GTP-binding protein of adenylate cyclase, G_i, by the β-adrenergic receptors in reconstituted phospholipid vesicles. J Biol Chem 259:9351-9354

Asano T, Pedersen SE, Scott CW, Ross EM (1984) Reconstitution of catecholamine-stimulated binding of guanosine 5'-O-(3-thiotriphosphate) to the stimulatory GTP-binding protein of adenylate cyclase. Biochemistry 23:5460-5467

Ashkenazi A, Winslow JW, Peralta EG, Peterson GL, Schimerlik MI, Capon DJ, Ramachandran J (1987) An M2 muscarinic receptor subtype coupled to both adenylyl cyclase and phosphoinositide turnover. Science 238:672-675

Bigay J, Deterre P, Pfister C, Chabre M (1985) Fluoroaluminates activate transducin-GDP by mimicking the γ-phosphate of GTP in its binding site. FEBS Lett 191:181-185

Bigay J, Deterre P, Pfister C, Chabre M (1987) Fluoride complexes of aluminum or beryllium act on G-proteins as reversibly bound analogues of the gamma phosphate of GTP. EMBO J 6:2907-2913

Brandt DR, Ross EM (1986) Catecholamine-stimulated GTPase cycle: Multiple sites of regulation by β-adrenergic receptor and Mg^{2+} studied in reconstituted receptor-G_s vesicles. J Biol Chem 261:1656-1664

Cerione RA, Gierschik P, Staniszewski C, Benovic JL, Codina J, Somers R, Birnbaumer L, Spiegel AM, Lefkowitz RJ, Caron MG (1987) Functional differences in the $\beta\gamma$ complexes of transducin and the inhibitory guanine nucleotide regulatory protein. Biochemistry 26:1485-1491

Cerione RA, Regan JW, Nakata H, Codina J, Benovic JL, Gierschik P, Somers RL, Speigel AM, Birnbaumer L, Lefkowitz RJ, Caron MG (1986) Functional reconstitution of the α_2-adrenergic receptor with guanine nucleotide regulatory proteins in phospholipid vesicles. J Biol Chem 261:3901-3909

Cerione RA, Staniszewski C, Benovic JL, Lefkowitz RJ, Caron MG, Gierschik P, Somers R, Spiegel AM, Codina J, Birnbaumer L (1985) Specificity of the functional interactions of the β-adrenergic receptor and rhodopsin with guanine nucleotide regulatory proteins reconstituted in phospholipid vesicles. J Biol Chem 260:1493-1500

Cerione RA, Staniszewski C, Gierschik P, Codina J, Somers RL, Birnbaumer L, Spiegel AM, Caron MG, Lefkowitz RJ (1986) Mechanism of guanine nucleotide regulatory protein-mediated inhibition of adenylate cyclase. Studies with isolated subunits of transducin in a reconstituted system. J Biol Chem 261:9514-9520

Codina J, Yatani A, Grenet D, Brown AM, Birnbaumer L (1987) The α subunit of the GTP binding protein G_k opens atrial potassium channels. Science 236:442-445

De Vos AM, Tong L, Milburn MV, Matias PM, Jancarik J, Noguchi S, Nishimura S, Miura K, Ohtsuka E, Kim S-H (1988) Three-dimensional structure of an oncogene protein: Catalytic domain of human c-H-ras p21. Science 239:888-893

Dixon RAF, Kobilka BK, Strader DJ, Benovic JL, Dohlman HG, Frielle T, Bolanowski MA, Bennett CD, Rands E, Diehl RE, Mumford RA, Slater EE, Sigal IS, Caron MG, Lefkowitz RJ, Strader CD (1986) Cloning of the gene and cDNA for mammalian β-adrenergic receptor and homology with rhodopsin. Nature 321:75-79

Dixon RAF, Sigal IS, Rands E, Register RB, Candelore MR, Blake AD, Strader CD (1987) Ligand binding to the β-adrenergic receptor involves its rhodopsin-like core. Nature 326:73-77

Dohlman HG, Caron MG, Lefkowitz RJ (1987) A family of receptors coupled to guanine nucleotide regulatory proteins. Biochemistry 26:2657-2664

Edelman AM (1987) Protein serine/threonine kinases. In: "Annual Review of Biochemistry", vol 56, CC Richardson, Ed, Annual Reviews, Inc, CA pp 567-613

Findlay JBC, Pappin DJC (1986) The opsin family of proteins. Biochem J 238:625-642

Florio VA, Sternweis PC (1985) Reconstitution of resolved muscarinic cholinergic receptors with purified GTP-binding proteins. J Biol Chem 260:3477-3483

Fong HKW, Amatruda, III TT, Birren BW, Simon MI (1987) Distinct forms of the β subunit of GTP-binding regulatory proteins identified by molecular cloning. Proc Natl Acad Sci USA 84:3792-3796

Fung BK-K (1983) Characterization of transducin from bovine retinal rod outer segments. I. Separation and reconstitution of subunits. J Biol Chem 258:10495-10502

Gilman AG (1987) G proteins: Transducers of receptor-generated signals. Ann Rev Biochem 56:615-649

Haga K, Haga T, Ichiyama A (1986) Reconstitution of the muscarinic acetylcholine receptor: Guanine nucleotide-sensitive high affinity binding of agonists to purified muscarinic receptors reconstituted with GTP-binding proteins (G_i and G_o). J Biol Chem 261:10133-10140

Hamm HE, Deretic D, Arendt A, Hargrave PA, Koenig B, Hofmann KP (1988) Site of G protein binding to rhodopsin mapped with synthetic peptides from the α subunit. Science 241:832-835

Hamm HE, Deretic D, Hofmann KP, Schleicher A, Kohl B (1987) Mechanism of action of monoclonal antibodies that block the light activation of the guanyl nucleotide-binding protein, transducin. J Biol Chem 262:10831-10838

Higashijima T, Uzu S, Nakajima T, Ross EM (1988) Mastoparan, a peptide toxin from wasp venom, mimics receptors by activating GTP-binding regulatory proteins (G proteins). J Biol Chem 263:6491-6494

Higashijima T, Wakamatsu K, Takemitsu M, Fujino M, Nakajima T, Miyazawa T (1983) Conformational change of mastoparan from wasp venom on binding with phospholipid membrane. FEBS Lett 152:227-230

Hildebrandt JD, Codina J, Rosenthal W, Birnbaumer L, Neer EJ, Yamazaki A, Bitensky MW (1985) Characterization by two-dimensional peptide mapping of the γ subunits of N_s and N_i, the regulatory proteins of adenylyl cyclase, and of transducin, the guanine nucleotide-binding protein of rod outer segments of the eye. J Biol Chem 260:14867-14872

Hingorani VN, Tobias DT, Henderson JT, Ho Y-K (1988) Chemical cross-linking of bovine retinal transducin and cGMP phosphodiesterase. J Biol Chem 263:6916-6926

Hirai Y, Yasuhara T, Yoshida H, Nakajima T, Fujino M, Kitada C (1979) A new mast cell degranulating peptide "Mastoparan" in the venom of Vespula lewisii. Chem Pharm Bull 27:1942-1944

Hurley JB, Fong HKW, Teplow DB, Dreyer WJ, Simon MI (1984) Isolation and characterization of a cDNA clone for the γ subunit of bovine retinal transducin. Proc Natl Acad Sci USA 81:6948-6952

Jakobs KH, Aktories K, Minuth M, Schultz G (1985) Inhibition of adenylate cyclase. Adv Cyclic Nucleotide Prot Phos Res 19:137-150

Jakobs KH, Schultz G (1983) Occurrence of a hormone-sensitive inhibitory coupling component of the adenylate cyclase in S49 lymphoma cyc⁻ variants. Proc Natl Acad Sci USA 80:3899-3902

Jelsema CL, Axelrod J (1987) Stimulation of phospholipase A2 activity in bovine rod outer segments by the beta gamma subunits of transducin and its inhibition by the alpha subunit. Proc Natl Acad Sci USA 84:3623-3627

Jurnak F (1985) Structure of the GDP domain of EF-Tu and location of the amino acids homologous to ras oncogene proteins. Science 230:32-36

Katada T, Bokoch GM, Smigel MD, Ui M, Gilman AG (1984) The inhibitory guanine nucleotide-binding regulatory component of adenylate cyclase. Subunit dissociation and the inhibition of adenylate cyclase in S49 lymphoma cyc⁻ and wild type membranes. J Biol Chem 259:3586-3595

Katada T, Northup JK, Bokoch GM, Ui M, Gilman AG (1984) The inhibitory guanine nucleotide-binding regulatory component of adenylate cyclase. Subunit dissociation and guanine nucleotide-dependent hormonal inhibition. J Biol Chem 259:3578-3585

Kataoka T, Broek D, Wigler M (1985) DNA sequence and characterization of the S. cerevisiae gene encoding adenylate cyclase. Cell 43:493-505

Kobilka BK, Kobilka TS, Daniel K, Regan JW, Caron MG, Lefkowitz RJ (1988) Chimeric α_2-, β_2-adrenergic receptors: Delineation of domains involved in effector coupling and ligand binding specificity. Science 240:1310-1316

Kuroda Y, Yoshioka M, Kumakura K, Kobayashi K, Nakajima T (1980) Effects of peptides on the release of catecholamines and adenine nucleotides from cultured adrenal chromaffin cells: Mastoparan-induced release. Proc Japan Acad, Ser B 56:660-664

Logothetis DE, Kim D, Northup JK, Neer EJ, Clapham DE (1988) Specificity of action of guanine nucleotide-binding regulatory protein subunits on the cardiac muscarinic K^+ channel. Proc Natl Acad Sci USA 85:5814-5818

Logothetis DE, Kurachi Y, Galper J, Neer EJ, Clapham DE (1987) The beta gamma subunits of GTP-binding proteins activate the muscarinic K^+ channel in heart. Nature 325:321-326

Masters SB, Sullivan KA, Miller RT, Beiderman B, Lopez NG, Ramachandran J, Bourne HR (1988) Carboxyl terminal domain of $G_{s\alpha}$ specifies coupling of receptors to stimulation of adenylyl cyclase. Science 241:448-451

Nathans J, Thomas D, Hogness DS (1986) Molecular genetics of human color vision: The genes encoding blue, green, and red pigments. Science 232:193-202

Neer EJ, Echeverria D, Knox S (1980) Increase in the size of soluble brain adenylate cyclase with activation by guanosine 5'-(β,γ-imino)triphosphate. J Biol Chem 255:9782-9789

Northup JK, Smigel MD, Sternweis PC, Gilman AG (1983) The subunits of the stimulatory regulatory component of adenylate cyclase. Resolution of the activated 45,000-dalton (α) subunit. J Biol Chem 258:11369-11376

Northup JK, Sternweis PC, Gilman AG (1983) The subunits of the stimulatory regulatory component of adenylate cyclase. Resolution, activity, and properties of the 35,000-dalton (β) subunit. J Biol Chem 258:11361-11368

Okajima F, Katada T, Ui M (1985) Coupling of the guanine nucleotide regulatory protein to chemotactic peptide receptors in neutrophil membranes and its uncoupling by islet-activating protein, pertussis toxin. J Biol Chem 260:6761-6768

Parker EM, Ross, EM (1989) In:Claudio T (ed) Current Topics in Membrane and Transport: Protein-Membrane Interactions. Academic Press in press

Peralta EG, Ashkenazi A, Winslow JW, Ramachandran J, Capon DJ (1988) Differential regulation of PI hydrolysis and adenylyl cyclase by muscarinic receptor subtypes. Nature 334:434-7

Pfeuffer T (1977) GTP-binding proteins in membranes and the control of adenylate cyclase activity. J Biol Chem 252:7224-7234

Pfeuffer T, Gaugler B, Metzger A (1983) Isolation of homologous and heterologous complexes between catalytic and regulatory components of adenylate cyclase by forskolin-Sepharose. FEBS Lett 164:154-160

Ross EM, Gilman AG (1980) Biochemical properties of hormone-sensitive adenylate cyclase. Ann Rev Biochem 49:533-564

Smigel MD, Katada T, Northup JK, Bokoch GM, Ui M, Gilman AG (1984) Mechanisms of guanine nucleotide-mediated regulation of adenylate cyclase activity. Adv Cyclic Nucleotide Res 17:1-18

Smigel MD, Ross EM, Gilman AG (1984) Role of the β-adrenergic receptor in the regulation of adenylate cyclase. In: Cell Membranes: Methods and Reviews, E L Elson, W A Frazier and L Glaser, Eds, Vol 2, Plenum Publishing Corp, New York, New York, pp 247-294

Stadel JM, De Lean A, Lefkowitz RJ (1982) Molecular mechanisms of coupling in hormone receptor-adenylate cyclase systems. Adv Enzymol 53:1-43

Sternweis PC (1986) The purified α subunits of G_o and G_i from bovine brain require $\beta\gamma$ for association with phospholipid vesicles. J Biol Chem 261:631-637

Sternweis PC, Gilman AG (1982) Aluminum: A requirement for activation of the regulatory component of adenylate cyclase by fluoride. Proc Natl Acad Sci USA 79:4888-4891

Strader CD, Dixon RAF, Cheung AH, Candelore MR, Blake AD, Sigal IS (1987) Mutations that uncouple the β-adrenergic receptor from G_s and increase agonist affinity. J Biol Chem 262:16439-16443

Strader CD, Sigal IS, Blake AD, Cheung AH, Register RB, Rands E, Zemcik BA, Candelore MR, Dixon RAF (1987) The carboxyl terminus of the hamster β-adrenergic receptor expressed in mouse L cells is not required for receptor sequestration. Cell 49:855-863

Stryer L (1985) Molecular design of an amplification cascade in vision. Biopolymers 24:29-47

Stryer L, Bourne HR (1986) G Proteins: A family of signal transducers. Ann Rev Cell Biol 2:391-419

Sullivan KA, Miller RT, Masters SB, Beiderman B, Heideman W, Bourne HR (1987) Identification of receptor contact site involved in receptor-G protein coupling. Nature 330:758-760

Van Dop C, Yamanaka G, Steinberg F, Sekura R, Manclark CR, Stryer L, Bourne HR (1984) ADP-ribosylation of transducin by pertussis toxin blocks the light-stimulated hydrolysis of GTP and cGMP in retinal photoreceptors. J Biol Chem 259:23-25

Wakamatsu K, Higashijima T, Fujino M, Nakajima T, Miyazawa T (1983) Transferred NOE analyses of conformations of peptides as bound to membrane bilayer of phospholipid; mastoparan X. FEBS Lett 162:123-126

West RE, Jr, Moss J, Vaughan M, Liu T, Liu T-Y (1985) Pertussis toxin-catalyzed ADP-ribosylation of transducin. J Biol Chem 260:14428-14430

Whiteway M (1988) In:Cold Spring Harbor Symp Quant Biol, vol 53 in press

Wong SK-F, Slaughter C, Ruoho AE, Ross EM (1988) The catecholamine binding site of the β-adrenergic receptor is formed by juxtaposed membrane-spanning domains. J Biol Chem 263:7925-7928

Yarden Y, Rodriguez H, Wong SK-F, Brandt DR, May DC, Burnier J, Harkins RN, Chen EY, Ramachandran J, Ullrich A, Ross EM (1986) The avian β-adrenergic receptor: Primary structure and membrane topology. Proc Natl Acad Sci USA 83:6795-6799

Yatani A, Codina J, Brown AM, Birnbaumer L (1987) Direct activation of mammalian atrial muscarinic potassium channels by GTP regulatory protein G_k. Science 235:207-211

REGULATION OF ADENYLATE CYCLASE IN MAMMALIAN CELLS AND SACCHAROMYCES CEREVISIAE

Alexander Levitzki
Department of Biological Chemistry
Institute of Life Sciences
Hebrew University of Jerusalem
Jerusalem Israel 91904

SUMMARY

A large number of transmembrane signalling systems transduce signals through heterotrimeric GTP binding proteins (G-proteins). In the most intensively studied system - adenylate cyclase, there are two distinct G-protein transducing systems which respond to stimulatory and inhibitory receptors respectively. The studies on the hormonally regulated adenylate cyclase have led to a detailed understanding of the molecular mechanism of signal transduction. This system therefore serves also as an archtype model for the study of other transmembrane signalling systems which possess heterotrimeric GTP binding proteins as transducer elements. One system in which the molecular mechanism of G-protein transduction is not known is the proliferation signal activated by the monomeric G-protein p21^{RAS}. The involvement of a RAS protein as a transducer in the activation in the yeast S. cerevisiae adenylate cyclase opens new avenues towards the understanding of RAS function in mammalian cells.

INTRODUCTION

Cyclic AMP, since its discovery more than 30 years ago, has played a pivotal role in understanding transmembrane signalling systems. In bacteria and in yeast S. cerevisiae, as well as in mammalian cells, the production of cAMP is regulated by external signals. In both yeast and mammalian cells, it was found that the control of adenylate cyclase is mediated by

NATO ASI Series, Vol. H29
Receptors, Membrane Transport and Signal Transduction
Edited by A.E. Evangelopoulos et al.
© Springer-Verlag Berlin Heidelberg 1989

GTP-binding proteins. Activation and inhibition of receptor regulated adenylate cyclase in mammalian cells are mediated by two unique heterotrimeric GTP-regulatory proteins (G-proteins), Gs and Gi respectively. The flow of chemical information is from the receptor through a heterotrimeric G-protein to the adenylate cyclase catalyst (Levitzki, 1987, 1988, for reviews). The hormonally regulated adenylate cyclase is a member of a growing family of transmembrane signalling systems in which G-proteins function as transducer elements. The common structural denominators of these transmembrane signalling systems is that the receptors are highly hydrophobic and that the G-protein are all heterotrimers $\alpha\beta$ where the α-subunits are homologous to each other and harbour the GTP binding site. The β subunits seem to be highly homologous or even identical and the -subunits very similar to each other. The high degree of homology between the β subunits make them functionally interchangeable. The findings that the monomeric protein product of the RAS gene, p21^{RAS} is localized to the inner leaflet of the membrane and that its homologue RAS2 in S. cerevisiae regulates adenylate cyclase suggests that RAS proteins function as transducer elements in yet unidentified receptor systems (Engelberg et al., 1988, for review).

1. HORMONALLY REGULATED ADENYLATE CYCLASE

Adenylate cyclase is activated by stimulatory receptors such as the β-adrenoceptor through the stimulatory G protein-Gs. The enzyme is inhibited by inhibitory receptors such as the α_2-adrenoceptor and the muscarinic receptor through a G-protein-Gi, in which the α_i-subunit is homologous to α_s. The β subunits in both proteins are identical. Sequence data shows that the stimulatory β-adrenoceptor is homologous to the inhibitory muscarinic receptor where unique sequences of these receptors most probably interact with the unique sequence in α_s ($G_s\alpha$) and α_i ($G\alpha_i$) respectively. The role of the receptor in both stimulatory and inhibitory pathways is to catalyze the GDP to GTP exchange at the Gα subunit of the G-protein. The GTP bound form of the G-protein activates the adenylate cyclase. The GTP activated state decays with the concomitant hydrolysis of GTP to GDP (Levitzki, 1987, 1988).

1.1 The Receptors

A number of the receptors which interact with G-proteins have been cloned and sequenced: cyclase stimulatory receptor, the β-adrenoceptor, cyclase inhibitory receptors - the muscarinic receptors, and the α_2-adrenergic and the S. cerevisiae pheromone receptor (Herskovitz and Marsh, 1987). In all cases the receptors are highly hydrophobic and possess 7 stretches of hydrophobic amino acids which can be arranged in 7 transmembrane spanning α-helices. Another feature which seems to be common is the finding that the ligand binding site resides in a hydrophobic sequence within the phospholipid bilayer (Strader et al., 1987). This finding agrees with observation that the potent β-adrenoceptor affinity label N-bromoacetyl-aminocyanopindolol (BAM-CYP) labels a unique glycolipid which resides proximal to the β-adrenoceptor binding site (Bar-Sinai et al., 1986).

1.2 GTPase activity of Gs

The direct demonstration of β-adrenoceptor dependent GTPase in the turkey erythrocyte provided a proof for the "on-off" cycle (Levitzki, 1987). The presence of hormone stimulated GTPase has since been demonstrated in other stimulatory receptors such as for PGE_1 and glucagon as well as for Gi which interacts with inhibitory receptors. The finding of hormone dependent GTPase also provided the basis for the understanding of Cholera toxin induced increase in adenylate cyclase activity. ADP-ribosyla-tion of $G_s\alpha$ induces its activation as a result of the direct inhibition of the GTPase step (k_{off}). The G_s protein in its ADP ribosylated form spends more time in its active GTP state, leading to increased activity of the enzyme.

1.3 GDP release

Release of the GDP from G_s is facilitated by the agonist-bound β-adrenoceptor. The agonist bound receptor interacts with the G-protein and induces an "open" conformational state which allows a facilitated GDP to GTP exchange, generating the active GTP-bound state of the G protein (Braun et al., 1982). The rate limiting step seems to be a conformational change within the G-protein which can be directly monitored by the release of [^{3}H]GDP or [^{3}H]-GPPNHP. The process of GDP to GTP exchange is much slower than the hydrolysis of GTP to GDP at the $G_s\alpha$ subunit. In native turkey erythrocyte membranes both kinetic constants were measured. The activation rate constant "k_{on}" is in the range of 0.5 to 1.5 min^{-1} whereas the rate constant of the GTPase turn-off reaction, (k_{off}) is 13 - 15 min^{-1}, at 37° C (Arad and Levitzki, 1979). Thus the rate limiting step in the on-off cycle is the GDP to GTP exchange reaction. Higashijima et al. (1987) measured directly the rate constant of GTP hydrolysis in pure G_o as well as the rate of GDP release from G_o. They have shown that at 20° C the overall rate of GTP hydrolysis is 0.4 min^{-1} for GDP bound G_o, identical to the rate of GDP-release from G_o. The initial rate of GTP hydrolysis by GDP-free G_o is 5-fold faster, demonstrating directly that the hydrolysis site per se is not the rate limiting step. The rate limiting step is either the GDP release itself or a conformational transition at the G-protein, which precedes the GDP-release as suggested for G_s in the native turkey erythrocyte $G_s C$ complex (Tolkovsky and Levitzki, 1978). The role of the agonist (H) bound receptor (H R) is to facilitate the GDP to GTP exchange at the α subunit of the G-protein. The overall process described in equation 1 is first order:

$$GTP + G^{GDP} \xrightarrow{H\ R} G^{GTP} + GDP \qquad\qquad (1)$$

where the pseudo-first order rate constant is given by:

$$k_{on} = k \frac{[H][R_T]}{K_H + [H]}$$ (2)

k is the intrinsic first order rate constant, $[R_T]$ is the total receptor concentration, K_H the hormone receptor-agonist dissociation constant and [H] the free hormone concentration.

1.4   Amplification of the hormonal signal

The first amplification step of the hormonal signal occurs at the level of receptor to G_s interaction. One hormone receptor can activate numerous G_s molecules as demonstrated in native membranes (Tolkovsky and Levitzki, 1978; Arad et al., 1981) and in β-adrenoceptor dependent adenylate cyclase reconstituted from purified components (Hekman et al., 1984; Feder et al., 1986). The second step of amplification occurs because the rate constant for GTP hydrolysis (k_{off}) at G_s (13 min^{-1} at 37°) is 100-times smaller than the rate constant of cAMP formation by the catalytic subunit of adenylate cyclase (1200 min^{-1}). The combined amplification at the receptor to G_s interaction is therefore in the range of 1,000. The mechanism where one receptor molecule catalyzes the activation of numerous G proteins is known as "collision coupling" (Levitzki, 1988; Tolkovsky and Levitzki, 1978; Arad et al., 1981). A completely analogous situation exists for the interaction of light activated rhodopsin with transducin which in turn activates cGMP phosphodiesterase. In that system it was shown that one activated rhodopsin can activate up to 300-500 G-protein molecules.

1.5 Role of β subunits

The ability of the β subunits to dissociate from the G-protein heterotrimer has suggested to some investigators that this dissociation plays a regulatory role. Gilman and his colleagues (1984) suggest that β dissociates from the G_s-protein upon its activation and loading of the α subunit with GTP. The naked but active α_s^{GTP} subunit then seeks the catalytic unit of adenylate cyclase, generating the active complex $\alpha_s^{GTP}C$. Concomitantly to GTP-hydrolysis, the α_s^{GDP} C dissociates from C and α_s^{GDP} recombines with β . According to this model (Figure 1), β and

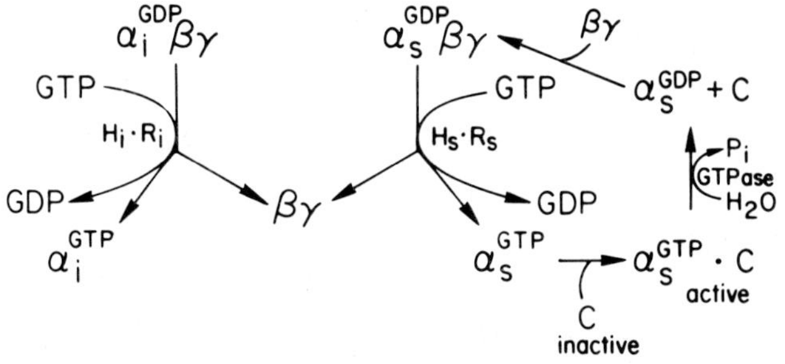

Figure 1. The β -dissociation model for adenylate cyclase regulation.

the catalyst C compete for α_s^{GTP}. Thus, if an extra supply of β subunits is provided, inhibition of adenylate cyclase results. The β dissociation hypothesis is the basis for the model for adenylate cyclase inhibition. When G_i is activated by an inhibitory receptor, it dissociates to α_i^{GTP} and β . Elevation of β levels within the membrane attenuates adenylate cyclase activity since they compete with α_s^{GTP} for C (Figure 1). This model contradicts "collision coupling" since its basic feature is that G_s and C are separate protein units. We have however demonstrated that the overall reaction kinetics predicted by dissociation models (Tolkovsky et al., 1982) is complex since the separate reactants: hormone-bound receptor, G-protein, and the catalyst interact in sequence. The kinetic features found experimentally are simple first order where the first order rate constant is linearly dependent on receptor concentration (Levitzki, 1987, 1988; Tolkovsky and Levitzki, 1978; Arad et al., 1981; Pedersen and Ross, 1982; Hekman et al., 1984). These features argue for a permanent association of G_s to C throughout the "on off" cycle. This assertion has been verified by direct biochemical experiments: G_sC can be purified as a complex in its GDP state as well as in its GPPNHP active state (Arad et al., 1984). If one however assumes that $\alpha_s(G_s\alpha)$ never separates from C but allows β to dissociate, the basic feature of the Gilman dissociation model can coexist (Levitzki, 1984, 1987) with the "collision coupling" mechanism (Figure 2).

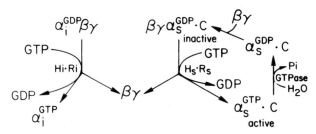

Figure 2. The partial β -dissociation model.

A serious objection against both molecular models are the findings that the β subunit co-purifies with the GPPNHP activated adenylate cyclase (Bar-Sinai et al., submitted). Other considerations also argue against the β -dissociation model. For example, activation of receptor systems which are linked to other G-proteins should lead to adenylate cyclase inhibition since all contain interchangeable β subunits; such non-specific phenomena have not been observed.
What is then the role of the β subunits? For one thing they provide an anchor for the $G\alpha$ subunits to the membrane. Localization of the α subunits to the phospholipid membrane is absolutely dependent on the presence of β subunits (Sternweis,

1986). Whether β subunits play an active regulatory role in addition is yet unresolved.

1.6 Other models for Gi action

The difficulties arising from invoking the β subunits as regulatory elements focused our attention on alternative molecular models for Gi action. These models must assume a physical interaction between Gi and the Gs-C system for which we recently published evidence (Marbach et al., 1988).

2. YEAST ADENYLATE CYCLASE

2.1 RAS Proteins as transducing elements

Mammalian RAS proteins are relatively low molecular weight (21 KD) GTP binding proteins localized to the inner leaflet of the bilayer and are essential for normal cell growth (for review see Lowy and Willumsen, 1986). Like heterotrimeric G-proteins RAS proteins possess slow GTPase activity. Single point mutations at position 12 from the amino terminal (for example Gly 12 --> Arg 12) convert the p21 to a transforming protein. Thus, in Harvey Sarcoma virus the oncogene which codes for $p21^{RASArg12}$ and transfection of NIH3T3 cell with DNA which codes for $p21^{RASArg12}$ induces permanent transformation of the cell. The mutated protein exhibits reduced GTPase activity and enhanced rate of GDP dissociation. The localization of $p21^{RAS}$ proteins to the inner leaflet of the plasma membrane makes them good candidates for transducer elements. However, neither a receptor nor an effector protein which interact with $p21^{RAS}$ proteins have thus far been identified.

It is noteworthy that mutations at position 12 which lead to reduced GTPase activity or enhanced GDP-dissociation rate, are associated with higher transforming potential. This correlation suggests that the active form of $p21^{RAS}$ is its GTP bound species. This correlation makes them very similar to the heterotrimeric G-proteins.

2.2 S. cerevisiae RAS and Adenylate cyclase

To the surprise of many it was recently found that the yeast S. cerevisiae possesses RAS proteins known as RAS1 and RAS2 which are highly homologous to mammalian p21^{RAS}. These two proteins, mainly RAS2, have been found to activate the GTP dependent yeast adenylate cyclase (see Engelberg et al., 1988, for review). Mammalian p21^{Ha-RAS} protein can reconstitute yeast adenylate cyclase in vitro, in membranes prepared from a S. cerevisiae in which the two RAS gene products are missing (Toda et al., 1985). This finding is corroborated by the observation that mammalian p21^{RAS} genes can "rescue" yeast cells which do not express their own RAS genes (because of mutations) and make them viable. These observations suggest that certain fundamental features of RAS proteins have been conserved through evolution. In mammalian cells, however, p21^{RAS} does not interact with the catalytic unit of adenylate cyclase (Beckner et al., 1985). Strikingly, the potency of p21^{RAS} mutant proteins to transform NIH3T3 in cells is proportional to their efficacy in activating S. cerevisiae cyclase, when expressed in RAS deficient yeast (Sigal et al., 1986). This finding suggests that target protein domains in S. cerevisiae and mammalian cells are extremely similar and therefore highly conserved. Another intriguing parallel is the behaviour of RAS mutant proteins in S. cerevisiae. The mutation Gly19 --> Val19 in yeast RAS2 protein which is homologous to the Gly12 --> Val12 mutation in mammalian RAS results in an "oncogenic" transformation. The yeast cells divide more rapidly, do not arrest at the G1 phase upon starvation and diploid cells fail to sporulate in sporulating medium (Toda et al., 1985). This observation supports the hypothesis that a protein domain in the yeast adenylate cyclase cascade system has been conserved through evolution and appears in mammalian cells in a different biochemical context.

2.3 The CDC25 Protein in S. cerevisiae - a receptor?

Mutations in S. cerevisiae which eliminate cAMP formation are lethal if a compensating mutation is not introduced. For example, a yeast cell which lacks adenylate cyclase can only grow in the presence of cAMP or if its cAMP dependent protein kinase is mutated such that it is constitutively active. Recently the gene CDC25 which is different from the adenylate cyclase structural gene (CDC35,CYR1) and the RAS genes has been recognized as a RAS regulator in the cAMP pathway. In the temperature sensitive S. cerevisiae mutant cdc25^{ts} it was found that at the permissive temperature (24° C) cells grow almost normally but at the non-permissive temperature (37° C) the cells stop growing and cAMP levels decline to zero within a half- life of 7 minutes after the temperature shift (Camonis et al., 1986). A detailed study of the adenylate cyclase from membranes isolated from these cells reveals that guanyl nucleotide regulation of the adenylate cyclase by RAS is attenuated whereas the intrinsic activity of the catalytic unit remains intact (Daniel et al., 1987).

The recent demonstration that a yeast cell which lacks the CDC25 gene altogether is viable only when its RAS2 protein is mutated to its activated "oncogenic" form RAS2^{Val19} and is made by the cell in many copies (Broek et al., 1987; Robinson et al., 1987), further strengthens the assertion that CDC25 catalyzes GDP/GTP exchange on RAS (see Engelberg et al., 1988, for review). Recent work from our laboratory shows that the activation of yeast cyclase by GPPNHP and GTPᴛS is a first order process similar to the activation of mammalian adenylate cyclase by these nucleotides. The first order rate constant of activation is higher in membranes prepared from cells which carry many copies of the CDC25 gene and lower in membranes prepared from cells which carry a defective CDC25 gene

(Engelberg and Levitzki, unpublished results). An intriguing question is whether there is a mammalian CDC25 homologue which regulates RAS function.

REFERENCES

Arad H and Levitzki A (1979) The mechanism of partial agonism of the beta-receptor dependent adenylate cyclase of turkey erythrocytes. Mol. Pharmacol. 16:749-756.

Arad H, Rimon G and Levitzki A (1981) The reversal of the GPP(NH)P-activated state of adenylate cyclase by GTP is by the "collision coupling" mechanism. J. Biol. Chem. 256:1593-1597.

Arad H, Rosenbusch J and Levitzki A (1984) Stimulatory GTP regulatory unit Ns and the catalytic unit of adenylate cyclase are tightly associated: Mechanistic consequences. Proc. Natl. Acad. Sci. USA 81:6579-6583.

Bar-Sinai A, Aldouby Y, Chorev M and Levitzki A (1986) Association of turkey erythrocyte β-adrenoceptors with a specific lipid component. EMBO J. 5:1175-1180.

Beckner SK, Hattori S and Shih T (1985) The ras oncogene product p21 is not a regulatory component of adenylate cyclase. Nature 317:71-72.

Braun S, Tolkovsky AM and Levitzki A (1982) Mechanism of control of the turkey erythrocyte β-adrenoceptor dependent adenylate cyclase by guanyl nucleotide: A minimum model. J. Cyclic. Nucl. Res. 8:133-147.

Broek D, Toda T, Michaeli T, Levin L, Birchmeier C, Zoller M, Powers S and Wigler M (1987) The S. cerevisiae CDC25 gene product regulates the RAS/adenylate cyclase pathway. Cell 48:789-799.

Camonis JH, Kalekin M, Bernard G, Garreca H, Boy-Marcotte E and Jacquet M (1986) Characterization cloning and sequence of the CDC25 gene which controls the cyclic AMP level of Saccharomyces cerevisiae. EMBO J. 5:375-380.

Daniel J, Becker J, Enari B and Levitzki A (1987) The activation of adenylate cyclase by guanyl nucleotides in S. cerevisiae is controlled by the CDC25 start gene product. Mol. Cell. Biol. 7:3857-3861.

Engelberg D, Perlman R and Levitzki A (1988) Transmembrane signalling in Saccharomyces cerevisiae. Cellul. Signal., in press.

Feder D, Im MJ, Klein HW, Hekman M, Dees C, Levitzki A, Helmreich EJM and Pfeuffer T (1986) Reconstitution of β_1-adrenoceptor dependent adenylate cyclase from purified components. EMBO J. 5:1509-1514.

Gilman AG (1984) G-proteins and dual control of adenylate cyclase. Cell 36:577-579.

Hekman M, Feder D, Gal A, Klein HW, Pfeufer T, Helmreich EJM and Levitzki A (1984) Reconstitution of β-adrenergic receptor with components of adenylate cyclase. EMBO J. 3(13):3339-3345.

Herskowitz I and Marsh L (1987) Conservation of a Receptor/Signal Transduction System. Cell 50:995-996.

Higashijima T, Ferguson KM, Smigel MD and Gilman AG (1987) The effect of GTP and Mg^{2+} on the GTPase activity and the fluorescent properties of Go. J. Biol. Chem., 262:757-761.

Levitzki A (1988) From Epinehprine to cAMP. Science, in press.

Levitzki A (1987) Regulation of adenylate cyclase by hormones and G-proteins. FEBS Lett. 211:113-118.

Levitzki, A (1984) Receptor to effector coupling in the receptor dependent adenylate cyclase system. J. Receptor Res. 4:399-409.

Lowy DR and Willumsen BM (1986) The ras gene family. Cancer Surveys 5:275-289.

Marbach I, Shiloach J and Levitzki, A. (1988) Gi affects the agonist-binding of the β-adrenoceptors in the presence of Gs. Eur. J. Biochem., 172:239-246.

Pedersen SE and Ross EM (1982) Functional reconstitution of the β-adrenergic receptors and the stimulatory GTP binding protein of adenylate cyclase. Proc. Natl. Acad. Sci. USA 79:7228-7232.

Sigal IS, Gibbs JB, D'Alonzo JS and Scolnick EM (1986) Mutant ras-encoded proteins with altered nucleotide binding exert dominant biological effects. Proc. Nat. Acad. Sci. 83: 4725-4729.

Sternweis PC (1986) The purified α subunits of Go and Gi from bovine brain require β for association with phospholipid vesicles. J. Biol. Chem. 261:631-637.

Strader CD, Sigal, IS, Register, RB, Candelore MR, Rands E and Dixon RAF (1987) Identification of residues required for ligand binding to the β-adrenergic receptor. Proc. Natl. Acad. Sci. USA 84:4384-4388.

Toda T, Uno I, Ishikawa T, Powers S, Kataoka T, Broek D, Cameron S, Broach J, Matsumoto K and Wigler M (1985) Yeast RAS proteins are controlling elements of adenylate cyclase. Cell 40:27-36.

Tolkovsky AM, Braun S and Levitzki A (1982) Kinetics of interaction between the β-receptor, the GTP regulatory protein and the catalytic unit of adenylate cyclase. Proc. Natl. Acad. Sci. USA 79:213-217.

Tolkovsky AM and Levitzki A (1978) Mode of coupling of the β-adrenergic receptor and adenylate cyclase in turkey erythrocytes. Biochemistry 51:3795-3810.

Acknowledgements

This work from the author's laboratory was supported by grants from the Israel Academy of Sciences, the U.S.-Israel Binational Research Foundation (BSF) Jerusalem and by NIH grant GM 37710.

PROTEIN KINASES, PROTEIN PHOSPHATASES AND
THE REGULATION OF GLYCOGEN METABOLISM

T.G. Sotiroudis and A.E. Evangelopoulos

The National Hellenic Research Foundation,
48 Vassileos Constantinou Avenue,
Athens   116 35,
Greece

Although it has been known for almost a hundred years that proteins contain covalently bound phosphorous, the importance of protein phosphorylation has only been realized since the discovery of enzyme regulation by this type of post-translational modification. The first enzyme found to be regulated by a phosphorylation-dephosphorylation mechanism was glycogen phosphorylase, an enzyme that had been known to exist in two intercovertible forms, phosphorylase b and a (Krebs, 1986). Phosphorylase b, the dephosphorylated form whose activity was dependent on the allosteric activator 5'-AMP could be converted to a phosphorylated a-form, largely active in the absence of 5'-AMP, through the action of a protein kinase. Phosphorylase kinase (PhK), the kinase involved in the above process was the second enzyme proved to be controlled by reversible phosphorylation, while a few years later it was determined that glycogen synthase, another key enzyme of the glycogen metabolism system also exists in intercovertible phosphorylated and non-phosphorylated forms (Krebs, 1985). Nevertheless, it was only after the discovery of cAMP-dependent protein kinase (cAMPdPK), also as a result of studies on the hormonal control of glycogen metabolism, that it was realized that the phosphorylation/dephosphorylation of cellular proteins constitutes a major process utilized in the control of diverse cellular activities such as the metabolism of lipids and carbohydrates, contractility, secretion, protein synthesis, cell growth, differentiation and communication (Krebs, 1985 & Cohen, 1982).

Protein phosphorylation systems involve a minimum of three proteins and two reactions:

NATO ASI Series, Vol. H29
Receptors, Membrane Transport and Signal Transduction
Edited by A. E. Evangelopoulos et al.
© Springer-Verlag Berlin Heidelberg 1989

$$\text{Protein} \quad + \quad n\text{NTP} \longrightarrow \text{Protein-Pn} + n\text{NDP} \quad (1)$$

$$\text{Protein-Pn} \quad + \quad n\text{H}_2\text{O} \longrightarrow \text{Protein} \quad + \quad n\text{Pi} \quad (2)$$

Reaction (1) is catalyzed by protein kinase(s) and reaction (2) by phosphoprotein phosphatase(s). In general NTP is ATP but several protein kinases are today known in which GTP is almost as effective as ATP (Krebs & Beavo, 1979). Most of the protein kinases catalyze the phosphorylation of serine and threonine residues while tyrosine-specific protein kinases represent another important group of protein kinases because their kinase activity has appeared so far to be intrinsic for the transforming proteins of certain retrovival oncogenes and the membrane receptors for certain cellular growth factors (Hunter & Cooper, 1985). The total number of protein kinases encoded by a mamalian genome is unexpectedly high (Hunter, 1987) but the protein phosphatases appear to be smaller in total number and they probably exhibit broader specificities (Krebs, 1986). In addition, most protein kinases are subject to control through their interaction with "second messengers" generated within cells in response to hormones, neurotransmitters and other extracellular signals. In contrast, most of the protein phosphatases, except for a calmodulin-dependent protein phosphatase appear to act independently of such modulators (Ballou & Fischer, 1986).

Studies on the control of glycogen metabolism have been of unique importance to our understanding of protein phosphorylation in enzyme regulation. It is today well established (in vitro or in vivo) that a number of enzymes and regulatory proteins involved in the control of this metabolic pathway can be posttranslationally modified by several phosphorylation processes (Table 1), involving a number of protein kinases and phosphatases and thus this system can be used as a model for the comparison with other cellular systems regulated by the same type of covalent modification.

In this overview article we summarize currently available information, on the role of protein kinases and protein phosphatases on the control of glycogen metabolism.

Table 1. Proteins of glycogen metabolism whose action
is regulated by phosphorylation

Glycogen phosphorylase

Glycogen synthase

Inhibitor -1

Inhibitor -2 (regulatory subunit of type-1 protein phosphatase

G-component (subunit of type-1 protein phosphatase associated
          with glycogen) / deinhibitor protein

DARPP-32

Type 1 protein phosphatase (catalytic subunit)

Protein serine/threonine kinases

a. cAMP-dependent protein kinase

Adrenergic stimulation of glycogenolysis results from an increase
in the intracellular concentration of cAMP which transmits the hormonal
signal by activation of cAMPdPK (Cohen, 1983). The activation reaction is
indicated by the equation:

$$R_2C_2 \text{ (inactive)} + 4cAMP \longrightarrow R_2(cAMP)_4 + 2 C \text{ (active)}.$$

The inactive tetramer of cAMPdPK is composed of two types of dissimilar
subunits: the regulatory (R) and the catalytic (C) subunit. Upon binding
of cAMP to the R subunits their affinity for the C subunits becomes lower
leading to the dissociation from the holoenzyme of two free C subunits
expressing phosphotransferase activity (Beebe & Corbin, 1986). The
cAMPdPK is represented by two different major types of isozymes which have
identical C subunits but are distinguished by containing either R_I or R_{II}
subunits which differ in several of their properties (Beebe and Corbin,
1986). This protein kinase appears to stimulate glycogenolysis by
phosphorylating four proteins, namely PhK, glycogen synthase, inhibitor 1
and the G subunit of type-1 protein phosphatase (Cohen, 1983; Stralfors et
al, 1985).

Phosphorylation of PhK by cAMPdPK is accompanied by the
modification of one major serine residue on the α subunit and one on the β
subunit (Cohen, 1983). Phosphorylation of both the α and β subunits
modulates PhK activity (Ramachandran et al, 1987). Phosphorylation of the

β subunit correlates with increase in enzyme activity, although both serines become phosphorylated in vivo in response to adrenalin (Cohen, 1983). In contrast to skeletal and liver type isosymes, PhK from chicken gizzard smooth muscle cannot be activated by phosphorylation with cAMPdPK or by autophosphorylation (Nikolaropoulos & Sotiroudis, 1985), a property also shared by dogfish PhK, suggesting that in some cases the hormonal control of glycogenolysis may not be exercised by a phosphorylation-induced activation of PhK.

Glycogen synthase (muscle isoenzyme) an excellent substrate for cAMPdPK, is phosphorylated mainly at sites -1a, -1b and -2 (seryl residues). The initial rate of phosphorylation of site -1a is 7- to 10-fold faster than site -2 and 15-20-fold faster than site-1b (Cohen, 1986). In this respect the studies demonstrated that site-2 and site-1a are both inactivating sites but phosphorylation of site-1b appears to have little or no effect on the activity (Cohen, 1986). Inhibitor-1, a thermostable protein inhibitor of protein phosphatase-1, can express its activity only after phosphorylation on a threonine residue by cAMPdPK (Cohen, 1982). In addition, DARPP-32 (dopamine- and cAMP-regulated phosphoprotein, 32 kDa) which was found only in nervous tissue, is an effective substrate for CAMPdPK and many of its physicochemical properties resemble those of phosphatase inhibitor -1 (Hemmings et al, 1986).

The G subunit of protein phosphatase -1 (protein phosphatase 1G) can be phosphorylated by cAMPdPK on a serine residue(s). Phosphorylation is rapid and stoichiometic and increases the rate of inactivation of protein phosphatase-1 by inhibitor-1 (Stralfors et al, 1985). cAMPdPK phosphorylates also inhibitor-2 but without affecting its activity (Ballou & Fischer, 1986).

Recently, it was repoted that protein phosphatase -1 (1:1 complex between the catalytic subunit and inhibitor -2) is potently inhibited by the regulatory subunit of type II cAMPdPK suggesting a new type of action of cAMP in glycogen metabolism (Jurgensen et al, 1985).

b. cGMP-dependent protein kinase

cGMP-dependent protein kinase (cGMPdPK) consists of a dimer of identical subunits, each of which contains a cGMP-binding domain and a catalytic domain. The activation mechanism is indicated by the equation:
$$E_2 \text{ (inactive)} + 4cGMP \longrightarrow E_2 \cdot cGMP_4 \text{ (active)} \quad \text{(Beebe \& Corbin, 1986).}$$

The cGMPdPK catalyzes the phosphorylation of both α and β subunits of PhK but in contrast to the cAMP-dependent enzyme, α-subunit phosphorylation is faster than that of the β subunit. It is not known if the phosphorylation sites are identical.

Glycogen synthase (skeletal muscle) is also a target for cGMPdPK; sites-1a, -1b and -2 are labeled as in the case of cAMPdPK (Cohen, 1986) but in this case the rate of phosphorylation is about 100-fold slower than that of cAMPdPK. The cGMPdPK phosphorylates in addition inhibitor-1, probably at the residue phosphorylated by the cAMP-dependent enzyme (Ballou & Fischer, 1986).

c. Phosphorylase Kinase

PhK plays a central role in the regulation of glycogen metabolism and catalyzes the phosphorylation of phosphorylase b and glycogen synthase leading to the activation of glycogenolysis and the intensive degradation of glycogen (Pickett-Giese and Walsh, 1986). It is a multisubunit, Ca^{2+}/calmodulin dependent enzyme composed of four different subunits with a stoichiometry $(αβγδ)_4$. The γ subunit contains a catalytic site and belongs to the family of protein kinases while the smallest δ subunit is calmodulin and serves as activator (in its Ca^{2+} saturated form) or as inhibitor (in its Ca^{2+} free form) (Pickett-Giese & Walsh, 1986). The major reaction thought to be catalyzed by PhK in vivo is the phosphorylation and activation of phosphorylase b. In this reaction the dimer of phosphorylase b is phosphorylated at each of two identical serine residues (ser-14). On the other hand the phosphorylation of glycogen synthase (muscle) occurs at site-2 and the rate of phosphorylation is comparable to that of glycogen phosphorylase (Cohen, 1986). Control of glycogen metabolism through a dual action of PhK on glycogen phosphorylase and glycogen synthase is an attractive idea. Cohen (1986), suggested that PhK may be responsible for the phosphorylation of glycogen synthase (site-2) in resting muscle in the presence of epinephrine, although studies with liver glycogen synthase in response to α-adrenergic agonists seem to raise doupt about its in vivo phosphorylation by PhK (Chan and Graves, 1984).

Recently, experimental evidence suggested that ADP-ribosylation could be an important regulatory mechanism for the control of glycogen metabolism in skeletal muscle (Tsuchiya et al, 1985; Soman and Graves, 1988). ADP-ribosylation of PhK suppressed both the autophosphorylation and the phosphorylation induced by cAMPdPK. This reduction of phosphorylation resulted in a decrease of the phosphorylation-dependent activation of PhK (Tsuchiya et al, 1985).

d. Multifunctional calmodulin-dependent protein kinases

The isolation and characterization of Ca^{2+}/calmodulin dependent protein kinases from a variety of tissues revealed the presence of an enzyme or class of enzymes distinct from the narrow substrate specificity kinases, PhK and myosin light chain kinase. This enzyme has been variably referred to as the multifunctional Ca^{2+}/calmodulin protein kinase or as Ca^{2+}/calmodulin dependent protein kinase II (Kin II) (Stull et al, 1986; Edelman et al, 1987). Kin II, a high Mr multisubunit enzyme consits of a 49-55 kDa subunit (α) and often a 60-58 kDa (β/β') doublet. It has a characteristic broad substrate specificity in vitro and is capable of phosphorylating a number of proteins including glycogen synthase at site-2 and site-1b (muscle). In the later case, the initial rate of phosphorylation of site-2 is 5-to 10-fold faster than site-1b (Cohen, 1986). Phosphorylase b and PhK are very poor substrates for kin II (Stull et al, 1986).

e. Protein Kinase C

Protein kinase C (PKC) is a multifunctional protein kinase identified by Nishizuka and co-workers as a Ca^{2+}-and phospholipid-dependent protein kinase that plays a crucial role in the signal transduction for a variety of biologically active substances involved in cellular function and proliferation. Although once considered as a single entity, enzymological and molecular cloning analysis has revealed that PKC exists as a family of multiple subspecies with subtle individual characteristics (Nishizuka, 1988). In the presence of limiting amounts of Ca^{2+} and phospholipids its activity is stimulated by sn-1,2-diacylglycerols or by phorbolesters (the enzyme has been shown to be the intracellular receptor for the tumor promoting phorbol esters) and the kinase phosphorylates a broad range of cellular proteins (Nishizuka, 1986). Glycogen synthase (muscle) can be phosphorylated in vitro by PKC to 1 mol/subunit and the major tryptic peptides that become labeled are N5-

Table 2. Independent protein kinases involved in the phosphorylation of proteins of glycogen metabolism

Names of kinases	Protein substrates in glycogen metabolism	Total Mr ($\times 10^3$) (subunit composition)	References
Glycogen synthase kinase-3, Factor F_A	Glycogen synthase:sites-3; Inhibitor-2; R_{II}(Ser 44,47); Autophosphorylation	51 (monomeric)	a-c
Glycogen synthase kinase-4, $PC_{0.4}$	Glycogen synthase:site-2	115 (unknown)	a
Glycogen synthase kinase-5, $PC_{0.7}$, casein kinase-II, Casein kinase-G, Casein kinase-TS	Glycogen synthase:site-5; R_{II}(Ser 74,76); Inhibitor-2; Autophosphorylation	130-180 ($\alpha_2\beta_2$) 43(α) 25(β)	a,d,e
Casein kinase -I	Glycogen synthase:sites-3,site-5; Phosphorylase kinase (α,β subunits); Inhibitor-2; Autophosphorylation; Calcineurin	35 (monomeric)	a, e-h
Glycogen synthase kinase-M	Glycogen synthase:site-2,sites-3; Inhibitor-2	62	i

(a) Cohen, 1986; (b) Aitken et al, 1984; (c) Hemmings et al, 1981; (d) Ballou & Fischer, 1986; (e) Ahmad et al, 1984; (f) Singh et al, 1982; (g) Singh & Wang, 1987; (h) Agostinis et al, 1987; (i) Hegazy et al, 1987

N38(containing site-2) and C85-C97 (containing site-1a) (Cohen, 1986). Although for the muscle enzyme phosphorylation was shown to be associated with inactivation, PKC phosphorylated liver glycogen synthase but was ineffective in causing inactivation (Nakabayashi et al, 1987). It has been suggested that the phorbol ester-induced inactivation of glycogen synthase observed in hepatocytes cannot be accounted for entirely by the activation of PKC (Nakabayashi et al, 1987; Bouscarel et al, 1988). In this case the effect of PKC could be indirect, i.e. through activation of another glycogen synthase kinase or inhibition of a protein phosphatase.

f. Protein kinases independent of regulation

According to the classification of protein kinases proposed by Krebs (Krebs, 1986), the so-called "independent" protein kinases represent a group of kinases that are not definitely known to be regulated through the interaction of a messenger(s) directly with the kinase. It is clear that glycogen synthase can be phosphorylated in vitro by at least 10 protein kinases, a large number of which belong to the group of "independent" protein kinases (Cohen, 1986). Though not all of these protein kinases may act physiologically, there is strong evidence that multiple phosphorylation of glycogen synthase does occur in muscle cells. Some important characteristics of "independent" protein kinases in relation to the regulation of glycogen metabolism are summarized in Table 2.

Protein tyrosine kinases

The protein tyrosine kinases have arbitrarily been divided into two groups: the hormone or growth factor-dependent enzymes and the independent tyrosine kinases for which no regulatory agents are known. In the first set of tyrosine kinases belong membrane receptors with tyrosine kinase activity such as EGF-, PDGF-, insulin-, IGF-, acidic FGF- and macrophage/CSF-I- receptors. The latter group of kinases comprises several oncogene products of RNA tumor viruses, their closely related cellular counterparts and some other tyrosine phosphorylating activities (Krebs, 1986; Feige and Chambaz, 1987).

Insulin increases the glycogen content of a wide variety of cells and it is well established that this hormone regulates glycogen synthase activity through a glucose-dependent and a glucose-independent pathway (Chan and Krebs, 1986). The glucose-independent pathway could act through a "mixed" phosphorylation/dephosphorylation cascade initiated by a tyrosine phosphorylation reaction catalyzed by the insulin receptor, but Cohen et al (1985) reported negative results using purified insulin receptors to phosphorylate glycogen synthase kinase-3 and the catalytic subunit of protein phosphatase-1 (Parker et al, 1983). In this respect it has been reported that the acute activation of glycogen synthase by insulin is associated with a decreased phosphorylation state of site-3. This result implicates an inhibition of glycogen synthase kinase-3 or an activation of phosphatases acting on site-3. Inhibitor-2 can be phosphorylated by the insulin receptor but does not exhibit changes in activity (Cohen et al, 1985), while recently Laurino et al (1988), presented experimental evidence on the in vitro phosphorylation of calmodulin by the insulin receptor tyrosine kinase. This later phosphorylation potentially represents an important post-translational modification altering calmodulin's ability to regulate a variety of enzymes.

Recent results have shown that the growth factors EGF, PDGF, and IGF-I, just as insulin, stimulate glycogen synthase activity (Chan and Krebs, 1986; Chan et al, 1987) and thus many laboratories are now looking for the "missing link" between tyrosine phosphorylation and serine/threonine phosphorylation, that is for a serine/threonine kinase or phosphatase which is a physiological substrate for the receptor kinase.

Concerning the retrovival oncogene products pocessing protein tyrosine kinase activity that have been linked to the control of cell growth in normal and transformed cells, there is a very interesting report by Johansen and Ingebritsen (1986) showing that the 37 kDa catalytic subunit of protein phosphatase-1 is an in vitro substrate for $pp60^{v-src}$ and that this phosphorylation results in a significant decrease in its activity. As a consequence, the phosphorylation of a set of cellular proteins on serine and threonine residues is likely to be increased through inhibition of their dephosphorylation. It is also important to cite a recent work by Mahrenholz et al (1988) revealing the phosphorylation of glycogen synthase by a bovine thymus protein tyrosine kinase, p40. This phosphorylation is specific for a single tyrosine residue and the inactivation of glycogen synthase is comparable to that observed for modification by cAMPdPK.

Protein phosphatases

Although numerous studies have focused on the regulation of kinase activities as a major factor in controlling glycogen metabolism, however, the phosphorylation level of the phosphoproteins involved in the regulation of this metabolic pathway is established in concert with the counterpoised dephosphorylation reaction carried out by a number of protein phosphatases.

According to the classification scheme proposed by Ingebritsen and Cohen(1983) the protein phosphatases(serine/threonine) can be grouped into two classes. The type 1 protein phosphatase selectively dephosphorylates the β subunit of PhK and is inhibited by two thermostable regulatory proteins, termed inhibitor-1 and inhibitor-2. Type 2 protein phosphatases (there are three type 2 enzymes: protein phosphatases 2A, 2B and 2C) selectively dephosphorylated the α subunit of PhK and are insensitive to inhibitor -2. Merlevede (1985), proposed a different classification scheme according to which four different types of protein phosphatases can be distinguished on the basis of their regulatory properties: ATP/Mg^{2+}-dependent (type 1), polycation-stimulated (type 2A), calcineurin (type 2B) and Mg^{2+}-dependent (type 2C).

A considerable number of phosphotyrosyl protein phosphatases from various tissues and cells have also been reported (Ballou & Fischer, 1986) but the preparations so far reported are not pure enough to clarify their amino acid specificity. In addition, no thorough investigation of the purified serine/threonine protein phosphatases has been performed in terms of their capacity for dephosphorylating the phosphotyrosine residues of proteins except for calcineurin which was found to possess a potent phosphotyrosyl protein phosphatase activity (Ballou & Fischer, 1986). Furthermore, a number of alkaline or acid phosphatases(orthophosphate-monoester phosphohydrolases with alkaline or acid pH optimum) are known to function also as protein phosphatases(serine or tyrosine protein phosphatases) (Ballou & Fischer, 1986), but they are not classified in Cohen's two type (Ingebritsen & Cohen, 1983) or Merlevede's four-class systems (Merlevede, 1985).

Protein phosphatase 1 (ATP, Mg^{2+}-dependent protein phosphatase)

This is the major phosphatase acting on the phosphorylation sites that alter the kinetic properties of the key enzymes of glycogen metabolism in skeletal muscle: serine-14 of glycogen phosphorylase, the β-subunit of PhK and sites 1a, 2 and (3a+3b+3c) of glycogen synthase (Cohen, 1983). Its enzymic activity is regulated by many extracellular stimuli like adrenaline and insulin and possibly also by neuronal stimulation (Ballou & Fischer, 1986). The 37 kDa catalytic (C) subunit exists in several stable inactive or active conformations and interacts with various regulatory subunits. Thus, protein phosphatase-1$_I$ can be purified as inactive complex composed of two proteins: an inactive catalytic subunit and inhibitor-2 (also called modulator). Activation involves phosphorylation of inhibitor-2 on a threonine residue by kinase F_A (glycogen synthase kinase 3) (Ballou & Fischer, 1986). In this activation process the autocatalytic dephosphorylation of the modulator subunit is an essential step largely dependent on Mg^{2+} (Vandenheede et al, 1987). The glycogen-bound form of protein phosphatase-1 (protein phosphatase-1$_G$) is a heterodimer composed of the catalytic subunit complexed to a 103-kDa G-subunit that anchors the enzyme to glycogen. The G-subunit is phosphorylated by cAMPdPK and the phosphorylation promotes the release of the phosphatase activity from glycogen and increases the rate at which protein phosphatase 1$_G$ is inactivated by inhibitor -1. The GC complex is less sensitive to inhibitor-1 and inhibitor-2, than the isolated C-subunit (Alemany et al, 1986). It has been proposed (Alemany et al, 1986) that Merlevede's deinhibitor protein (Merlevede et al, 1984) may be a proteolytic fragment of the G-subunit that retains its ability to interact with the C-subunit.

Recently two other forms of protein phosphatase-1 have been identified in skeletal muscle. A form tightly bound to myosin (termend protein phosphatase 1$_M$ (Chisholm & Cohen, 1988) and a membrane bound form obtained from microsome-contaminated glycogen particles (Villa-Moruzzi & Heilmeyer, 1987). The presence of the same catalytic subunit at various cell locations suggests a new mechanism for protein phosphatase-1 regulation.

Protein phosphatase 2A (Polycation stimulated phosphatase)

Protein phosphatases 2A can be recovered from tissue extracts in three subtypes termed 2A$_O$, 2A$_1$ and 2A$_2$ in order of their elution from DEAE-cellulose. These phosphatases are multisubunit enzymes consisting of a 36 KDa catalytic subunit bound to various other components. The molecular mass of the subunits was estimated by SDS-PAGE as 60,54 and 36 KDa for $2A_O (AB'C_2)$; 60,55 and 36 KDa for $2A_1 (ABC_2)$ and 60 and 36 KDa for $2A_2 (AC)$ (Ballou & Fischer 1986). In contrast to the type 1 phosphatase whose activity towards phosphorylase a is inhibited by micromolar concentration of basic proteins or polyamines, the 2A phosphatases are stimulated 5- to 10-fold and this effect of polycations seems to be primarily phosphatase-directed (Ballou & Fischer, 1986). Waelkens et al (1987) have recently purified four types of polycation-stimulated (PCS) phosphorylase phosphatases from rabbit skeletal muscle. They were called $PCS_H (390$ KDa), $PCS_M (250$ KDa) and PCS_L (200 KDa). The PCS_H could be resolved into a 3-subunit (PCS_{H1}) and a 2-subunit (PCS_{H2}) enzyme which both display deinhibitor phosphatase activity. PCS_{H1} and PCS_L phosphatases are probably identical to $2A_1$ and $2A_2$ respectively. Protein phosphatase 2A is the major PhK (α subunit) phosphatase in resting muscle (in the absence of Ca^{2+}), it accounts for an appreciable percentage of the glycogen synthase phosphatase and for a small proportion of the phosphorylase phosphatase, while it efficiently dephosphorylates inhibitor-1 and -2 (Cohen, 1983; Ballou & Fischer, 1986). Determination of the primary structure of the catalytic subunit of phosphatase 2A by molecular cloning has revealed the existence of two isotypes α and β with a 98% degree of homology between their amino acid sequences (Stone et al, 1987; DaCruz e Silva & Cohen, 1987) while type-1 protein phosphatase catalytic subunit was strikingly homologous to protein phosphatase 2A (Berndt et al, 1987).

Protein phosphatase 2B (calcineurin)

Calcineurin is a calmodulin-dependent protein phosphatase originally purified as a major Ca^{2+} and calmodulin binding protein from bovine brain. The enzyme is found to consist of two subunits having different molecular masses: subunit A (61 KDa), interacts with calmodulin and contains the catalytic site while the subunit B (192 KDa) binds up to four Ca^{2+} with high affinity (Ballou & Fischer, 1986). Although physiological substrates for calcineurin have not been established the enzyme is considered to have a very narrow substrate specificity compared to other protein phosphatases. Under optimal conditions it is the major PhK (α subunit), phosphatase and inhibitor-1 phosphatase in skeletal muscle, but it does not dephosphorylate glycogen phosphorylase, glycogen synthase or the β subunit of PhK (Cohen, 1983; Ballou & Fischer, 1986).

Protein phosphatase 2C (Mg^{2+}-dependent protein phosphatase)

Protein phosphatase 2C like protein phosphatases 1 and 2A has a broad substrate specificity, but although it dephosphorylates many proteins that are substrates for protein phosphatases 1 and 2A it is distinguished by its requerement for Mg^{2+} and by its very low phosphorylase phosphatase activity. Good substrates include the α-subunit of PhK and glycogen synthase (it dephosphorylates site 2 more rapidly than site 1a or sites (3a+3b+3c) (Cohen, 1982; Ballou & Fischer, 1986). McGowan and Cohen (1987) have recently demonstrated the existence of two isoenzymes of protein phosphatase 2C in rabbit skeletal muscle and liver termed protein phosphatase $2C_1$ and $2C_2$ with molecular masses 44 and 42 kDa respectively.

Insulin mediators and the control of glycogen metabolism

Our present knowledge of the mechanism of insulin action is still limited (Rosen, 1987), thus the molecular events linking the occupation of the insulin receptor to the effects of this hormone on glycogen metabolism remain obscure. Nevertheless, much has been learned about the early actions of insulin during the last years and there is direct evidence that the protein kinase activity of the insulin receptor is essential for six post-receptor effects, including glycogen synthesis (Rosen, 1987). It is thought that postbinding effects of insulin may involve the regulation of certain multifunctional protein kinase(s) and/or phosphatase(s) which in term control activation/inactivation reactions of key proteins of glycogen metabolism. In this respect there is now evidence that a group of low Mr compounds generated in response to insulin are able to act as "second messengers" or "mediators" by modulating the activities of various intracellular enzymes. Recent studies indicate that one or more of these mediators are inositol phosphate-glycans generated by hydrolysis of glycanphosphoinositide(s) in the plasma membrane (Mato et al 1987; Armstrong & Newman 1987). Concerning glycogen metabolism these substances were reported to activate glycogen synthase phosphatase (Armstrong & Newman, 1987) to inhibit cAMPdPK (Villalba et al, 1988) and to antagonize glucagon-dependent activation of glycogen phosphorylase in isolated rat hepatocytes (Alvarez et al, 1987), supporting the hypothesis that this class of compounds mediates at least some insulin actions on glycogen metabolism.

Concluding remarks

The study of phosphorylation of proteins involved in the control of glycogen metabolism continues to give us valuable informations concerning the mechanism of action of a large number of protein kinases and phosphatases. The evidence summarized in this article suggests that almost all the types of protein kinases and phosphatases are able, at least in vitro, to affect the phosphorylation state of a wide range of enzymes and regulatory proteins integrated in this intracellular process. In this respect further investigation of the phosphorylation-dephosphorylation mechanisms of glycogen metabolism may help to explain the molecular basis for the mode of action of several hormones and growth factors by understanding the relationship between the expression of receptor tyrosin kinase activity and serine/threonine phosphorylation of physiological substrates.

References

Agostinis,P. , Vandenheede,J.R. ,Goris,J. , Meggio,F. , Pinna,L. and Merlevede,W. (1987) The ATP,Mg-dependent protein phosphatase:regulation by casein kinase-1. FEBS Lett.224, 385-390.

Ahmad,Z. , Camici,M. , DePaoli-Roach,A.A. and Roach,P.J. (1984) Glycogen synthase kinases. Classification of a rabbit liver casein and glycogen synthase kinase(casein kinase-1) as a distinct enzyme. J.Biol.Chem.259, 3420-3428.

Aitken,A. , Holmes,C.F.B. , Campbell, D.G. , Resink,T.J. , Cohen,P. , Leung,C.T.W. and Williames,D.H. (1984) Amino acid sequence at the site on protein phosphatase inhibitor-2 phosphorylated by glycogen synthase kinase-3. Biochim.Biophys.Acta,790, 288-291.

Alemany,S. , Pelech,S. , Brierley,C.H. and Cohen,P. (1986). The protein phosphatases involved in cellular regulation. Evidence that dephosphorylation of glycogen phosphorylase and glycogen synthase in the glycogen and microsomal fractions of rat liver are catalysed by the same enzyme: protein phosphatase-1. Eur.J.Biochem.156, 101-110.

Alvarez,J.F. , Cabello, M.A. , Feliu,J.E. and Mato,J.M. (1987) A phospho-oligosaccharide mimics insulin action on glycogen phosphorylase and pyruvate kinase activities in isolated rat hepatocytes. Biochem. Biophys.Res.Commun. 147, 765-771.

Armstrong,J.McD. and Newman,J.D. (1987) Insulin mediators and protein phosphatase activity. Adv.Prot.Phosphatases 4, 329-350.

Ballou. L.M. and Fischer, E.H. Phosphorprotein phosphatases, In: The Enzymes (Boyer, P.D. and Krebs,E.G. eds) Vol.17, 311-361,Academic Press, New York.

Beebe, S.J. and Corbin, J.D. Cyclic nucleotide-dependent protein kinases, In:The Enzymes (Boyer,P.D. and Krebs,E.G. ,eds)Vol.17,43-111, Academic Press, New York.

Berndt,N. , Campbell,D.G. , Caudwell,F.B. , Cohen,P. , Da Cruz e Silva,E.F. , Da Cruz e Silva,O.B. and Cohen,P.T.W. (1987) Isolation and sequence analysis of a cDNA clone encoding a type-1 protein phosphatase catalytic subunit: homology with protein phosphatase 2A. FEBS Lett.223, 340-346.

Bouscarel,B. , Meurer,K. , Decker,C. and Exton, J.H. (1988) The role of protein kinase C in the inactivation of hepatic glycogen synthase by calcium-mobilizing agonists. Biochem.J.251, 47-53.

Chan,K.-F.J. and Graves,D.J. (1984) Molecular properties of phosphorylase kinase. In: Calcium and Cell Function (Cheung,W.Y.ed)vol.5, 1-31 Academic Press, New York.

Chan,C.P. and Krebs,E.G. (1986) Effects of growth factors on carbohydrate metabolism in cultured mammalian cells. In: Mechanisms of insulin action (Belfrage,P.et al, eds) 13-31 Elsevier/North-Holland, Amsterdam.

Chan,C.P. , Bowen-Pope,D.F. , Ross,R. and Krebs,E.G. (1987) Regulation of glycogen synthase activity by growth factors. Relationship between synthase activation and receptor occupancy. J.Biol.Chem.262, 276-281.

Chisholm,A.A.K. and Cohen,P. (1988) Identification of a third form of protein phosphatase 1 in rabbit skeletal muscle that is associated with myosin. Biochim.Biophys.Acta 968, 392-400.

Cohen, P. (1982) The role of protein phosphorylation in neural and hormonal control of cellular activity. Nature, 296, 613-620.

Cohen, P. (1983) Protein phosphorylation and the control of glycogen metabolism in skeletal muscle. Phil.Trans.R.Soc.Lond. B302. 13-25.

Cohen,P. , Parker,P.J. and Woodgett,J.R. (1985) The molecular mechanism by which insulin activates glycogen synthase in mammalian skeletal muscle. In: Molecular basis of insulin action (Czech.M.ed.) 213-233,Plenum Press, New York.

Cohen,P. Muscle glycogen synthase, In: The Enzymes (Boyer,P.D. and
    Krebs,E.G.eds) vol.17,461-497, Academic Press, New York.
DaCruz e Silva,O.B. and Cohen,P.T.W. (1987) A second catalytic subunit of
    type-2A protein phosphatase from rabbit skeletal muscle. FEBS Lett.226,
    176-178.
Edelman,A.M. , Blumenthal,D.K. and Krebs,E.G. (1987) Protein serine/threonine
    kinases. Ann.Rev.Biochem.56, 567-613.
Feige,J.-J. and Chambaz,E.M. (1987) Membrane receptors with protein-tyrosine
    kinase activity. Biochimie,69, 379-385.
Hegazy,M.G. , Thysseril,T.J. , Schlender,K.K. and Reimann, E.M. (1987)
    Characterization of GSK-M, a glycogen synthase kinase from rat skeletal
    muscle. Arch.Biochem.Biophys.258, 470-481.
Hemmings,B.A. , Yellowlees,D. , Kernohan,J.C. and Cohen,P. (1981) Purification
    of glycogen synthase kinase-3 from rabbit skeletal muscle. Copurifica-
    tion with the activating factor(F$_A$) of the (Mg-ATP)dependent Protein
    phosphatase. Eur.J.Biochem.119, 443-451.
Hemmings,H.C. , Nairn,A.C. and Greengard,P. (1986) Protein kinases and
    phosphoproteins in the nervous system. In: Neuropeptides in neurologic
    and phychiatric disease(Martin,J.B. and Barchas,J.D.eds) 47-69, Raven
    Press, New York.
Hunter, T. and Cooper, J.A. (1985) Protein-tyrosine kinases.
    Ann.Rev.Biochem.54, 897-930.
Hunter, T. (1987) A thousand and one protein kinases. Cell,50, 823-829.
Ingebritsen,T.S. and Cohen,P. (1983) Protein phosphatases: Properties and
    role in cellular regulation. Science, 221, 331-338.
Johansen,J.W. and Ingebritsen,T.S. (1986) Phosphorylation and inactivation
    of protein phosphatase-1 by pp60^{v-src}. Proc.Natl.Acad.Sci.USA, 83,
    207-211.
Jurgensen,S.R. , Chock, P.B. , Taylor,S. , Vandenheede,J.R. and Merlevede,W.
    (1985) Inhibition of the Mg(II) ATP-dependent phosphoprotein phos-
    phatase by the regulatory subunit of cAMP-dependent protein kinase.
    Proc.Natl.Acad.Sci.USA 82, 7565-7569.
Krebs, E.G. and Beavo, J.A. (1979) Phosphorylation-dephosphorylation of
    enzymes. Ann.Rev.Biochem.48, 923-959.
Krebs, E.G. (1985) The phosphorylation of proteins: a major mechanism for
    biological regulation. Biochem.Soc.Trans.13, 813-820.
Krebs, E.G. (1986) The enzymology of control by phosphorylation. In: The
    Enzymes (Boyer, P.D. and Krebs, E.G. eds)vol. 17, pp.3-20, Academic
    Press, New York.
Laurino,J.P. , Colca,J.R. , Pearson,J.D. , DeWald,D.B. and McDonald,J.M.
    (1988) The in vitro phosphorylation of calmodulin by the insulin
    receptor tyrosine kinase. Arch.Biochem.Biophys. 265, 8-21.
Mato,J.M. , Kelly,K.L. , Abler,A. , Jarett,L. , Corkey,B.E. , Cashel,J.A. and
    Zopf,D. (1987) Partial structure of an insulin-sensitive glycophos-
    pholipid. Biochem.Biophys.Res.Commun. 146, 764-770.
Mahrenholz,A.M. , Votaw,P. , Roach,P.J. , Depaoli-Roach,A.A. , Zioncheck,T.F. ,
    Harrison,M.L. and Geahlen,R.L. (1988) Phosphorylation of glycogen
    synthase by a bovine thymus protein-tyrosine kinase, p40. Biochem.
    Biophys.Res.Commun. 155, 52-58.
McGowan,C.H. and Cohen,P. (1987) Identification of two isoenzymes of protein
    phosphatase 2C in both rabbit skeletal muscle and liver. Eur.J.Biochem.
    166, 713-722.
Merlevede,W. , Vandenheede,J.R. , Goris,J. and Yang,S.-D. (1984) Regulation of
    ATP-Mg-dependent protein phosphatase. Curr.Top.Cell.Regul.23, 177-215.
Merlevede,W. (1985) Protein phosphates and the protein phosphatases.
    Landmarks in an eventful century. Adv.Protein Phosphatases 1, 1-18.

Nakabayashi,H. , Chan,K.-F.J. and Huang, K.-P. (1987) Role of protein kinase
    C in the regulation of rat liver glycogen synthase. Arch.Biochem.
    Biophys.252, 81-90.
Nikolaropoulos,S. and Sotiroudis,T.G. (1985) Phosphorylase kinase from
    chichen gizzard. Partial purification and characterization. Eur.J.
    Biochem.151, 467-473.
Nishizuka,Y. (1986) Studies and persectives of protein kinase C. Science,233,
    305-312.
Nishizuka,Y. (1988) The heterogeneity and differential expression of
    multiple species of the protein kinase C family. Biofactors, 1, 17-20.
Parker,P.J. , Caudwell,F.B. and Cohen,P. (1983) Glycogen synthase from rabbit
    skeletal muscle; effect of insulin on the state of phosphorylation of
    the seven phosphoserine residues in vivo. Eur.J.Biochem.130, 227-234.
Pickett-Giese,C.A. and Walsh,D.A. (1986) Phosphorylase kinase: In: The
    Enzymes (Boyer,P.D. and Krebs,E.G.eds)vol.17, 395-459, Academic
    Press,New York.
Ramachandran,C. , Goris,J. , Waelkens,E. , Merlevede,W. and Walsh,D.A. (1987)
    The interrelationship between cAMP-dependent α and β subunit phospho-
    rylation in the regulation of phosphorylase kinase activity. Studies
    using specific phosphatases. J.Biol.Chem.262, 3210-3218.
Rosen,O.M. (1987) After insulin binds. Science 237, 1452-1458.
Singh,T.J. , Akatsuka,A. and Huang, K.-P. (1982) Phosphorylation and
    activation of rabbit skeletal muscle phosphorylase kinase by a cyclic
    nucleotide-and Ca^{2+}-independent protein kinase, J.Biol.Chem.257,
    13379-13384.
Singh,T.J. and Wang,J.H. (1987) Phosphorylation of calcineurin by glycogen
    synthase(casein) kinase-1. Biochem.Cell Biol.65, 917-921.
Soman,G. and Graves,D.J. (1988)Endogenous ADP-ribosylation in skeletal
    muscle membranes. Arch.Biochem.Biophys.260, 56-66.
Stinson,R.A and Chan,J.R.A. (1987) Alkaline phosphatase and its function as
    a protein phosphatase. Adv.Protein phosphatases 4, 127-151.
Stone,S.R. , Hofsteenge,J. and Hemmings,B.A. (1987) Molecular cloning of
    cDNAs encoding two isoforms of the catalytic subunit of protein
    phosphatase 2A. Biochemistry 26, 7215-7220
Stralfors,P. ,Hiraga,A. and Cohen,P. (1985) The protein phosphatases involved
    in cellular regulation. Purification and characterisation of the
    glycogen-bound form of protein phosphatase-1 from rabbit skeletal muscle.
    Eur.J.Biochem.149, 295-303.
Stull, J.T. , Nunnally,M.H. andf Michnoff,C.H. (1986) Calmodulin-dependent
    protein kinases. In: The Enzymes (Boyer,P.D. and Krebs,E.G.eds)vol.17,
    113-166, Academic Press, New York.
Tsuchiya,M. , Tanigawa,Y. , Ushiroyama,T. , Matsuura,R. and Shimoyama,M. (1985)
    ADP-ribosylation of phosphorylase kinase and block of phosphate
    incorporation into the enzyme. Eur.J.Biochem.147, 33-40.
Vandenheede,J.R. , Vanden Abeele, C. and Merlevede,W. (1987) On the dephospho-
    rylation of the ATP, Mg-dependent protein phosphatase modulator. FEBS
    Lett. 216, 291-294.
Villalba,M. , Kelly,K.L. and Mato,J.M. (1988) Inhibition of cyclic AMP-
    dependent protein kinase by the polar head group of an insulin-sensitive
    glycophospholipid. Biochim.Biophys.Acta 968, 69-76.
Villa-Moruzzi,E. and Heilmeyer,L.M.G.Jr. (1987) Phosphorylase phosphatase
    from skeletal muscle membranes. Eur.J.Biochem.169, 659-667.
Waelkens,E. , Goris,J. and Merlevede,W. (1987) Purification and properties of
    polycation-stimulated phosphorylase phosphatases from rabbit skeletal
    muscle. J.Biol.Chem.262, 1049-1059.

PHOSPHORYLASE KINASE AND PROTEIN KINASE C: FUNCTIONAL SIMILARITIES

T. G. Sotiroudis, S. M. Kyriakidis, L. G. Baltas,
T. B. Ktenas, V. G. Zevgolis and A. E. Evangelopoulos

The National Hellenic Research Foundation,
48 Vassileos Constantinou Avenue,
Athens 116 35
Greece

Introduction

Protein serine and threonine kinases can be classified into individual groups or subclasses on the basis of the type of regulation of their activities (Krebs, 1986). Two of the most intensively studied groups are Ca^{2+}-regulated, i.e. the Ca^{2+}/calmodulin (CaM)-dependent and the Ca^{2+}-phospholipid (diacylglycerol)-dependent protein kinases. Of the enzymes belonging in the category of Ca^{2+}/ CaM-dependent kinases, myosin light chain kinases (MLCK) are distinguished by their high degree of substrate specificity and CaM dependency (Edelman et al, 1987). Phosphorylase kinase (PhK) another member of the same group is characterized by a broader substrate specificity. Its primary substrate is phosphorylase b but the enzyme may catalyze the phosphorylation of other proteins (Chan & Graves, 1984). In addition, a number of Ca^{2+}/CaM-dependent multifunctional protein kinases identified in a variety of tissues shows a broad substrate specificity suggesting that such a group of CaM-dependent protein kinases may play important roles in the control of different cellular processes (Shenolikar et al, 1986). On the other hand, protein kinase C (PKC) is a multifunctional protein kinase identified by Nishizuka and co-workers as a Ca^{2+}- and phospholipid-dependent protein kinase that plays a crucial role in the signal transduction for a variety of biologically active substances involved in cellular function and proliferation (Nishizuka, 1984). In the presence of limiting amounts of Ca^{2+} and phospholipids its activity is stimulated by sn-1,2-diacylglycerols or by phorbol esters (Nishizuka, 1984) and it

NATO ASI Series, Vol. H29
Receptors, Membrane Transport and Signal Transduction
Edited by A. E. Evangelopoulos et al.
© Springer-Verlag Berlin Heidelberg 1989

phosphorylates a broad range of cellular proteins (Kikkawa and Nishizuka, 1986).

During the last years a model has evolved which predicts a mechanism for enzyme activation by CaM (Gietzen et al, 1981; Kyriakidis et al, 1986a). This model is based primarily on the basic similarity of CaM-dependent enzymes in their inhibition by hydrophobic drugs and their activation by acidic amphiphiles and limited proteolysis, three features that also characterize PKC, although this enzyme is known to be not responsive to CaM (Takai et al, 1979). On the basis of this hypothetical model it would be reasonable to assume that PKC obeys to the same general mechanism of hydrophobic activation and inhibition of CaM-dependent enzymes.

Studies in our laboratory, were focused on the understanding of the hydrophobic properties of cytoplasmic PhK in an effort to test the hypothetical model by which most types of activation and inhibition phenomena of Ca^{2+}/CaM- and Ca^{2+}/phospholipid-regulated protein kinases can be ascribed to a common general mechanism involving similar hydrophobic and ionic interactions. This type of interactions may lead under certain intracellular conditions to the transformation of the above Ca^{2+}-dependent kinases from a "soluble" to a membrane-bound form. We describe below the experimental evidence suggesting functional similarities between PhK and PKC.

Limited proteolysis

PhK from rabbit skeletal muscle, a hexadecamer composed of four dissimilar subunits $\alpha_4\beta_4\gamma_4\delta_4$, has been shown to exist in a non-activated and an activated form at pH 6.8. The non-activated kinase can be converted to the activated form by covalent modification processes including protein phosphorylation and limited proteolysis (Chan & Graves, 1984; Pickett-Giese & Walsh, 1986). In most cases, initial proteolytic activation is concomitant with a marked degradation of the α and β subunits, whereas the γ and δ subunits remain intact. Limited proteolysis eliminates the absolute requirement of PhK for Ca^{2+} and the binding by extrinsic CaM but increases sensitivity to Ca^{2+} activation via the intrinsic δ-subunit (Pickett-Giese & Walsh, 1986). PhK from chicken gizzard (Nikolaropoulos & Sotiroudis, 1985) and bovine stomach (Zevgolis et al, in prep.) smooth muscle, a type of Ca^{2+}-CaM protein kinase, lacking α and β subunits (one low Mr protein band was shown upon SDS-PAGE), has also been found to be effectively activated after limited proteolysis by trypsin. PKC can also be irreversibly activated by proteolysis and the proteolytically activated form (called PK-M by the original authors) was catalytically active in the absence of Ca^{2+} and phospholipid (Kishimoto et al, 1983).

Effect of naturally occuring lipids

Another means by which Ca^{2+} might alter the activity or subcellular distribution of a protein kinase is to facilitate the association of hydrophobic ligands with the protein. One such enzyme is PKC which depends on Ca^{2+} as well as phospholipid for its activation (Nishizuka, 1984). Diacylglycerol, which is generated by hormone-induced turnover of phosphatidylinositol in the plasma membrane dramatically increases reaction velocity and decreases the activation constant of phospholipid as well as of Ca^{2+} in such a way that the enzyme can be activated at physiological Ca^{2+} concentrations (0.1 μM) (Kikkawa & Nishizuka, 1986; Castagna et al, 1985). Among various phospholipids tested the order of potency for phospholipid in supporting enzyme activity is as follows: phosphatidylserine > phosphatidic acid > phosphatidylinositol > phosphatidylethanolamine > sphingomyelin > phosphatidylcholine (Castagna et al, 1985). Tumor-promoting phorbol esters, have also been shown to bind to and stimulate PKC, by substituting for diacylglycerol in the activation of the enzyme (Kikkawa & Nishizuka, 1986). Another type of PKC activation reported recently, involves the interaction of the kinase with cis-fatty acids, in the absence of Ca^{2+} and phospholipids (Murakami et al, 1986), while in a recent paper (Sakai et al, 1987) the authors presented evidence suggesting that purified PKC phosphorylates clupein sulfate in a phosphatidylserine -dependent but Ca^{2+}-independent process; in the later case, the effect of diolein was not significant. Moreover, PKC-dependent phosphorylation of profilin has been shown to be specifically stimulated by phosphatidylinositol biphosphate, whereas phosphatidylserine was ineffective (Hansson et al, 1988). Other hydrophobic activators of PKC which are known to substitute for phosphatidylserine include E.coli lipid X (Wightman & Raetz, 1984), the lipophilic muramyltripeptide MTP-PE (Meyer et al, 1986) and the sythetic naphthalenesulfonamide derivative SC-9 (Ito et al, 1986). It has been also observed that sphingosine (Hannun et al, 1986), lysosphingolipids (Hannun & Bell, 1987) and gangliosides (Kreutter et al, 1987), a specific type of lipids ubiquitous in eukaryotes, potently and reversibly inhibited PKC activity (in presence of phosphatidylserine and diacylglycerol) while, ganglioside GM3 (Mamoi, 1986) and cerebroside sulfate (Fujiki et al, 1986) were able to activate PKC as substitutes for phosphatidylserine. It was proposed that sphingosine inhibition of PKC may have physiological significance acting as a negative effector of the enzyme (Hannun et al, 1986) and that lysosphingolipids represent the functional missing link between the accumulation of sphingolipids and the pathogenesis of sphingolipidoses (Hannum & Bell, 1987). More interestingly, a recent communication suggests that phosphatidylinositol biphosphate may effectively antecede diacylglycerol as activator of PKC, reducing the Ca^{2+}-requirement of this kinase, and that phosphatidylinositol biphosphate rather diacylglycerol is the primary pre-activator of PKC (Chauhan & Brockerhoff, 1988).

Although initial studies in other laboratories have shown that hydrophobic organic solvents and unsaturated fatty acids stimulated PhK activity at neutral pH, they failed to find any significant effect of phospholipids on phosphorylase b to $\underline{a}$ conversion (Singh and Wang, 1979; Negami et al, 1986). In this respect, we were able to demonstrate that the activity of non-activated PhK at pH 6.8 can be stimulated by a variety of phospholipids and certain anionic amphiphiles, the activating effect being largely dependend on the size of lipid vesicles which in turn is connected with the procedure of vesicle preparation (Kyriakidis et al, 1986a). In parallel, both acidic phospholipids and SDS drastically increase autophosphorylation of PhK (Negami et al, 1986; Kyriakidis et al, 1986a) while in presence of mixed acidic phospholipids, the autoactivation of PhK, at pH 6.8, is enhanced (Negami et al, 1986). Diacylglycerol was also found to slightly stimulate phospholipid activation of phosphorylase b to $\underline{a}$ conversion while PhK was unable to bind phorbol esters (unpublished results).

In our effort to gain better insight into the hydrophobic behaviour of PhK and to compare its lipid-binding properties with those of PKC we have tested a number of lysosphingolipids and gangliosides on

several reactions catalyzed by PhK (Baltas et al, in prep.). We found
that sphingosine and galactosylsphingosine (phsychosine) potently
inhibited phosphorylase b to a conversion. Psychosine was also able to
stimulate the autophosphorylation of PK and this stimulation was
accompanied with a sharp decrease of the rate of autoactivation at pH 6.8.
Moreover the activity of nonactivated PhK at pH 6.8 could be stimulated by
a number of brain gangliosides. Among the individual gangliosides tested
the activation potency of $GD1_\alpha$ and $GT1_b$ was higher than that of GM_1. Most
important, $GD1_a$ dramatically increases the affinity of the enzyme for
Ca^{2+}.

Effect of other hydrophobic compounds

Several hydrophobic compounds interacting with phospholipids were
found to inhibit PKC to various extents. These include psychotic drugs,
local anesthetics, W-7, polyamines, melittin, heparin, polymyxin B and
some other drugs (Kikkawa & Nishizuka, 1986). In addition, the hydrophobic
flavonoid drug quercetin, known to affect membrane-linked activities and
to inhibit cAMP-independent kinases have also been shown to inhibit
Ca^{2+}/phospholipid dependent activity (Gschwendt et al, 1983). Most of the
above compounds have been shown to inhibit a number of Ca^{2+}/CaM dependent
regulatory systems and therefore they have become known as anticalmodulin
agents or CaM inhibitors.

Concerning PhK, a number of studies in this and other laboratories
have shown that many anticalmodulin agents effectively alter the catalytic
and structural properties of the enzyme. Thus, we found that polymyxin B
inhibited both cytoplasmic (Ktenas et al, 1985) and sarcoplasmic reticulum
(SR)-bound PhK (Ktenas et al, in prep.). In the first case, we observed
also that this cyclic polycationic peptide greatly stimulated
autophospholytation and autoactivation of the kinase at pH 6.8. Moreover,
when SR membranes were ^{32}P labelled by endogenous kinase(s), in presence
of exogenous calmodulin, it was revealed that polymyxin B increased ^{32}P
incorporation into a protein band with a similar Mr to that of a subunit
of PhK (Ktenas et al, in prep.), suggesting that the drug is able to
stimulate also autophosphorylation of membrane-bound kinase. In another
series of experiments we have examined the interaction of flavonoids with

skeletal muscle PhK (Kyriakidis et al, 1986b). From 14 flavonoids tested, the flavones quercetin and fisetin were proved to be efficient inhibitors while the flavanone hesperetin stimulated PhK activity. Quercetin inhibited also the autophosphorylation of the kinase while this flavonoid was found to be a competitive inhibitor of ATP for the phosphorylation of phosphorylase b. Shenolikar et al (1979) studying the role of calmodulin in the structure and regulation of PhK have observed that the antipsychotic drug trifluoperazine completely abolished calmodulin-stimulated PhK activity. On the other hand Hessova et al (1985) have shown that heparin reversibly stimulated Ca^{2+}-independent pH 7.0 activity with concomitant inhibition of Ca^{2+}-dependent, pH 8.2 activity. At the same time the aggregation state of the enzyme was drastically changed.

Association with membranes

The most important property of PKC concerning signal transduction mechanisms, is its ability to be rapidly and reversibly distributed between soluble and membrane-bound forms (Kraft & Anderson, 1983). The translocation process is closely related to the activation of the enzyme itself, since its physiological regulation requires, as a prerequisite, a specific interaction with membrane phospholipids (Nishizuka, 1986). To this point, PCK represents the best known example of an "amphitropic" protein kinase, the term "amphitropic" being recently introduced by Burn (1988) for the proteins which can exist both in a cytoplasmic soluble form as well as in a membrane embedded form. It is thus important to emphasize that a prerequisite for a Ca^{2+}/CaM- dependent protein kinase to be characterized as a functional equivalent of PKC must be the capacity of the kinase for a reversible distribution between soluble and membrane-bound forms.

Concerning PhK, it is known that although cytosolic enzyme, this kinase is also associated with SR or plasma membranes (Hörl et al, 1978; Dombradi et al, 1984). Recently Thieleczek et al (1987) have localized molecular structures related to phosphorylase kinase at the SR of rabbit skeletal muscle employing polyclonal antibodies against the holoenzyme as well as monoclonal antibodies specific for its α-, β-, or γ-subunits. In our effort to further investigate the relation between cytoplasmic and membrane bound PhK we have chosen as a membrane system the inside-out human erythrocyte vesicles a system also used for the study of the mechanism of PKC-membrane interaction (Wolf et al, 1985). We found (Kyriakidis et al, 1988) that at pH 7.0 PhK binds to the inner face of the erythrocyte membrane in a Ca^{2+} and Mg^{2+}-dependent manner and the sharpest increase of this association occurs between 70 and 550 nM free Ca^{2+} when 3 mM Mg^{2+} is present. CaM decreases the original binding capacity about 50% suggesting that this Ca^{2+} binding protein may block some sites on the kinase molecule responsible for enzyme-membrane association. These sites are possibly located on exposed areas of α or/and β subunits which are susceptible to proteolysis, a hypothesis compatible with a highly reduced capacity of trypsinolyzed PhK for erythrocyte membrane binding (Kyriakidis et al, 1988). It was found also that several proteins of the membrane can be labelled with ^{32}P by PhK, a 93 kDa band being the most prominent phosphorylated protein and possibly represents band 3 polypeptide of red blood cell membrane. In contrast, we could not find any significant Ca^{2+}-dependent binding of PhK to SR either in presence or in absence of glycogen, suggesting that endogenous PhK molecules associated very tightly with SR (Thieleczek et al, 1987) do not permit exogenous PhK to interact with specific SR membrane binding sites.

Activation by lanthanide ions and Cd^{2+}

Although both Ca^{2+}/CaM-and Ca^{2+}/phospholipid-dependent kinase activation seem to be specific for Ca^{2+} (Pickett-Giese & Walsh, 1986; Stull et al, 1986; Kikkawa & Nishizuka, 1986), experimental evidence suggests that trivalent lanthanide ions (Ln^{3+}) and Cd^{2+} are able to substitute for Ca^{2+} in regulating the activity of the above kinase systems. Thus, PKC although less sensitive to Ca^{2+} substitution by Ln^{3+} shows an enhanced ability to be activated by suboptimal Ca^{2+}, when these metal ions are present (Mazzei et al, 1983). Similarly, we observed that Ln^{3+} effectively mimic the stimulatory action of Ca^{2+} on PhK (Sotiroudis, 1986) but in contrast to the effect on PKC which is biphasic (stimulation followed by inhibition with increasing metal cation concentration), Ln^{3+} exhibited only a stimulatory action on PhK. Cd^{2+} was also found to mimic effectively, potentiate and antagonize the stimulatory action of Ca^{2+} on

PKC in a biphasic manner (Mazzei et al, 1984). We obtained similar results when Cd^{2+} was substituted for Ca^{2+} in PhK activity assays both at pH 6.8 and 8.2 (Sotiroudis, 1986). It must also be noted that both Ln^{3+} and Cd^{2+} were able to replace Ca^{2+} required for the stimulation of PhK by exogenous CaM (Sotiroudis, 1986).

According to the hypothesis of the existence of two classes of sites for divalent and trivalent cations in CaM (capital and auxiliary) (Cox, 1988), Ca^{2+} binds only to capital sites but Ln^{3+} and Cd^{2+} are able to occupy both type of sites. Concerning PKC, the location of Ca^{2+} binding site(s) is unclear and there is no typical E-F hand structure present that would provide a CaM like Ca^{2+}-binding site. Moreover there is experimental evidence that diacylglycerol and phospholipid binding sites are located within this regulatory domain (Parker & Ullrich, 1987). Recently, evidence has been presented (Murakami et al, 1987) that PKC possesses high and low affinity Ca^{2+}-binding sites and that at least one Zn^{2+}-binding site (auxiliary site ?) exists which is dinstict from Ca^{2+}-binding sites.

Conclusions

Many hormones and extracellular messengers regulate cell function, in part at least, by inducing an increase in Ca^{2+} concentration of the cell cytosol. Ca^{2+}/CaM and Ca^{2+}/phospholipid-dependent protein kinases represent a very important group of Ca^{2+}-receptor proteins which is responsible for the propagation and amplification of the signal formed by a variety of physiological stimuli.

An essential phase in the mechanism of activation of enzymes by CaM is a conformational change induced in CaM by binding of Ca^{2+} and as a consequence a hydrophobic site becomes exposed on the surface of the molecule (Cox, 1988). Peptides modeled on the CaM-binding domains of several enzymes have demonstrated both Ca^{2+}-dependent CaM-binding and the pontential to form amphiphilic helices. Overall, studies with MLCK, PhK (γ subunit) (Lucas et al, 1986) and Ca^{2+}/CaM-dependent protein kinase II (Hanley et al, 1988) provide a model for CaM binding domains in structurally diverse CaM binding proteins that contain clusters of basic residues within potential amphiphilic α-helical structures. On the other hand from the amino acid sequence of bovine PKC one can predict amphiphatic helices that could provide a site for hydrophobic interaction, although no single hydrophobic stretch was identified within the regulatory domain (Parker and Ullrich, 1987). In this context, experimental evidence from this and other laboratories permits us to develop a hypothesis according to which the hydrophobic activation and inhibition of both PhK and PKC can be described by the assumption of similar hydrophobic domains, which possibly determine in a similar way the responsiveness of these kinases to transmembrane signaling. Moreover, the data available argue against the existence of specific (e.g. phosphatidylserine) phospholipid-activated kinases indicating instead that the Ca^{2+}-dependent stimulation of the above protein kinases reflects the more general actions of CaM and phospholipids as hydrophobic probes (Juskevich et al , 1983). As far as the inability of different research groups to identify a CaM-induced activation of PKC there are several explanations: Thus, CaM may substitute for phospholipid for the phosphorylation of specific substrates (Hansson et al, 1988), or that PKC may lose its sensitivity to be activated by CaM, during purification (Juskevich et al, 1983). Accordingly, the CaM-induced inhibition of phosphatidylserine-dependent PKC activity (Albert et al, 1984) may be due to a direct competition between CaM and phospholipid for the same hydrophobic site on the kinase.

Finally, we suggest that PhK and PKC are amphitropic protein kinases which under mobilizing Ca^{2+} conditions may translocate to membrane compartments where they serve for a specialized function not necessarily connected with the corresponding cytosolic catalytic action. This translocation may lead under certain conditions to an insertion of the kinase into the membrane (Bazzi & Nelsestuen, 1988) and thus the integral membrane form of the kinase may function as a long-term cell regulator.

References

Albert,K.A., Wu,W.C-S., Nairn,A.C. and Greengard,P. (1984) Inhibition of
calcium/phospholipid-dependent protein phosphorylation. Proc.Natl.
Acad.Sci. USA 81, 3622-3625

Baltas,L.G., Zevgolis,V.G., Kyriakidis,S.M., Sotiroudis,T.G. and
Evangelopoulos, A.E. in preparation

Bazzi,M.D. and Nelsestuen,G.L. (1988) Constitutive activity of membrane-
inserted protein kinase C. Biochem.Biophys.Res.Commun. 152, 336-343

Burn,P. (1988) Amphitropic proteins: A new class of membrane proteins.
Trends Biochem.Sci., 13, 79-83

Castagna,M., Pavone,C., Bazgar,S., Couturier,A., Chevalier,M. and
Fiszman,M. (1985) Phospholipid/Ca^{2+}-dependent protein kinase, cell
differentiation and tumor promotion. In: Hormones and Cell Regulation (Dumont,J.E. et al, eds) Vol.9, 185-206, Elsevier Science
Publishers BV

Chan,K.-F.J. & Graves,D.J. (1984) Molecular properties of phosphorylase
kinase. In: Calcium & Cell Function (Cheung,W.Y.,ed) Vol.5, 1-31,
Academic Press, New York

Chauhan,V.P.S. and Brockerhoff,H. (1988) Phosphatidylinositol,-4-5 biphosphate antecede diacylglycerol as activator of protein kinase C.
FASEB J. 2, A349

Cox,J.A. (1988) Interactive properties of calmodulin. Biochem.J. 249,
621-629

Dombradi,V.K., Silberman,S.R., Lee,E.Y.C., Caswell,A.H. & Brandt,N.R.
(1984) The association of phosphorylase kinase with rabbit muscle
T-tubules. Arch.Biochem.Biophys. 230, 615-630

Edelman,A.M., Blumenthal,D.K. and Krebs,E.G. (1987) Protein serine-
threonine kinases. Ann.Rev.Biochem.56, 567-613

Fujiki,H., Yamashita,K., Suganuma,M., Horiuchi,T., Taniguchi,N. and
Makita,A. (1986) Involvement of sulfatide in activation of protein
kinase C by tumor promoters Biochem.Biophys.Res.Commun.138,153-158

Gietzen,K., Sadorf,I. and Bader,H. (1981) A model for the regulation of
the calmodulin-dependent enzymes erythrocyte Ca^{2+}-transport
ATPase and brain phosphodiesterase by activators and inhibitors.
Biochem. J. 207, 541-548

Gschwendt,M., Horn,F., Kittstein,W.and Marks,F. (1983) Inhibition of the
calcium- and phospholipid-dependent protein kinase activity from
mouse brain cytosol by quercetin. Biochem.Biophys. Res. Commun.
117, 444-447

Hanley,R.M., Means,A.R., Kemp,B.E. and Shenolikar,S. (1988) Mapping of
calmodulin-binding domain of Ca^{2+}/calmodulin-dependent protein
kinase II from rat brain. Biochem.Biophys.Res.Commun. 152, 122-128

Hannun,Y.A., Loomis,C.R., Merrill,A.H.Jr and Bell,R.M. (1986) Sphingosine
inhibition of protein kinase C activity and of phorbol dibutyrate
binding in vitro and in human platelets. J.Biol.Chem. 261, 12604-
12609

Hannun,Y.A. and Bell,R.M. (1987) Lysosphingolipids inhibit protein kinase
C: Implications for the sphingolipidoses. Science, 235, 670-674

Hansson,A., Skoglund,G., Lassing,I., Lindberg,U. and Ingelman-Sundberg,M.
(1988) Protein kinase C-dependent phosphorylation of profilin is
specifically stimulated by phosphatidylinositol biphosphate
(PIP$_2$). Biochem.Biophys.Res.Commun. 150, 526-531

Hessova,Z., Varsanyi,M. & Heilmeyer,L.M.G.,Jr. (1985) Dual function of
calmodulin (δ) in phosphorylase kinase. Eur.J.Biochem.146, 107-115

Hörl,W.H., Jennissen,H.B. and Heilmeyer,L.M.G.,Jr. (1978) Evidence for the
participation of a Ca^{2+}-dependend protein kinase and a protein
phosphatase in the regulation of the Ca^{2+}-transport ATPase of

the sarcoplasmic reticulum. 1.Effect of inhibitors of the Ca^{2+}-dependent protein kinase and protein phosphatase. Biochemistry 17, 759-766

Ito,M., Tanaka,T., Inagaki,M., Nakanishi,K. and Hidaka,H. (1986) N-(6-Phenylhexyl)-5-chloro-1-Naphthalenesulfonamide. A novel activator of protein kinase C. Biochemistry 25, 4179-4184

Juskevich,J.C. Kuhn,D .M. and Lovenberg,W. (1983) Phosphorylation of brain cytosol proteins. Effects of phospholipids and calmodulin. J.Biol. Chem. 258, 1950-1953

Kikkawa,V. and Nishizuka,Y. (1986) Protein kinase C. In: The Enzymes (Boyer,P and Krebs,E.G.eds) Vol.17, 167-189, Academic Press, New York.

Kishimoto,A., Kajikawa,N., Siota,M. and Nishizuka,Y. (1983) Proteolytic activation of calmodulin-activated, phospholipid-dependent protein kinase by calcium-dependent neutral protease. J.Biol.Chem. 258, 1156-1164

Kraft,A.S. and Anderson,W.B. (1983) Phorbol esters increase the amount of Ca^{2+}, phospholipid-dependent protein kinase associated with plasma membrane. Nature (London) 301, 621-623

Krebs,E.G. (1986) The enzymology of control by phosphorylation. In: The Enzymes (Boyer,P. and Krebs,E.G. eds) Vol.17, 3-20, Academic Press New York

Kreutter,D., Kim,J.Y.H., Goldenring,J.R., Rasmussen,H. Ukomadu,C., DeLorenzo,R.J. and Yu,R.K. (1987) Regulation of protein kinase C activity by gangliosides. J.Biol.Chem. 262, 1633-1637

Ktenas,T.B., Sotiroudis,T.G., Nikolaropoulos,S. and Evangelopoulos,A.E. (1985) Interaction of phosphorylase kinase with polymixins Biochem.Biophys.Res.Commun. 133, 891-896

Ktenas,T.B., Sotiroudis,T.G. and Evangelopoulos,A.E. in preparation

Kyriakidis,S.M., Sotiroudis,T.G. & Evangelopoulos,A.E. (1986a) Stimulation of glycogen phosphorylase kinase with phospholipids. Biochem. Inter. 13, 853-861

Kyriakidis,S.M., Sotiroudis,T.G. & Evangelopoulos,A.E. (1986b) Interaction of flavonoids with rabbit muscle phosphorylase kinase. Biochim. Biophys.Acta 871, 121-129

Kyriakidis,S.M., Sotiroudis,T.G. and Evangelopoulos,A.E, (1988) Ca^{2+} and Mg^{2+} dependent association of phosphorylase kinase with human erythrocyte membranes. Submitted for publication

Lucas, T.J., Burgess, W.H., Prendergast, F.G., Lau, W. and Watterson, D.M. (1986) Calmodulin binding domains: Characterization of a phosphorylating and calmodulin binding site from myosin light chain kinase. Biochemistry, 25, 1458-1464

Mamoi,T. (1986) Activation of protein kinase C by ganglioside GM3 in the presence of calcium and 12-o-tetradecanoylphorbol-13-acetate Biochem.Biophys.Res.Commun. 138, 865-871

Mazzei,G.J., Qi,D.-F., Schatzman,R.C., Raynor,R.L., Turner,R.S. and Kuo,J.F. (1983) Comparative abilities of lanthanide ions La^{3+} and Tb^{3+} to substitute for Ca^{2+} in regulating phospholipid-sensitive Ca^{2+}-dependent kinase and myosin light chain kinase. Life Sci. 33, 119-129

Mazzei,G.J., Girrard,P. and Kuo, J.F. (1984) Environmental pollutant Cd^{2+} biphasically and differentially regulates myosin light chain kinase and phospholipid/Ca^{2+}-dependent protein kinase FEBS Lett. 173, 124-128

Meyer,T., Fabro,D., Eppenberger,U. and Matter,A. (1986) The lipophilic muramyltripeptide MTP-PE, a biological response modifier, is an activator of protein kinase C.Biochem.Biophys. Res.Commun. 140, 1043-1050

Murakami,K. , Chan,S.Y. and Routtenberg,A. (1986) Protein kinase C activation by cis-fatty acid in the absence of Ca^{2+} and phospholipids. J.Biol.Chem. 261, 15424-15429

Murakami,K. , Whitley,M.K. and Routtenberg,A. (1987) Regulation of protein kinase C activity by cooperative interaction of Zn^{2+} and Ca^{2+}. J.Biol.Chem. 262, 13902-13906

Negami,A.I. , Sasaki,H. and Yamamura,H. (1986) Activation of phosphorylase kinase through autophosphorylation by membrane component phospholipids. Eur.J.Biochem. 157, 597-603

Nikolaropoulos,S. and Sotiroudis,T.G. (1985) Phosphorylase kinase from chicken gizzard. Partial purification and characterization. Eur.J. Biochem. 151, 467-473

Nishizuka,Y. (1984) The role of protein kinase C in cell-surface signal transduction and tumor promotion. Nature 308, 693-698

Nishizuka,Y. (1986) Studies and perspectives of protein kinase C. Science 233, 305-312

Parker,P.J. and Ullrich,A. (1987) Protein kinase C. J.Cell.Physiol.Suppl.5 53-56

Pickett-Giese,C.A. & Walsh,D.A. (1986) Phosphorylase Kinase. In: The Enzymes (Boyer,P. & Krebs,E.G.,eds) Vol.17, 395-459, Academic Press, New York

Sakai,K. , Kobayashi,T. , Komuvo,T. , Nakamura,S. , Mizuta,K. , Sakanoue, Y. , Hashimoto,E. and Yamamura,H. (1987) Non-requirement of calcium on protamine phosphorylation by calcium-activated, phospholipid dependent protein kinase. Biochem.Inter. 14, 63-70

Shenolikar, S. , Cohen, P.T.W. , Cohen, P. , Nairn, A.C. and Perry,S.V. (1979) Role of calmodulin in the structure and regulation of phosphosphorylase kinase from rabbit skeletal muscle. Eur.J.Biochem. 100, 329-337

Shenolikar,S. , Lickteig,R. , Hardie,D.G. , Soderling,T.R. , Hanley,R.M. and Kelly,P.T. (1986) Calmodulin-dependent multifunctional protein kinases. Evidence for isoenzyme forms in mammalian tissues Eur.J. Biochem. 161, 739-747

Singh,T.J. & Wang,J.H. (1979) Stimulation of glycogen phosphorylase kinase from rabbit skeletal muscle by organic solvents. J.Biol.Chem. 254, 8466-8472

Sotiroudis,T.G. (1986) Lanthanide ions and Cd^{2+} are able to substitute for Ca^{2+} in regulating phosphorylase kinase. Biochem.Inter. 13, 59-64

Stull,J.T. , Nunnally,M.H. and Michnoff,C.H. (1986) Calmodulin-dependent protein kinases. In: The Enzymes (Boyer,P. and Krebs,E.G.,eds) vol. 17, 113-166, Academic Press, New York

Takai,Y. ,Kishimoto,A. , Iwasa,Y. , Kawahara,Y. , Mori,T. and Nishizuka,Y. (1979) Calcium-dependent activation of a multifunctional protein kinase by membrane phospholipids. J.Biol.Chem. 254, 3692-3695

Thieleczek,R. , Behle,G. , Messer,A. , Varsanyi,M. , Heilmeyer.L.M.G.,Jr. & Drenckhahn,D. (1987) Localization of phosphorylase kinase subunits at the sarcoplasmic reticulum of rabbit skeletal muscle by monoclonal and polyclonal antibodies. Eur.J.Cell Biol. 44, 333-340

Wightman,P.D. and Raetz,C.R.H. (1984) The activation of protein kinase C by biologically active lipid moietes of lipopolysaccharide. J.Biol. Chem. 259, 10048-10052

Wolf,M. , LeVine III,H. , May,S.,Jr, Cuatrecasas,P. and Sahyoun,N. (1985) A model for intracellular translocation of protein kinase C involving synergism between Ca^{2+} and phorbol esters. Nature 317, 546-549

Zevgolis,V.G. , Sotiroudis,T.G. and Evangelopoulos,A.E. in preparation

THE USE OF SPECIFIC ANTISERA TO LOCATE FUNCTIONAL DOMAINS OF GUANINE NUCLEOTIDE BINDING PROTEINS

F.R. McKenzie and G. Milligan
Departments of Biochemistry and Pharmacology
University of Glasgow
Glasgow G12 8QQ
Scotland, U.K.

In many signal transduction systems, the transfer of information from a plasma membrane receptor to an effector system is mediated by members of a family of guanine nucleotide binding proteins (G-proteins) (Gilman, 1987). Members of this family include Gs and Gi, which serve to couple receptors to adenylate cyclase in either a stimulatory or inhibitory fashion, both rod and cone specific forms of transducin which couple the photon receptor rhodopsin to forms of cyclic GMP phosphodiesterase, Go, a G-protein which may be involved in Ca^{2+} channel regulation, Gk, which has a specific role in the activation of K^{+} channels and Gp, a G-protein involved in the stimulation of phospholipase C-mediated inositol phospholipid breakdown.

All of these G-proteins are heterotrimers consisting of α (52-39kDa), β (36-35kDa) and γ (10kDa) subunits. The α subunit is the site of guanine nucleotide binding and interacts with both receptor and effector entities. As such it is this subunit which defines the individuality and is likely to confer specificity to the G-protein (Masters et al. 1986). The function of the β/γ subunits, which are permanently bound to one another, is currently unclear, although they may play a role in mediating the inhibition of adenylate cyclase (Gilman, 1987) and are required for anchorage of α-subunits to the plasma membrane (Sternweis 1986). The individual G-proteins share a mechanism of action whereby the binding of agonist to receptor promotes the loss of bound GDP from, and the binding of GTP to, the G-proteins α subunit. The α subunit can now dissociate from the β/γ subunits in a Mg^{2+} dependent manner and is now able to interact with the effector system. To deactivate the G-protein, bound GTP is hydrolysed to GDP by a GTPase activity which is intrinsic to the α subunit. The inactivated α subunit can now recombine with the β/γ complex. Therefore agonist-induced stimulation of GTPase activity is indicative that a receptor interacts with a G-protein (Klee et al. 1985).

The recent isolation and analysis of cDNA clones from a variety of tissues has demonstrated that a number of "Gi-like" G-proteins exist,

NATO ASI Series, Vol. H29
Receptors, Membrane Transport and Signal Transduction
Edited by A. E. Evangelopoulos et al.
© Springer-Verlag Berlin Heidelberg 1989

including Gi1, Gi2, Gi3, transducin and Go (Jones and Reed, 1987). These G-proteins have been shown to be highly homologous in primary structure and in addition, all of these proteins are substrates for an exotoxin produced by <u>Bordetella pertussis</u>, called simply pertussis toxin. Pertussis toxin catalyses the mono-ADP-ribosylation of the α subunit of these G-proteins at a cysteine residue four amino-acids from the C-terminus. The effect of this covalent modification is to prevent interaction of G-protein and receptor. This can be assayed by the loss of receptor stimulated high affinity GTPase activity in membranes produced from cells which have been pretreated with pertussis toxin. This observation is consistent with the idea that the C-terminal region of the G-protein α subunit might play a specific role in the interaction with receptor.

More definitive evidence in support this hypothesis came from work performed on the S49 murine lymphoma cell line. In membranes produced from wild type cells, it is possible to stimulate adenylate cyclase by agonists that act at β-adrenergic receptors, presumably through the activation of Gs. However, this ability is absent in the unc (uncoupled) mutant of this cell line, even though β-adrenergic receptors, Gs and adenylate cyclase are all present. The recent isolation of cDNA's encoding the α subunit of Gs from both wild type S49 cells and the unc mutant (Sullivan <u>et al</u>. 1987; Rall and Harris, 1987) has demonstrated that there is a single base substitution in the mutant which results in a single amino-acid change. This alteration substitutes a proline residue for an arginine residue 6 amino acids from the C-terminus of the α subunit and provides an elegant rationale for the inability of the G-protein to interact with receptor. It further explains why it had previously been noted that unc Gs α migrates as a more acidic spot on 2-dimensional electrophoresis than does wild type Gs α (Schleifer et al. 1980). This work suggests that the α-subunit C-terminus might play a specific role in receptor coupling in Gs, as well as in the "Gi-like" G-proteins.

In order to analyse the mechanism of G-protein interaction with other components of signal transduction pathways, we have generated a range of antipeptide antisera directed against either regions of variance between the highly homologous "Gi-like" G-proteins, or against putative regions of the G-protein which represent functional domains (Goldsmith <u>et al</u>. 1987, Mullaney <u>et al</u>. 1988). It was thought that these antisera would potentially be more specific than antisera raised against purified G-proteins as a high degree of epitope conservation is likely amongst the "Gi-like" G-proteins. Antiserum AS7 was raised against a synthetic peptide corresponding to the C-terminal decapeptide of the α-subunit of rod transducin (Goldsmith et al 1987). As there is only one amino-acid difference in the corresponding region of Gi and the change (from glutamic acid (transducin) to asparagine (Gi)) is a conservative one, it is not surprising to note that the antiserum recognises both transducin and Gi. However this antiserum does not recognise Go as the equivalent region of Go contains changes in five amino-acids present in the peptide used. As transducin in only present in phoreceptor containing tissues, this allows the use of antiserum AS7 as a specific probe for Gi in all other tissues.

Antiserum OC1 was raised against a synthetic peptide corresponding to the C-terminal 10 amino-acids of Go. This antiserum recognised Go. However it fails to recognise Gi or transducin, which is in agreement with the points noted above. It is therefore possible to use OC1 as a specific probe for Go (Milligan, 1988).

The neuroblastoma x glioma hybrid cell line, NG 108-15, has been of great use in the analysis of signal transduction systems (Klee et al 1985). This is because it possesses a variety of receptors which are known to couple to either adenylate cyclase, the stimulation of phospholipase C or the regulation of ion channels. The cell line also expresses a range of G-proteins including Gi, Go and Gs (Milligan et al, 1986). Both a δ opioid receptor and a poorly characterised growth factor receptor are present on NG108-15 cells. Ligands which are agonists at these receptors, namely synthetic enkephalins at the opioid receptor and foetal calf serum at the growth factor receptor, stimulated high affinity GTPase activity in membranes produced from these cells. These stimulations were abolished in membranes produced from cells which had been pretreated with pertussis toxin (McKenzie et al. 1988a).

To determine whether both receptors were functioning through activation of the same pertussis-toxin sensitive G-protein, adenylate cyclase activity was examined. Opioid peptides produced an inhibition of adenylate cyclase activity in NG108-15 membranes, however foetal calf serum did not, thus suggesting that the opioid receptor was functioning through activation of Gi, and the growth factor receptor was functioning through activation of a separate pertussis toxin-sensitive G-protein.

Moreover, when GTPase activity was determined in NG108-15 membranes, saturating concentrations of both opioid peptides and foetal calf serum produced fully additive GTPase responses, thus indicating that the receptors for each agonist were activating distinct pools of G-protein and hence by extension interacting with different G-proteins (McKenzie et al. 1988a,b).

To further analyse the nature of the G-protein(s) which interacts with the δ opioid receptor in the NG108-15 cell line we employed both antisera AS7 and OC1. Membranes from NG108-15 cells were pre-incubated with each antiserum diluted with normal rabbit serum for 60 minutes, in the presence of components required to assay high affinity GTPase activity except [γ -^{32}P]GTP and agonist. After the pre-incubation [γ -^{32}P]GTP and suitable agonist were added, and the assay extended. Under these conditions, opioid peptides were unable to stimulate GTPase activity in membranes which had been pre-incubated with antiserum AS7 (McKenzie et al. 1988 a,b) (Table 1), whilst stimulation was apparent in membranes incubated with normal rabbit serum or with antiserum OC1 (Table 1). In contrast, there was only a small reduction in GTPase stimulation produced by foetal calf serum in membranes incubated with AS7 when compared to the foetal calf serum stimulation produced in membranes incubated with normal rabbit serum, however even here pre-incubation with antiserum OC1 had no effect (Table 1).

TABLE 1 - Attenuation of opioid-receptor stimulated high-affinity GTPase activity in membranes of NG108-15 cells by affinity purified antibodies from antiserum AS7.

Addition to preincubation	Basal high-affinity GTPase activity (pmol/min/mg)	Receptor-stimulated high-affinity GTPase activity (pmol/min/mg)	
		DADLE (1uM)	Foetal calf serum (10% v/v)
Water	14.25 + 1.08	7.80 + 0.23	17.70 + 0.75
Normal rabbit serum	12.90 + 0.83	6.60 + 0.35	15.75 + 1.07
AS7 antibodies	16.50 + 0.69	0.03 + 0.01	12.60 + 0.63
OC1 antibodies	12.93 + 0.26	7.00 + 0.33	14.87 + 0.57

NG108-15 membranes were incubated with water, normal rabbit serum or a 1:100 dilution in normal rabbit serum of either antiserum OC1 or of antibodies which had been affinity-purified from antiserum AS7 for 1h at 37°C in the presence of all the GTPase assay reagents except [γ -^{32}P]GTP. After this time the appropriate receptor ligand and [γ -^{32}P]GTP were added, and incubation continued for a further 20min. Samples were then processed as previously described (McKenzie et al. 1988a). Data which are means + S.D. are from a single experiment, representative of five experiments performed. Receptor-stimulated GTPase represents agonist-stimulation above basal GTPase levels.

If the C-terminal region of G-proteins is indeed important for interaction with receptor, then it might be postulated that any alteration in the conformation of this region will modify the coupling of G-protein to receptor. As such, antibodies which are able to interact with the C-terminal region, might be expected to interfere with G-protein-receptor coupling. Table 1 shows that preincubation of NG108-15 membranes with antibodies derived from antiserum AS7 abolished productive coupling between the δ -opioid receptor and the G-protein signalling system, as determined by the loss of stimulation of GTP hydrolysis. Since antiserum AS7 will only recognise Gi in NG108-15 membranes, it was concluded that the δ -opioid receptor functions through the activation of Gi and not Go (McKenzie et al. 1988a,b).

As well as producing an inhibition of adenylate cyclase, both opioid peptides and α 2 adrenergic receptor agonists are able to inhibit Ca^{2+} currents across the membrane of NG108-15 cells. This can conveniently be measured using patch clamp techniques. This inhibition can be attenuated by prior treatment of the cells with pertussis toxin (Hescheler et al. 1987). Using this approach, Heschler et al. (1987) have successfully reconstituted opioid receptor mediated inhibition of the Ca^{2+} current across the plasma membrane of pertussis toxin-treated NG108-15 cells by introducing a purified preparation of Go. This process could be mimicked to a lower degree of efficiency by the addition of a purified preparation of Gi. In contrast to our use of selective antisera in the GTPase assay, we have noted that introduction of antiserum OC1, but not antiserum AS7 into a patch derived from NG108-15 cells leads to the attentuation of α 2 adrenergic receptor-mediated inhibition of the Ca^{2+} channel current (I.McFadzen, G. Milligan and D.Brown, in preparation). In an analagous series of experiments Harris-Warwick et al. (1988) have demonstrated that a polyclonal anti-Go antiserum is able to prevent dopaminergic inhibition of Ca^{2+} current in identified neurons of a snail.

The explanation of these apparently contradictory results may lie in the nature of the cells used for electrophysiological experiments. Measurement of Ca^{2+} currents across the membrane of NG108-15 cells is only possible in cells which have been morphologically "differentiated" by treatment with agents which raise intracellular concentration of cyclic AMP. When NG108-15 cells are differentiated in this manner, then membranes derived from these cells have some 4-10 greater amounts of the α -subunit of Go as compared membranes from untreated cells (Mullaney et al. 1988; Mullaney and Milligan, in preparation). There is indeed only very low levels of Go in the untreated cells. As such, the observation that opioid peptide-mediated stimulation of GTPase activity was completely abolished in membranes of undifferentiated NG108-15 by antiserum AS7 may simply reflect that the low levels of Go present in undifferentiated cells can provide very little measurable GTPase activity. If the δ-opioid receptor in NG108-15 cells can indeed interact with both Gi and Go, then it could be hypothesised that the receptor-mediated inhibition of adenylate cyclase would occur through activation of Gi whilst inhibition of the Ca^{2+} current would be mediated by receptor activation of Go.

A considerable range of other experiments have recently been performed using similar approaches to our own. In the visual transduction system of vertebrates, the detection of photons by the light receptor rhodopsin promotes the activation of transducin which then stimulates cyclic GMP phosphodiesterases (cGMP PDE). The functional reconstitution of purified transducin, rhodopsin and cGMP PDE into phospholipid vesicles by a number of research groups has allowed the direct analysis of interactions between transducin and rhodopsin (Fung, 1985).

Using this system, it has recently been shown that preincubation of purified transducin α-subunit with antiserum AS7 before incorporation into phospholipid vesicles, leads to a diminished ability of rhodopsin to stimulate GTPase activity, whereas preincubation with non-specific rabbit IgG had no effect (Cerione et al. 1988). Maximal inhibition of the rhodopsin-stimulated GTPase activity occurred at a molar ratio of transducin α-subunit to AS7 of 1:1. Since the presence of transducin β/γ subunits during the preincubation did not alter the effect of AS7, it was concluded that the binding of AS7 to transducin did not alter the ability of the α-subunit of transducin to interact with β/γ. Further, the intrinsic basal GTPase activity of transducin α was not reduced by preincubation with AS7. A further point obtained from the above experiments was that preincubation of antiserum AS7 with the α-subunit of transducin which had been activated by the addition of a non-hydrolysable analogue of GTP, guanosine 5-0-(3-thiotriphosphate) (GTP γ S), enhanced the ability of transducin to interact with the cGMP PDE.

When rod outer segment membranes are preincubated with monoclonal antibodies to transducin α-subunit which identify a region close to the C-terminus, then light activation of transducin by rhodopsin is blocked (Hamm et al. 1987, Deretic and Hamm, 1987). As the antibody which produced inhibition was able to immunoprecipitate the holomeric transducin $\alpha\beta\gamma$ complex, it is unlikely that the effect was produced by preventing the interaction of transducin α with β/γ subunits.

Whilst the interaction of rhodopsin and transducin is easily amenable to study, similar experiments have attempted to assess the interaction of receptors with Gk. The activation of a subset of potassium channels in isolated membrane patches of guinea pig atrial cells by muscarinic receptor activation is another event which is attenuated by pre-exposure of the system to pertussis toxin. Recently, the G-protein responsible for this effect has been isolated from human red blood cells and termed Gk (Codina et al. 1988). The same series of monoclonal antisera which were used to assess the interaction of rhodopsin and transducin, cross react with Gk and were applied to the solution bathing an inside out patch of atrial membrane. The antibody irreversibly blocked muscarinic receptor modulation of K^+ currents. The addition of a second monoclonal antibody which had been produced in a similar fashion and also cross-reacted with Gk, but had been immunochemically mapped to a site on the α-subunit of transducin further removed from the C-terminus was found to be totally ineffective (Yatani et al. 1988). When the antibody which inhibited receptor-mediated opening of the potassium channels was added to membrane patches where Gk had been activated by the addition of GTP γ S, no inhibition of the potassium channels was recorded. This suggests that the effect of the antibody was to prevent the interaction of receptor and Gk.

In the last year a considerable amount of evidence has been produced to demonstrate that in both membrane and reconstitution systems it is possible to use specific antisera to identify regions of G-protein α-subunits which are likely to represent functional domains for interaction between the α-subunit and other protein components of signal transduction systems. Good evidence now indicates that the extreme C-terminal region of a range of different G-proteins represents a key domain in G-protein-receptor interactions. It is likely that similar strategies will lead to a more detailed understasnding of the mechanism of interaction of G-proteins with their effector systems.

REFERENCES

Cerione RA, Kroll S. Rajaram R, Unson C, Goldsmith P and Spiegel AM
(1988) An antibody directed against the carboxy-terminal decapeptide
of the α -subunit of the retinal GTP-binding protein, transducin.
Effects on transducin function. J Biol Chem 263:9345-9352

Codina J, Olate J, Abramowitz J, Mattera R, Cook RG and Birnbaumer L
(1988) α i-3cDNA encodes the α -subunit of Gk, the stimulatory G
protein of receptor-regulated K^+ channels. J Biol Chem
263:6746-6750

Deretic D and Hamm HE (1987) Topographic analysis of antigenic
determinants recognized by monoclonal antibodies to the photoreceptor
guanyl nucleotide-binding protein, transducin. J Biol Chem
262:10839-10847

Fung B (1985) The light-activated cyclic GMP phosphodiesterase system in
retinal rods. Molecular Mechanisms of Transmembrane Signalling
4:184-214 (eds Cohen P and Houslay MD, Elsevier Press, Amsterdam)

Gilman AG (1987) G-proteins: transducers of receptor-generated signals.
Ann Rev Biochem 56:615-649

Goldsmith P, Gierschik P, Milligan G, Unson C, Vinitsky R, Malech H and
Spiegel AM (1987) Antibodies directed against synthetic peptides
distinguish between GTP-binding proteins in neutrophil and brain.
J Biol Chem 262:14683-14688.

Hamm HE, Deretic D, Hofmann KP, Schleicher A and Kohl B (1987) Mechanism
of action of monoclonal antibodies that block the light activation of
the guanyl nucleotide-binding protein, transducin. J Biol Chem
262:10831-10838

Harris-Warrick J, Hammond C, Paupardin-Tritsch D, Homburger V, Ruout B,
Bockaert J and Gerschenfeld HM (1988) The α -subunit of a GTP-binding
protein homologous to mammalian Go mediates a dopamine-induced
decrease of Ca^{2+} current in snail neurons. Neuron 1:27-34

Heschler J, Rosenthal W, Trautwein W and Schultz G (1987) The GTP-binding
protein, Go, regulates neuronal calcium channels. Nature 325:445-447

Jones DT and Reed RR (1987) Molecular cloning of five GTP-binding protein
cDNA species from rat olfactory neuroepithelium. J Biol Chem
262:14241-14249

Klee WA, Milligan G, Simonds WF and Tocque B (1985) Opiate receptors in
neuroblastoma x glioma hybrid cells: a system for investigating Ni/Ns
interactions. Molecular Mechanisms of Transmembrane Signalling
4:119-127 (eds Cohen P and Houslay MD, Elsevier Press, Amsterdam)

Masters SB, Stroud RM and Bourne HR (1986) Family of G-protein α chains:
amphipathic analysis and predicted structure of functional domains.
Protein Engineering 1:47-54

McKenzie FR, Kelly ECH, Unson C, Spiegel AM and Milligan G (1988a)
Antibodies which recognize the C-terminus of the inhibitory
guanine-nucleotide-binding protein (Gi) demonstrate that opioid
peptides and foetal-calf serum stimulate the high-affinity GTPase
activity of two separate pertussis-toxin substrates. Biochem. J
249:653-659

McKenzie FR, Mullaney I, Unson C, Spiegel AM and Milligan G (1988b). The
use of anti-peptide antisera to probe interactions between receptors
and guanine nucleotide binding proteins. Biochem Soc Trans 16:434-437

Milligan G (1988) Techniques used in the identification and analysis of
function of pertussis toxin-sensitive guanine nucleotide binding
proteins. Biochem J 255:1-13

Milligan G, Gierschik P, Spiegel AM and Klee WA (1986) The GTP-binding
regulatory proteins of neuroblastoma x glioma, NG108-15, and glioma,
C6, cells. FEBS Lett. 195, 225-230.

Mullaney I, Magee AI, Unson CG and Milligan G (1988) Differential
    regulation of amounts of the guanine-nucleotide-binding proteins Gi
    and Go in neuroblastoma x glioma hybrid cells in response to
    dibuturyl cyclic AMP. Biochem J 256:649-656.
Rall T and Harris BA (1987) Identification of the lesion in the
    stimulatory GTP-binding protein of the uncoupled S49 lymphoma. FEBS
    Lett 224:365-370
Schleifer LS, Garrison JC, Sternweis PC, Northup JK and Gilman AG (1980)
    The regulatory component of adenylate cyclase from uncoupled S49
    lymphoma cells differs in charge from the wild type protein. J Biol
    Chem 255:2641-2644
Sternweis PC (1986) The purified α subunits of Go and Gi from bovine
    brain require β/γ for association with phospholipid vesicles. J Biol
    Chem 261:631-637
Sullivan KA, Miller RT, Masters SB, Beiderman B, Heideman W and Bourne HR
    (1987) Identification of receptor contact site involved in receptor-G
    protein coupling. Nature 330:758-760
Yatani A, Hamm HE, Codina J, Mazzoni MR, Birnbaumer L and Brown AM (1988)
    A monoclonal antibody to the α subunit of Gk blocks muscarinic
    activation of atrial K^+ channels. Science 241:828-831

CALCIUM INHIBITS GTP-BINDING PROTEINS IN SQUID PHOTORECEPTORS

Jenny Baverstock, Julie Fyles and Helen Saibil
Department of Zoology
Oxford University
South Parks Road
Oxford OX1 3PS
United Kingdom

INTRODUCTION

Visual transduction in vertebrate rods commences with photoexcitation of rhodopsin, in which the 11-*cis*-retinal chromophore is photoisomerized to the all-*trans* isomeric form. Photoexcited rhodopsin catalyses guanine nucleotide exchange on the GTP-binding protein transducin. The α-subunit binds GTP and dissociates from the ßγ part of the trimer to activate cyclic GMP phosphodiesterase (cGMP-PDE), which hydrolyses cGMP. The subsequent reduction in cGMP concentration results in cation channel closure and membrane hyperpolarisation. By GTP hydrolysis, the α-subunit reverts to its GDP bound form (the resting state) and reassociates with the ßγ complex until reactivation by the same process (Stryer, 1986).

As yet, not all the activation steps in the invertebrate photoreceptor system have been identified. The basic structure of rhodopsin is conserved (Ovchinnikov et al., 1988) and, as in the vertebrate system, there is light-stimulated GTPase activity (Calhoon et al., 1980; Vandenberg & Montal, 1984; Saibil & Michel-Villaz, 1984; Blumenfeld et al., 1985; Paulsen & Bentrop, 1986). In squid photoreceptor membranes there is a 45 kD polypeptide substrate for ADP-ribosylation by cholera toxin which requires light stimulation (Vandenberg & Montal, 1984). A 40 kD component can be labelled in the dark using pertussis toxin, in isolated octopus photoreceptor membranes (Tsuda et al., 1986). It is thought that invertebrate photoreceptor G-proteins activate phospholipase C (PLC) (Payne, 1986). Guanine nucleotides have been shown to stimulate light-activated

NATO ASI Series, Vol. H29
Receptors, Membrane Transport and Signal Transduction
Edited by A. E. Evangelopoulos et al.
© Springer-Verlag Berlin Heidelberg 1989

PLC in squid and fly photoreceptors (Baer & Saibil, 1988; Devary et al., 1987). There is also evidence that cyclic GMP may act as a second messenger in invertebrate visual transduction (Saibil, 1984; Johnson et al., 1986). However recent work on *Drosophila* has shown that mutants lacking PLC expression in photoreceptors show no receptor potential (Bloomquist et al., 1988). Light induced changes in cGMP concentration may be part of a chemical chain of events linked to PLC activation. Upon light stimulation there is an increase in inositol(1,3,4)trisphosphate (InsP$_3$) in squid photoreceptor cells (Szuts et al., 1986, Brown et al., 1987). This is proposed to cause the release of calcium from internal stores. A rise in calcium concentration has been shown to mimic light excitation and adaptation in *Limulus* ventral photoreceptors. Increased calcium has also been shown to correlate with light adaptation in squid (Pinto and Brown, 1977).

In our studies, cholera and pertussis toxins have been used as tools to identify GTP-binding proteins and to investigate the effects of calcium. ADP-ribosylation and GTPase assays have been used to characterise two light-dependent, calcium sensitive, GTP-binding proteins in squid photoreceptor membranes.

EXPERIMENTAL PROCEDURES

Materials

Cholera toxin was obtained from Sigma Chemical Company Ltd., Poole, U.K. It was activated by incubation with 20 mM dithiothreitol in 100 mM Tris, pH 7.5 at 24°C for 30 minutes. Activated pertussis toxin (suspended in a 50% glycerol / 50% 0.05 M phosphate, 0.5 M NaCl buffer pH 7.2) was obtained from Porton Products Ltd, Salisbury, U.K. ^{32}P-nicotinamide adenine dinucleotide was obtained from Dupont, New England Nuclear. ATP, and GTP were obtained from Boehringer Mannheim U.K.. Thymidine and arginine were obtained from Sigma.

Membrane Preparation

Small squid, *Alloteuthis subulata*, were obtained from the Marine Biological Association, Plymouth, U.K.. Retinal photoreceptor membranes were isolated by the method of Baer and Saibil (1988). Whole frozen eyes were broken and thawed in isolation buffer: 0.4 M NaCl, 10 mM EGTA, 50 mM MOPS, 2.5 mM MgCl$_2$, 1 mM dithiothreitol, pH 6.9. Pieces of retina were shaken in the same buffer to detach the photoreceptor layer. The membranes were isolated by washing and 40% sucrose flotation and suspended in 0.2 M NaCl assay buffer (in which all other chemical concentrations and pH were the same as in isolation buffer).

Apart from the GTPase experiments, which were performed in dim red light, experimental procedures were carried out using infra-red illumination (Kodak filter 87c) and an infra-red viewer. The rhodopsin concentration was measured from the absorbance change at 494 nm after illumination at pH 10.

GTPase assays

GTPase activity of photoreceptor membranes, incubated with 0.45 µM γ^{32}P-GTP and two different free calcium concentrations (0.01 and 1 µM), was determined by charcoal adsorbtion of nucleotides and molybdate separation of inorganic phosphate (Whaler & Wollenberger, 1958).

Toxin-catalysed radiolabelling

Both toxins were incubated at 24°C for 30 minutes, with membranes containing 10 µg rhodopsin in the ratios of 0.5 µg pertussis toxin:1 µg rhodopsin; and 0.33 µg cholera toxin:1 µg rhodopsin. The incubation was performed in the light and dark at 24°C for 30 minutes, in assay buffer containing 7µM ^{32}P-NAD, 10 mM thymidine, 12 mM arginine, 1 mM ATP, and 100 µM GTP (with cholera toxin). The reaction was terminated by washing with excess sample buffer. The membranes were dissolved in SDS/mercaptoethanol, loaded on a 10% polyacrylamide gel and electrophoresed. Gels were stained with Coomassie Blue, vacuum dried and exposed to Kodak X-Omat 100 film with an intensifying screen.

Effect of calcium on toxin catalysed radiolabelling

The procedure was followed as above but this time pertussis toxin and cholera toxin catalysed labelling were performed in the dark and light respectively. Calcium chloride was added to the reaction mixture to give 1 µM free calcium and controls had no added calcium.   After washing in assay buffer, the photoreceptor membranes were centrifuged and the resulting pellet dissolved in SDS/mercaptoethanol and loaded on to a 10% polyacrylamide gel.   The wash supernatant was treated with cold acetone in order to precipitate any released protein.

Results

GTPase assay

Table 1. Effects of calcium and toxin treatment on GTPase activity. Slopes $\pm$ SD from linear regression analysis of light-stimulated GTP hydrolysis, expressed as pmol hydrolysed/mg rhodopsin/min.   Each rate is determined from 3 independent data sets measured at 4 time points over a period of 12 minutes.

Free calcium concentration	Control	Pertussis toxin	Cholera toxin	Pertussis and Cholera toxin
10^{-8}M	2.40$\pm$0.23	1.06$\pm$0.02	0.70$\pm$0.05	0.07$\pm$0.03
10^{-6}M	0.73$\pm$0.72	0.08$\pm$0.04	0.11$\pm$0.003	0.08$\pm$0.00007

Pertussis and cholera toxin treatments were used to selectively study GTPase activity and the results are summarized in Table 1.   Pretreatment with pertussis toxin gives a 55% decrease in light-stimulated activity and cholera toxin decreases light activity by 71% compared with non toxin-treated membranes.   An increase of calcium concentration to 1 µM results in a decrease of 71% for light stimulated GTPase activity in non toxin-treated membranes.   1 µM calcium reduces the GTPase activity by 92% in pertussis treated membranes and by 84% in cholera toxin treated membranes. Membranes treated with both toxins are no longer capable of GTP hydrolysis.

Effect of light on toxin-catalysed ADP-ribosylation

In the dark, pertussis toxin caused radiolabelling of a component, in this case a doublet of molecular weight around 40 kD. However, this component is often seen as a single labelled band. Upon illumination, toxin-catalysed ribosylation of the 40 kD component was reduced (Fig. 1). This is similar to the result obtained by Tsuda et al (1986) for octopus photoreceptor membranes. Cholera toxin catalysed the ribosylation of several proteins due to its lower substrate specificity (Fig. 1). However, a band of molecular weight 45 kD was labelled more strongly in the light than in the dark. Another band of low molecular weight was labelled more in the light, but the presence of this band appears to be associated with cholera toxin treatment, and it is not seen in gels of untreated membranes. The 45 kD component is likely to be the one identified by Vandenberg and Montal (1984).

Effect of calcium on toxin-catalysed ADP-ribosylation

1 μM calcium reduces labelling catalysed by both pertussis toxin and cholera toxin of the components described above (Fig. 2). After cold acetone treatment, 5% of the total label was retrieved in the supernatant (data not shown). There was no apparent difference in the amount and position of labelled proteins between precipitated supernatants obtained from controls and calcium treated samples.

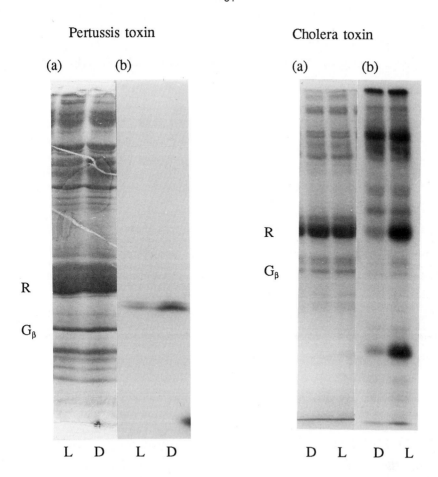

Figure 1. Toxin-catalysed, light-dependent radiolabelling of squid photoreceptor membranes. a) Coomassie Blue stained gels and b) autoradiographs of photoreceptors incubated with 7 μM ^{32}P-NAD, 10 mM thymidine, 12 mM arginine, 1 mM ATP and either pertussis toxin (2:1 rhodopsin:pertussis) or cholera toxin (3:1 rhodopsin:cholera) with 100 μM GTP. Photoreceptors were incubated at 24°C for 30 minutes with toxin both in light and darkness. Pertussis toxin catalyses increased labelling of a doublet (MW = 40kD) in the dark. Cholera toxin has a low substrate specificity and can ADP-ribosylate several proteins. A component of MW = 45kD is labelled more strongly in the light. Another protein of low molecular weight is also more strongly labelled in the light. This protein appears to be correlated with cholera toxin treatment. It is only visible in Coomassie Blue stained gels of membranes treated with cholera toxin.

Pertussis toxin

    (a)   (b)

Cholera toxin

    (a)   (b)

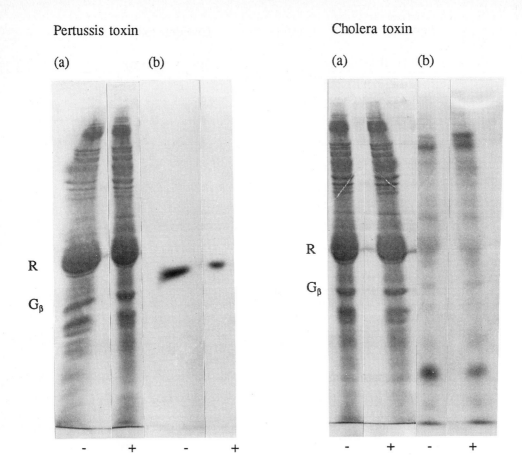

Figure 2. The effect of calcium on ADP-ribosylation of squid photoreceptor membranes. Photoreceptor membranes were incubated with ^{32}P-NAD and pertussis toxin (in the dark) and cholera toxin (in the light) in the presence of EGTA or 1 μM free calcium. Coomassie Blue stained gels (a) and autoradiographs (b) show ADP-ribosylation in the absence (-) and presence (+) of added calcium. Calcium inhibits radiolabelling catalysed by both toxins. The light-dependent cholera toxin substrate (see Fig 1) is very weakly labelled but labelling is further reduced by calcium.

Discussion

Our results show that squid photoreceptor membranes contain at least two G-proteins. In this respect they resemble hormonal systems more than vertebrate rod outer segment membranes. These results provide the first evidence of a large effect of calcium ions at physiological concentrations on G-proteins. High affinity GTPase activity is strongly inhibited by 1 μM free calcium but not by 0.01 μM. The light- induced GTPase activity of transducin in vertebrate rods is not inhibited by calcium under identical conditions (Fyles & Saibil, unpublished observations). The calcium increase after light stimulation in squid retina (Pinto & Brown, 1977) may provide a negative feedback pathway, turning off G-proteins and inhibiting further release of calcium via the PI pathway. Payne et al. (1988) have shown that the light-induced calcium increase in *Limulus* photoreceptors causes a decrease in sensitivity to light and to InsP$_3$ (adaptation). The effect of calcium on G-proteins in squid may be part of the negative feedback pathway of adaptation. Calmodulin has been shown to bind to the $\beta\gamma$-subunit of G$_i$ (Katada et al 1987), but there does not appear to be enough calmodulin in the squid photoreceptor membrane preparation to account for the inhibitory effect of calcium reported here (N. Ryba, unpublished observations). Further work is required to determine the mechanism by which calcium inhibits invertebrate photoreceptor G-proteins.

References

Baer KM, Saibil HR (1988) Light- and GTP-activated hydrolysis of phosphatidylinositol bisphosphate in squid photoreceptor membranes. J. Biol. Chem. 263:17-20

Bloomquist BT, Shortridge RD, Schneuwly S, Perdew M, Montell C, Steller H, Rubin G, Pak WL (1988) Isolation of a putative phospholipase C gene of *Drosophila*, *norpA*, and its role in phototransduction. Cell 54:723-733

Blumenfeld A, Eurusalimsky J, Heichal O, Selinger Z, Minke B (1986) Light-activated guanosine triphosphatase in *Musca* eye membranes resembles the prolonged depolarizing after potential in photoreceptor cells. Proc. Natl. Acad. Sci. USA 82:7116-7120

Brown JE, Watkins DC, Malbon CC (1987) Light-induced changes in the content of inositol phosphates in squid (*Loligo pealei*) retina. Biochem. J. 247:293-297

Calhoon R, Tsuda M, Ebrey TG (1980) A light-activated GTPase from octopus photoreceptors. Biochem. Biophys. Res. Commun. 94:1452-1457

Devary O, Heichal O, Blumenfeld A, Cassel D, Suss E, Barash S, Rubinstein CT, Minke B, Selinger Z (1987) Coupling of photoexcited rhodopsin to phosphoinositide hydrolysis in fly photoreceptors. Proc. Natl. Acad. Sci. USA 84:6939-6943

Johnson EC, Robinson PR, Lisman JE, (1986) Cyclic GMP is involved in the excitation of invertebrate photoreceptors. Nature 324:469-470

Katada T, Kusakabe K, Oinuma M, Ui M (1987) A novel mechanism for the inhibition of adenylate cyclase via inhibitory GTP-binding proteins. J. Biol. Chem. 262:11897-11900

Ovchinikov YA, Abdulaev NG, Zolotarev AS, Artamonov ID, Bespalov IA, Dergachev AE, Tsuda M (1988) Octopus rhodopsin. Amino acid sequence deduced from cDNA. FEBS Lett 232:69-72

Paulsen R, Bentrop J (1986) Light-modulated events in fly photoreceptors. Prog Zool. 33:299-319

Payne R, (1986) Phototransduction by microvillar photoreceptors of invertebrates: mediation of a visual cascade by inositol trisphosphate. Photobiochem. Photobiophys. 13:373-397

Payne R, Walz B, Levy S, Fein A (1988) The localisation of calcium release by inositol trisphosphate in *Limulus* photoreceptors and its control by negative feedback. Phil. Trans. R. Soc. Lond. B 320:359-379

Pinto LH, Brown JE (1977) Intracellular recordings from photoreceptors of the squid (*Loligo pealii*). J. Comp. Physiol. 122:241-250

Saibil HR (1984) A light-stimulated increase in cyclic GMP in squid photoreceptors. FEBS Lett 168:213-216

Saibil HR, Michel-Villaz M (1984) Squid rhodopsin and GTP-binding protein crossreact with vertebrate photoreceptor enzymes. Proc. Natl. Acad. Sci. USA 81:5111-5115

Stryer L (1986) Cyclic GMP cascade of vision. Ann. Rev. Neurosci. 9:87-119

Szuts EZ, Wood SF, Reid MA, Fein A (1986) Light stimulates the rapid formation of inositol trisphosphate in squid retinae. Biochem. J. 240:929-932

Tsuda M, Tsuda T, Terayama Y, Fukada Y, Akino T, Yamanaka, Stryer L, Katada T, Ui M, Ebrey T (1986) Kinship of cephalopod photoreceptor G-protein with vertebrate transducin. FEBS Lett. 198:5-10

Vandenberg CA, Montal M (1984) Light-regulated biochemical events in invertebrate photoreceptors. 1. Light-activated guanosine triphosphatase, guanine nucleotide binding and cholera toxin catalyzed labeling of squid photoreceptor membranes. Biochem 23:2339-2347

Whaler BE, Wollenberger A (1958) Zur Bestimmung des Orthophosphats neben säuremolybdat-labilen Phosphorsäureverbindungen. Biochemische Zeitschrift 329:508-520

DEGRADATION OF THE INVASIVE ADENYLATE CYCLASE TOXIN OF BORDETELLA PERTUSSIS
BY THE EUKARYOTIC TARGET CELL-LYSATE

Anat Gilboa-Ron and Emanuel Hanski
Department of Hormone Research
The Weizmann Institute of Science
Rehovot 76100, Israel

INTRODUCTION

Bordetella pertussis, the causative agent of whooping cough, produces several factors which have been suggested to play a role in the pathogenesis of the disease. One of these is an adenylate cyclase (AC) which has several unique properties. The location of the enzyme in the bacterium is largely extracytoplasmic and a small amount of it is secreted into the culture medium during exponential growth (Hewlett et al., 1976). It is activated up to 1000-fold by the Ca^{2+}-binding protein calmodulin (CaM) (Hewlett et al., 1976; Wolff et al., 1980) which does not exist in the bacteria. Confer and Eaton (1982) and Hanski and Farfel (1985) have shown that B. pertussis AC can penetrate various eukaryotic cells and generates high levels of intracellular cAMP. In immune effector cells the increase of cAMP impairs functions such as chemotaxis, phagocytosis, superoxide generation and microbial killing, leading to evasion of host defences (Weiss and Hewlett, 1986). It was, therefore, suggested that B. pertussis AC acts as a toxin which suppresses immune responses and assists the survival of the bacterium (Weiss and Hewlett, 1986). The importance of B. pertussis AC as a virulence factor was demonstrated by using mutants deficient in AC which were avirulent (Weiss et al., 1984).

Recent purification studies of B. pertussis AC have indicated that multiple heterologous forms of the enzyme are produced by the native bacterium. The molecular weights reported for the extracellular enzyme range between 43 and 50 kDa (Shattuck et al., 1985; Ladant et al., 1986), whereas a much larger form of 200 kDa was isolated from extracts of concentrated bacterial cells (Friedman, 1987; Rogel et al., 1988). Rogel et al. (1988) have recently shown that all froms of B. pertussis AC are immunologically

----

ABBREVIATIONS
CaM - calmodulin; PMSF - phenylmethansulfonyl fluoride; TPCK - L-1-tosyl-amido-2-phenylethyl chloromethyl ketone; App(NH)p - adenosine-5-[β-γ-imido]triphosphate; AC - adenylate cyclase; PBS - phosphate-buffered saline; CHAPS - 3-[(3-cholamidopropyl)dimethylamino]- 1-propanesulfonate; SDS - sodium dodecyl sulfate; PAGE - polyacrylamide gel electrophoresis.

related. Glaser et al., (1988) have cloned and sequenced B. pertussis gene for AC and showed that it has the potential to encode a polypeptide of 1706 amino acids which corresponds to calculated M_r of 177 kDa. Thus, the enzyme appears to be synthesized as a precursor of 200 kDa which is processed in the bacterium and secreted at lower M_r forms.

The structure of the toxic form of the enzyme, i.e. the from of the enzyme which is capable of penetrating eukaryotic cell has not been completely elucidated. The toxic form of the enzyme present in bacterial extract, migrated on gel filtration as a distinct form associated with a minor peak of AC activity with apparent size of 190 kDa (Hanski and Farfel, 1985). Immunoblot analysis of this form revealed the presence of a catalyst of 200 kDa, and in some preparations, additional breakdown products of lower molecular weight (Rogel et al., 1988). The mechanism by which B. pertussis AC penetrates mammalian cells is not known. It seems that the enzyme penetrates the plasma membrane of the target cell directly and does not enter the cytosol via endocytic vesicles. Agents that inhibit endocytosis by acidification of endosomes did not interfere with B. pertussis AC penetration (Hanski and Farfel, 1985; Gordon et al. 1988).

Hanski and Farfel (1985) developed a procedure which made it possible to demonstrate the dose-time and temperature-dependent accumulation of the invasive enzyme itself in human lymphocytes. This penetration led to rapid and extensive production of intracellular cAMP. The accumulation of the activity of the bacterial enzyme within the host cell was rapid and proceeded without any noticeable lag-time. After 10-30 min of cell exposure a constant level was reached and maintained provided that the invasive enzyme was present in the incubation medium (Friedman et al., 1987; Farfel et al., 1987). Transfer of the cells into toxin-free medium led to a rapid loss ($t_{1/2}$ ~20 minutes) of intracellular AC activity (Friedman et al., 1987). This decrease reflected intracellular inactivation of the bacterial enzyme and did not result from the release of the invasive enzyme back to the incubation medium. The inactivation process which occurs within the cell was dependent on cell integrity; when the treated cells were disrupted and the lysate was reincubated at 36°C there was no loss of invasive AC activity (Fig. 1) (Friedman et al., 1987). We tested whether or not metabolic energy is required for the process of B. pertussis AC intracellular inactivation. Here we demonstrated (Gilboa-Ron et al., 1988) that the decay of B. pertussis AC activity within human lymphocytes was completely prevented

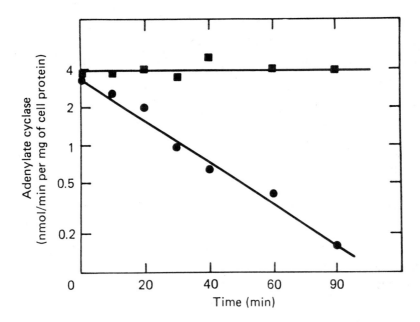

Fig. 1: Intracellular decrease of adenylate cyclase activity. (After Friedman et al., 1987). Human lymphocytes ($4 \cdot 10^7$ cells) were incubated in PBS with the invasive AC. After 6 min, cells were transferred to 4°C, washed and trypsin treated to remove enzyme that adhered to external cell surface. Treated cells were resuspended in 8 ml of PBS, and divided into two equal pools; one pool was reincubated at 36°C. At the indicated times, 0.5 ml samples were withdrawn, transferred to ice and intracellular cyclase activity was determined (●). The second pool was centrifuged, disrupted by freezing and thawing, and the lysate obtained was reincubated at 36°C. At the indicated times cyclase activity (■) was determined.

by depletion of cellular energy. Broken-cell preparations proteolytically inactivated B. pertussis AC in an ATP-dependent manner. These results suggest that the invasive AC is inactivated in the host cell by proteolysis.

METHODS

Culture of organism and purification of B. pertussis AC

B. pertussis strain 165 bacteria were grown as described by Hanski and Farfel (1985). The partially purified invasive form of AC was obtained by chromatography of crude dialyzed urea extract of the bacteria on ultrogel AcA 34 column as previously described (Hanski and Farfel, 1985; Rogel et al., 1988). A 200 kDa catalyst was purified from this form according to the procedure described by Rogel et al. (1988) and Gilboa-Ron et al. (1988).

Human lymphocytes and lymphocyte lysates

Human lymphocytes were isolated from whole blood as previously described by Hanski and Farfel (1985). Cells were suspended in lysing solution containing: 20 mM Hepes, pH 7.5, 2 mM $MgCl_2$, 2 mM EDTA, 2 mM leupeptin and 0.15 mM pepstatin. Then the cells were disrupted by freezing and thawing twice using liquid nitrogen followed by homogenization. Nuclei and unbroken cells were removed by centrifugation at 900 x g for 10 minutes.

Determination of the invasive-enzyme activity

Lymphocytes were incubated with the invasive form of B. pertussis AC and the activity of the enzyme within the cells was determined as previously described (Hanski and Farfel, 1985; Friedman et al., 1987). The intracellular decay in the activity of the invasive enzyme was monitored as explained by us before (Friedman et al., 1987; Farfel et al., 1987).

Inactivation of B. pertussis AC by lymphocyte-lysates

The inactivation reaction was as described by Gilboa-Ron et al. (1988). Essentially, the reaction was performed at 36°C in final volume of 50 μl and included the following ingredients: 0.2-0.4 mg·ml^{-1} cell lysates, 1 μM CaM, 0.1 mM $CaCl_2$, 2 mM $MgCl_2$, 2 mM EDTA and (when included) 3 mM ATP. In the absence of ATP, an ATP trap consisting of 10 mM glucose and 1 unit of hexokinase was added. When lysates of non-treated cells were used - B. pertussis AC or ^{125}I-labeled 200-kDa catalyst were added to the inactivation reaction. At various times, samples of 5-10 μl were withdrawn, diluted 10-fold and assayed for AC activity as described (Gilboa-Ron et al., 1988). Visualization of the degradation was performed by subjecting the withdrawn samples which contained ^{125}I-labeled enzyme to SDS-PAGE followed by autoradiography.

Iodination and PAGE

Iodination of B. pertussis AC was done according to the method devised by Ladant (1988), and PAGE was performed as described by Laemmli (1970). For detection of AC activity in a gel, gel lanes were cut into 0.4 cm slices, and the slices were treated with 2% CHAPS solution as described by Masure et al. (1988). After renaturation, AC activity and radioactivity were determined. Other procedures were as described by Gilboa-Ron et al. (1988).

RESULTS AND DISCUSSION

The inactivation of the invasive AC of B. pertussis requires metabolic energy.

We tested whether metabolic inhibitors will affect the intracellular inactivation of B. pertussis AC. As shown in Fig. 2. human lymphocytes which were depleted of their metabolic energy after being exposed to the invasive enzyme lost the ability to inactivate the invasive AC (Gilboa-Ron et al., 1988). To ensure complete depletion of metabolic energy the cells were treated with a combination of two metabolic inhibitors. 2-deoxyglucose, which inhibits glycolysis and the uncoupler of oxidative phosphorylation, NaN_3. This treatment resulted in depletion of the cellular ATP content. Transfer of cells into medium containing glucose and adenine restored the intracellular ATP level and the inactivation of the invasive enzyme (Gilboa-Ron et al., 1988).

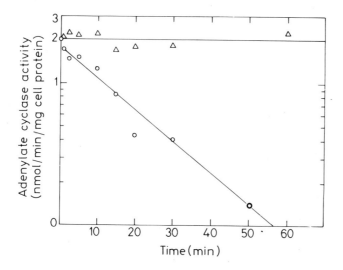

Fig. 2: The inactivation of B. pertussis AC in intact cells requires cellular energy. Human lymphocytes ($2 \cdot 10^7$) were incubated for 25 minutes at 36°C in 8 ml PBS with 500 μg of the invasive form of B. pertussis AC. Then the cells were divided into two equal pools. One pool (Δ-Δ) received 10 mM NaN_3 and 50 mM 2-deoxyglucose and both pools were further incubated for 10 minutes. Then the cells (of the two pools) were washed extensively and reincubated in toxin-free PBS in the presence (Δ-Δ) or in the absence (O-O) of metabolic inhibitors. At the indicated times the intracellular AC activity was determined. Methods were as described by Gilboa-Ron et al. (1988).

ATP could also promote inactivation of B. pertussis AC in lysates of lymphocytes pre-exposed to the invasive enzyme, while in the absence of ATP there was no inactivation (Fig. 3, see also Fig. 1).

We have tested proteolysis as a possible mechanism for the inactivation of B. pertussis AC. For that purpose the enzyme was radiolabeled in a way that preserved its catalytic activity. This was achieved by performing iodination of the enzyme while being bound to CaM-agarose column. The ^{125}I-labeled and active AC was subjected to lymphocyte lysate in the presence and absence of ATP. As shown in Fig. 4B the ^{125}I-labeled AC was degraded both in the absence and in the presence of ATP. In the presence of ATP there was rapid inactivation (Fig. 4A) associated with complete degradation of the enzyme (Fig. 4A and B). In the absence of ATP there was an increase to about 150 to 200% of the original enzyme-activity (Fig. 4A).

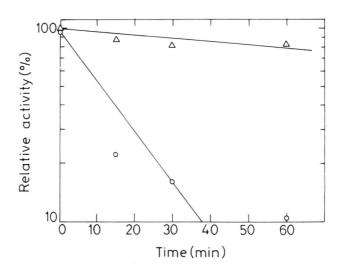

Fig. 3: The inactivation of B. pertussis AC in cell-lysates requires ATP. Human lymphocytes ($3 \cdot 10^7$) were incubated in 5 ml PBS with the invasive B. pertussis AC. After 20 minutes cells were washed and cell lysates were prepared. The lysates were reincubated at 36°C in the presence (O-O) or absence (Δ-Δ) of 2 mM ATP. At the indicated times the AC activity was determined. 100% represents the activity which was 4.5 nmol·min^{-1}·mg^{-1}. Methods were as described by Gilboa-Ron et al. (1988).

The autoradiography revealed accumulation of a 42-kDa fragment which was resistant to further proteolysis (Fig. 4B). This 42 kDa fragment was found to be catalytically active: This was demonstrated by renaturation of CaM-dependent AC activity of 42 kDa fragment after SDS-PAGE (Gilboa-Ron et al., 1988).

ADP, AMP, GTP and CTP did not support degradation and inactivation of the enzyme. The degradation was promoted by non-hydralyzable ATP analogs such as App(NH)p, α,β methylene ATP and β,γ methylene ATP. In addition, non-hydrolyzable GTP analogs or cAMP did not support inactivation and degradation of the enzyme (Gilboa-Ron et al., 1988). Further analysis of the degradation reaction of B. pertussis AC had shown that it did not require free Mg^{2+} and in fact it could occur in the presence of an excess of uncomplexed ATP. Breakdown of proteins by ATP-dependent proteases requires hydrolysis of ATP and an excess of free Mg^{2+} over ATP (Goldberg and St. John, 1976; Hersko and Ciechanover, 1982). Thus, the above result

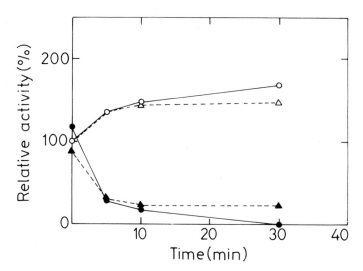

Fig. 4: Inactivation and degradation of ^{125}I-AC by lymphocyte-lysates. A. The 200-kDa form of B. pertussis AC was purified as described by Rogel et al. (1988). A solid-phase iodination was performed according to the method of Ladant (1988). The labeled catalyst was incubated with lymphocyte-lysates either in the presence (full symbols) or absence (empty symbols) of ATP. At the indicated times the AC activity was determined. The above was repeated in the presence ($\blacktriangle$-$\blacktriangle$, $\triangle$-$\triangle$) of the following protease inhibitors: PMSF (1 mM), leupeptin (2 mM), pepstatin (0.2 mM), TPCK (0.1 mM). The lysates were pretreated for 1 h at 4°C with 5 mM N-ethylmaleimide which was quenched with 5.5 mM 2-mercaptoethanol. 100% represents the activity of ^{125}I-labeled 200-kDa catalyst at time zero of the inactivation reaction.

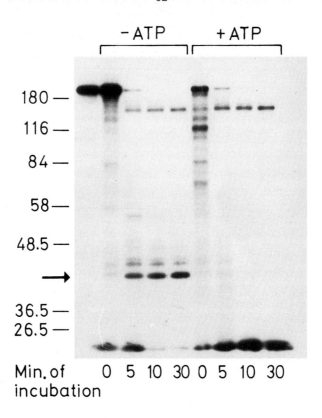

- ATP   + ATP

180 —

116 —

84 —

58 —

48.5 —

→

36.5 —
26.5 —

Min. of   0 5 10 30 0 5 10 30
incubation

B. Identical samples as in A were run on 9% PAGE in SDS and the gel was autoradiographed. The left lane represents the ^{125}I-labeled 200-kDa catalyst which was incubated with ATP and CaM under the conditions of the inactivation reaction but in the absence of lymphocyte-lysates. The molecular weight standards with their masses in kDa were; α-Macroglobulin (180), β-galactosidase (116), Fructose-6-phosphate Kinase (84), Pyruvate Kinase (58), Fumarase (48.5), Lactic Dehydrogenase (36.5) and Triose phosphate Isomerase (26.5). The arrow indicates the position of the 42-kDa fragment which was generated in the absence of ATP. Methods were as described by Gilboa-Ron et al. (1988).

suggests that ATP binds to B. pertussis AC and renders the enzyme susceptible to complete degradation, rather than acting by stimulation of ATP-dependent proteolysis.

Is the ATP-dependent proteolysis demonstrated by us in cell lysates related to the inactivation of the invasive enzyme observed within intact cells? The above notion is suggested by the following findings: Depletion of ATP by metabolic inhibitors prevented the inactivation of the invasive

AC in intact human lymphocyte and addition of exogenous ATP to lymphocyte-lysates restored it. In addition, the kinetics of the enzyme inactivation both in lysate and intact cells fitted a first order reaction with half-life of 8 to 15 minutes. Direct demonstration that the invasive enzyme in intact cells is inactivated by proteolysis must wait until an invasive and labeled preparation of B. pertussis AC is available.

REFERENCES

Confer DL, Eaton JW (1982) Phagocyte impotence caused by an invasive bacterial adenylate cyclase. Science 217:948-950

Farfel Z, Friedman E, Hanski E (1987) The invasive adenylate cyclase of Bordetella pertussis. Intracellular localization and kinetics of penetration into various cells. Biochem J 243:153-158

Friedman RL (1987) Bordetella pertussis adenylate cyclase: Isolation and purification by calmodulin-sepharose 4B chromatography. Infect Immun 55:129-134

Friedman E, Farfel Z, Hanski E (1987) The invasive adenylate cyclase of Bordetella pertussis. Properties and penetration kinetics. Biochem J 243:145-151

Gilboa-Ron A, Rogel A, Hanski E (1988) Lysate of host cell proteolytically inactivates Bordetella pertussis adenylate cyclase. Submitted for publication.

Glaser P, Ladant D, Sezer O, Pichot F, Ullmann A, Danchin A (1988) The calmodulin-sensitive adenylate cyclase of Bordetella pertussis: Cloning and expression in Eschrichia coli. Molec Microbiol 2:19-30

Goldberg AL, St John AC (1976) Intracellular protein degradation in mammalian and bacterial cells: Part 2. Annu Rev Biochem 45:747-803

Gordon VM, Leppla SH, Hewlett EL (1988) Inhibitors of receptor-mediated endocytosis block the entry of Bacillus anthracis adenylate cyclase toxin but not that of Bordetella pertussis adenylate cyclase toxin. Infect Immun 56:1066-1069

Hanski E, Farfel Z (1985) Bordetella pertussis invasive adenylate cyclase. Partial resolution and properties of its cellular penetration. J Biol Chem 260:5526-5532

Hershko A, Ciechanover A (1982) Mechanisms of intracellular protein breakdown. Annu Rev Biochem 51:335-364

Hewlett EL, Urban MA, Manclark CR, Wolff J (1976) Extracytoplasmic adenylate cyclase of Bordetella pertussis. Proc Natl Acad Sci USA 73:1926-1930

Ladant D (1988) Interaction of Bordetella pertussis adenylate cyclase with calmodulin. Identification of two separated calmodulin-binding domains. J Biol Chem 263:2612-2618

Ladant D, Brezin C, Alonso JM, Crenon I, Guiso N (1986) Bordetella pertussis adenylate cyclase. Purification, characterization and radioimmunoassay. J Biol Chem 261:16264-16269

Laemmli UK (1970) Cleavage of structural proteins during the assembly of the head of bacteriophage T4. Nature 227:680-685

Masure HR, Oldenburg DJ, Donovan MG, Shattuck RL, Storm DR (1988) The interaction of Ca^{2+} with the calmodulin-sensitive adenylate cyclase from Bordetella pertussis. J Biol Chem 263:6933-6940

Rogel A, Farfel Z, Goldschmidt S, Shiloach J, Hanski E (1988) Bordetella pertussis adenylate cyclase. Identification of multiple forms of the enzyme by antibodies. J Biol Chem 263:13310-13316

Shattuck RL, Oldenburg DJ, Storm DR (1985) Purification and characterization of a calmodulin-sensitive adenylate cyclase from Bordetella pertussis. Biochemistry 24:6356-6362

Weiss AA, Hewlett EL (1986) Virulence factors of Bordetella pertussis. Annu Rev Microbiol 40:661-686

Weiss AA, Hewlett EL, Myers, GA, Falkow S (1984) Pertussis toxin and extracytoplasmic adenylate cyclase as virulence factors of Bordetella pertussis J Infect Dis 150:219-222

Wolff J, Cook G, Goldhammer A, Berkowitz S (1980) Calmodulin activates prokaryotic adenylate cyclase. Proc Natl Acad Sci USA 77:3841-3844

IDENTIFICATION AND CHARACTERIZATION OF ADENYLATE CYCLASES IN VARIOUS TISSUES BY MONOCLONAL ANTIBODIES

Stefan Mollner, Ulrich Heinz and Thomas Pfeuffer
Department of Physiological Chemistry
University of Würzburg
Koellikerstr. 2
D-8700 Würzburg
Federal Republic of Germany

Introduction

In vertebrate cells, the adenylate cyclase is regulated by
hormones or neurotransmitters. Adenylate cyclase of eucaryo-
tic organisms is a membrane-bound multi-component enzyme.
The complex contains hormone receptor (R), guanine nucleotide
binding proteins (G_s and G_i) and the catalytic unit (C).
Recently the catalyst, a calmodulin-insensitive enzyme from
myocardium and a calmodulin-sensitive enzyme from brain has
been purified to homogeneity by forskolin-Sepharose affinity
chromatography (Pfeuffer et al., 1985a; Pfeuffer et al. 1985b;
Smigel, 1986).
Brostrom et al. (1975) first reported on the existence of two
forms of adenylate cyclase activity in the neuronal system of
higher organisms, one which is activated by calmodulin at low
concentrations of Ca^{2+} and the other which has a calmodulin-
independent adenylate-cyclase activity.
Very little is known about the calmodulin-insensitive enzyme
from the brain since it has not been purified till now. More
insights in the relationships between the different adenylate
cyclases in brain could be obtained by antibodies directed
against diverse sites of the enzyme.
In this study we developed monoclonal antibodies against the
catalytic unit of the purified calmodulin-sensitive enzyme
from bovine brain cortex. Two monoclonal antibodies were obtained with a strict specificity to the calmodulin-sensitive
enzyme of bovine brain. Two other clones produced antibodies
which showed in addition cross-reaction with the calmodulin-

NATO ASI Series, Vol. H29
Receptors, Membrane Transport and Signal Transduction
Edited by A. E. Evangelopoulos et al.
© Springer-Verlag Berlin Heidelberg 1989

insensitive enzyme from brain, heart and lung. The cross-
reacting antibodies also recognized C in other species like
rabbit, rat and turkey.

Results

Hybridoma cells of BALB/c mice which have been immunized with
pure calmodulin-sensitive bovine enzyme were initially
screened in an ELISA applying the homogenous 115-kDa antigen. In Fig. 1 four monoclonal antibodies were tested for
their ability to react with the purified calmodulin-sensitive enzyme (115-kDa) in an ELISA. The titration curve shows
that the antibodies BBC-2 and BBC-4 have a lower affinity to
the antigen compared with BBC-1 and BBC-3 and do not attain
the maximal absorbance of the latter.

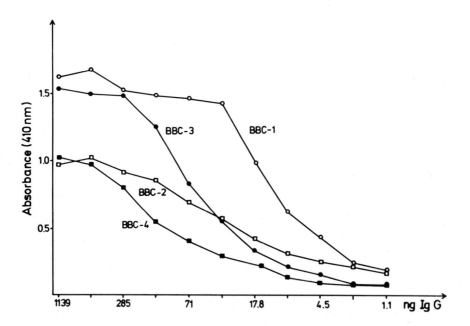

Fig.1: ELISA of adenylate cyclase from bovine brain cortex.
A microtiter plate was coated with pure bovine brain adenylate cyclase und incubated with various dilutions of BBC-1 to BBC-4 as
described by Mollner and Pfeuffer (1988).

To determine whether the monoclonal antibodies were directed
to different sites of the antigen an epitope analysis was performed (Fig.2). This experiment in which an antibody competed
against an other for a given recognition site on pure bovine
brain cortex adenylate cyclase suggested that BBC-1 and BBC-3
as well as the pair BBC-2 and BBC-4 respectively bind to an
overlapping, if not identical epitope.

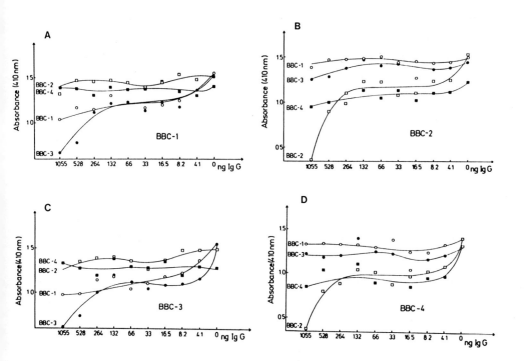

Fig.2: Analysis of monoclonal antibody epitopes by ELISA. The
wells of a microtiter plate were coated with a monoclonal antibody
and incubated with biotinylated adenylate cyclase and decreasing
dilutions of BBC-1 to BBC-4. (A) coated with BBC-1, (B) BBC-2,
(C) BBC-3, (D) BBC-4. Preincubation of the biotinylated adenylate
cyclase with: (O) BBC-1, (□) BBC-2, (●) BBC-3, (■) BBC-4.

Best results in identification of adenylate cyclases in Western blots were achieved when antibodies radioiodinated by

the chloramine T method were used. To verify whether the adenylate cyclase is recognized in membranes immunoblots of
solubilized striatal membranes with all four monoclonal antibodies were performed (Fig.3A). BBC-1 and BBC-3 recognize the
115-kDa antigen only, whereas BBC-2 and BBC-4 bind in addition
to a 150-kDa adenylate cyclase which seems to be more abundant in brain striatum than the smaller protein. Also more
than one antigen is recognized in membranes from bovine brain
cortex by the latter antibodies (Fig.3B). However, in brain
cortex the 115-kDa species is the most dominant.

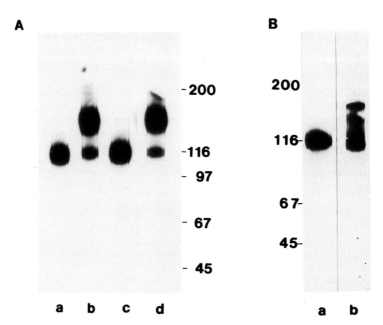

Fig.3: Recognition of different adenylate cyclases by monoclo-
       nal antibodies in bovine brain tissues. Bovine striatum
       (A) and bovine brain cortex (B) membranes were Lubrol-PX-solubi-
       lized, electrophoresed on a 5 - 15% SDS/polyacrylamide gel and
       blotted onto nitrocellulose. Antigens were probed with [^{125}I]
       iodinated antibodies BBC-1 (a), BBC-2 (b), BBC-3 (c), BBC-4 (d).

In Fig. 4 the species and the tissues specificity was investigated in immunoblots with the two types of monoclonal
antibodies (represented by BBC-1 and BBC-2). Antibody BBC-1
(lane a) is strictly specific for the 115-kDa enzyme from

bovine brain (cortex and striatum). In addition BBC-2 (lane b)
is recognizing the brain enzymes from other vertebrates like
rabbit, rat and turkey (Fig.4A). All these brain adenylate
cyclases show a molecular mass of 115-kDa with exception of
the turkey enzyme (135-kDa). The 60-kDa band in the preparation from the rat is most likely a product of limited proteolysis. In comparison to solubilized membranes the partially
forskolin-Sepharose purified preparation from bovine brain
cortex lacks the 150-kDa band and is reduced in the preparation of the bovine brain striatal cyclase (compare Fig.2
lanes b with Fig.4) due to the washing procedure during the
forskolin-Sepharose chromatography. BBC-2 also crossreacts
with the cyclase from bovine lung and heart and with that
from the rabbit myocardium (Fig.4B).

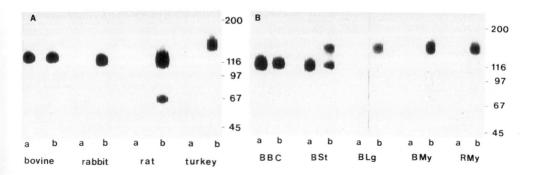

Fig.4: Immunoreactivity of monoclonal antibodies BBC-1 and
BBC-2 against brain adenylate cyclase from different
species (A) and adenylate cyclases from various tissues
(B). Enzyme preparations, partially purified on forskolin-Sepharose, were run on 5 - 15% SDS/polyacrylamide gels, blotted onto
nitrocellulose and incubated with radioiodinated monoclonal antibodies. BBC-1 (a); BBC-2 (b); BBC, bovine brain cortex; BSt, bovine striatum; BLg, bovine lung; BMy, bovine heart; RMy, rabbit heart.

Two adenylate cyclase activities from bovine striatum can be
separated by chromatography on calmodulin-Sepharose (Fig.5).
As demonstrated by immunoblotting only the 115-kDa form of

the cyclase was retained on the column and eluated with 2 mM
EGTA. The affinity purified enzyme is fully activated by Ca^{2+}/
calmodulin.

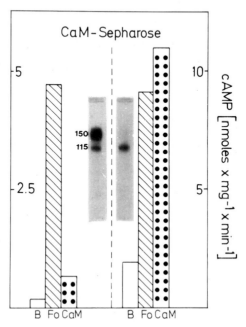

Fig.5: Separation of two striatal adenylate cyclases by calmodulin-Sepharose. Crude solubilized bovine striatal adenylate
cyclase was depleted of endogenous calmodulin according to
Westcott et al. (1979), bound to calmodulin-Sepharose and after
the resin was washed, calmodulin-bound proteins were released after addition of EGTA. Adenylate cyclase activities were determined in the absense (B, basal) and presence of forskolin (Fo) or
calmodulin (CaM). Immunoblots were performed with iodinated monoclonal antibody BBC-2.

Conclusions

Here we demonstrated the utility of monoclonal antibodies
against a mammalian adenylate cyclase for identifying the
enzyme in various tissues of several higher vertebrates.
Two types of antibodies were applied: one type (BBC-1 and
BBC-3) indicates that at least one epitope is restricted to

the calmodulin-sensitive enzyme from bovine brain. Adversely two other antibodies BBC-2 and BBC-4 are directed to a highly conserved epitope of the cyclase molecule, since this epitope is shared by the calmodulin-dependent enzymes from several higher vertebrates like cow, rat, rabbit, turkey and it is also present on the polypeptide of the calmodulin-insensitive enzyme from different tissues and species. The cross-reacting antibody BBC-2 clearly demonstrates that in brain tissues indeed distinct adenylate cyclases are existing: a calmodulin/Ca^{2+}-sensitive (115-kDa) and a calmodulin/Ca^{2+}-insensitive form (150-kDa) which cannot result from one precursor by limited proteolysis (Fig.3).
In non-neuronal tissues like heart and lung no 115-kDa enzyme (calmodulin-sensitive) could be detected. These data are in contradiction to reports that detergent-solubilized membranes of rat heart contain the calmodulin-sensitive form of the adenylate cyclase (Sulimovici et al., 1984; Panchenko and Tkachuck, 1984).

References

Brostrom, C.O., Huang, Y., Breckenridge, B.McL. and Wolff, D.J. (1975)
    Proc. Natl. Acad. Sci. USA 72, 64-68.
Mollner, S. and Pfeuffer T. (1988) Eur. J. Biochem. 171, 265-271.
Panchenko, M.P. and Tkachuck, V.A. (1984) FEBS Lett. 174, 50-54.
Pfeuffer, E., Dreher, R.M., Metzger, H. and Pfeuffer, T. (1985a)
    Proc. Natl. Acad. Sci. USA 82, 3086-3090.
Pfeuffer, E., Mollner, S. and Pfeuffer, T. (1985b) EMBO J. 4, 3675-3679.
Smigel, M.D. (1986) J. Biol. Chem. 261, 1976-1982.
Sulimovici, S., Pinkus, L.M., Susser, F.I., and Roginsky, M.S. (1984)
    Arch. Biochem. Biophys. 234, 434-441.
Westcott, K.R. LaPorte, D.C., and Storm, D.R. (1979) Proc. Natl. Acad.
    Sci. USA 76, 204-208.

THE ROLE OF G-PROTEINS IN EXOCYTOSIS

Jane Stutchfield, Blandine Geny[1] and Shamshad Cockcroft.
Department of Experimental Pathology,
University College London,
University Street,
London WC1E 6JJ, U.K.

ABSTRACT

Stimulation of many cell types with their agonists leads to the generation of two intracellular messengers, diacylglycerol and inositol trisphosphate (IP_3) derived from hydrolysis of phosphatidylinositol bisphosphate by polyphosphoinositide phosphodiesterase (PPI-pde) in the membrane. PPI-pde is coupled to receptors by a putative G-protein, G_p. PPI-pde activity and β-glucuronidase secretion can be measured simultaneously in permeabilized undifferentiated HL60 cells stimulated by Ca^{2+} and guanine nucleotides. The effect of GTPγS on secretion could be explained by its ability to generate IP_3 and diacylglycerol, an activator of protein kinase C. However, when increased intracellular Ca^{2+} mobilized by IP_3 is buffered, the phorbol ester, phorbol 12-myristate 13-acetate (PMA), a potent and stable activator of protein kinase C, does not fully substitute for GTPγS. The effects of PMA are (1) to increase the sensitivity to Ca^{2+}, (2) to dramatically increase secretion in the presence of GTPγS at nanomolar concentrations of Ca^{2+}, (3) to promote some secretion in the presence of Ca^{2+} alone and (4) to inhibit PPI-pde activity. PPI-pde activity stimulated by both GTPγS and Ca^{2+} alone was inhibited by pretreatment of the cells with PMA. Under these conditions, secretion of β-glucuronidase was increased. The dissociation between PPI-pde activity and secretion implicates the existence of a further G-protein, G_e, in the stimulus-secretion pathway.

---

[1]INSERM Unite 204, Centre Hayem, Hopital Saint Louis, Paris Cedex 10, France.

NATO ASI Series, Vol. H29
Receptors, Membrane Transport and Signal Transduction
Edited by A. E. Evangelopoulos et al.
© Springer-Verlag Berlin Heidelberg 1989

INTRODUCTION

The chemotactic peptide, fMetLeuPhe, stimulates secretion in neutrophils and HL60 cells; this is thought to be triggered by the two intracellular messengers, diacylglycerol which stimulates protein kinase C and inositol trisphosphate (IP_3) which mobilises intracellular Ca^{2+}. The generation of these two second messengers arises from the hydrolysis of phosphatidylinositol bisphosphate (PIP_2) by the enzyme known as polyphosphoinositide phosphodiesterase (PPI-pde) or phospholipase C (PLC). The membrane receptors for such agonists as fMetLeuPhe have been suggested to be coupled to the PPI-pde by a putative G-protein, G_p (Cockcroft & Gomperts, 1985).

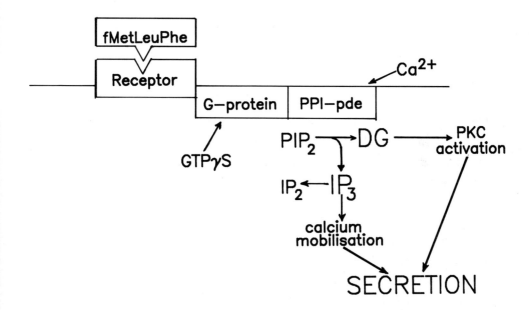

Figure 1 A proposed model for the activation of PPI-pde.

Figure 1 shows the proposed scheme for activation of G_p (for review see Cockcroft & Stutchfield, 1988a). The agonist (fMetLeuPhe in this case) interacts with the receptor which catalyses the exchange of bound GDP on the G-protein for GTP; this leads to dissociation of the α-subunit. The released α-subunit activates PPI-pde which catalyses the hydrolysis of PIP_2 to diacylglycerol and IP_3, the two intracellular messengers. The intrinsic GTPase activity of the α-subunit hydrolyses the bound GTP thus terminating the PPI-pde activity. Non-hydrolysable analogues of GTP, e.g. GTPγS, and fluoride maintain the activity by stimulating the G-protein directly. Stimulation of secretion and PPI-pde activity via the G-protein can be competitively inhibited by GDP. Millimolar concentrations of Ca^{2+} can directly stimulate PPI-pde activity without involving the G-protein (see figure 2) (Geny et al, 1988).

METHODS

In order to address the question of whether Ca^{2+} and protein kinase C activation are sufficient to elicit secretion, we have used cells permeabilized with streptolysin-O (Howell & Gomperts, 1987; Stutchfield & Cockcroft, 1988a). Streptolysin-O is a bacterial cytolysin which forms pores in the plasma membrane (of approximately 12 nm) which allow the introduction of Ca^{2+} (buffered with EGTA between pCa8 and pCa5) and GTPγS. GTPγS can directly interact with G_p and activate PPI-pde so we have also introduced GTPγS into the cell as an alternative way of raising endogenous diacylglycerol. PMA was used as an activator of protein kinase C.

Experiments were conducted in which secretion (of β-glucuronidase) and PPI-pde activity (formation of IP_2 and IP_3) were simultaneously measured in permeabilized HL60 cells. HL60 cells are a human promyelocytic cell-line that can be differentiated by dibutyryl cAMP to neutrophil-like cells (Chaplinski & Niedel, 1982). We have observed that the undifferentiated cells, although lacking receptors for fMetLeuPhe, have all the intracellular machinery required for the secretion of β-glucuronidase (Stutchfield & Cockcroft 1988a).

The cells were cultured, permeabilized and labelled as described previously (Stutchfield & Cockcroft, 1988a;1988b). After a 10 minute incubation period, the cells were quenched and samples were taken from the supernatant for measurement of β-glucuronidase and inositol phosphates as described previously (Stutchfield & Cockcroft 1988b). The membrane enriched preparation was made as described previously (Cockcroft & Stutchfield, 1988b). Pretreatment with phorbol 12-myristate 13-acetate (PMA) was carried out on intact cells for 60 minutes at 10nM PMA (Geny et al, 1988).

PPI-pde activity is expressed as % hydrolysis of inositol lipids; this is d.p.m. in IP_2 and IP_3 as a percentage of the radioactivity in total inositol-containing lipids.

RESULTS & DISCUSSION

Pretreatment of the cells with 10nM PMA for 60 minutes inhibited PPI-pde activity (Geny et al, 1988). In membranes, PPI-pde activity can be directly stimulated with millimolar concentrations of Ca^{2+} (this effect is not seen in cells permeabilized with streptolysin-O probably due to the presence of Ca^{2+}-dependent proteases). Inhibition of PPI-pde activity induced by Ca^{2+} (which directly stimulates PPI-pde) demonstrates that the effect of PMA-pretreatment is at the level of the enzyme (figure 2).

Using HL60 cells permeabilized with streptolysin-O, secretion of β-glucuronidase (figure 3, top panel) and PPI-pde activity (figure 3, lower panel) were measured simultaneously. Introduction of buffered Ca^{2+} alone (open bars) was not sufficient to support secretion whereas addition of GTPγS (hatched bars), in the presence of Ca^{2+}, stimulated secretion maximally. Pretreatment of the cells with PMA produced a modest increase in the level of secretion (stippled bars) but was markedly less effective than GTPγS. When GTPγS was also added (filled bars) secretion was stimulated markedly, particularly in the presence of nanomolar concentrations of Ca^{2+}. The effect of pretreatment with 10nM PMA on secretion is similar to direct addition of 100nM PMA (see Stutchfield & Cockcroft, 1988a). The effect of GTPγS on secretion is not solely due to generation of IP_3 or diacylglycerol. In these experiments, the effect of IP_3 on intracellular Ca^{2+} mobilization can be disregarded because Ca^{2+} is already provided in the form of Ca^{2+} buffers. Activation of protein kinase C by PMA cannot fully substitute for GTPγS in stimulating secretion, suggesting that maximal secretion requires a further G-protein dependent step.

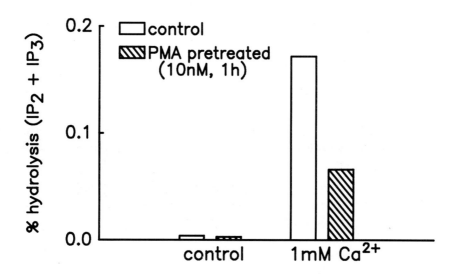

Figure 2 PPI-pde activity in HL60 membranes stimulated with 1mM Ca^{2+}: Effect of PMA-pretreatment.
^{3}H-inositol-labelled HL60 cells were pretreated with 10nM PMA for 1h and the cells homogenized to make a membrane-enriched fraction. The membranes were incubated with 1mM Ca^{2+} for 10 min at 37°C.

In the control cells, the effects of Ca^{2+} and GTPγS on PPI-pde activity were similar to those observed on secretion. GTPγS

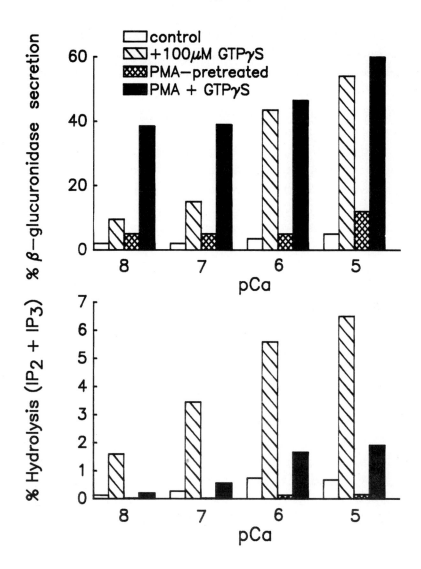

Figure 3 The upper panel shows β-glucuronidase secretion and
the lower panel PPI-pde activity in permeabilized
cells in the presence of 10nM, 100nM, 1μM and 10μM Ca^{2+}.
After pretreatment of the cells with 10nM PMA for 1h,
pretreated and control cells were permeabilized by
streptolysin-O in the presence of 1mM Mg.ATP and 10mM Li$^+$. The
cells were then incubated in the absence and presence of 100μM
GTPγS at the indicated Ca^{2+} concentration for 10 min at 37°C.

greatly potentiated both secretion and PPI-pde activity; both responses showed a similar sensitivity to Ca^{2+}. However, in the PMA-pretreated cells secretion and PPI-pde activity were dissociated. PMA inhibited both the Ca^{2+}-induced PPI-pde activity and the GTPγS-induced PPI-pde activity by 70-90% (figure 3, lower panel). Thus, in the presence of a maximal concentration of GTPγS (100μM) and optimal concentrations of Ca^{2+}, secretion from cells pretreated with PMA was dramatically increased whereas PPI-pde activity was markedly inhibited.

From these observations, we conclude that stimulation of secretion is dependent on two effectors, Ca^{2+} and GTPγS. The role of GTPγS in secretion, additional to its effect on PPI-pde, has led us to consider the possibility that GTPγS activates a second G-protein (designated G_e) that is responsible for triggering an effector system providing additional signals.

Dissociation of secretion and PPI-pde activity has been observed in several permeabilized cell types. In mast cells, neomycin which inhibits PPI-pde activity does not inhibit secretion stimulated by GTPγS and Ca^{2+} (Cockcroft et al, 1987). At extremely low Ca^{2+} concentrations which inhibit PPI-pde activity, GTPγS-stimulated secretion is observed in neutrophils (Barrowman et al, 1986), RINm5F cells (Vallar et al, 1987) and adrenal chromaffin cells (Bittner et al, 1986). The evidence for a further G-protein regulated step in secretion is accumulating although we cannot ignore the possibility that G_p itself may be coupled to the other effector(s).

REFERENCES

Barrowman MM, Cockcroft S & Gomperts BD. (1986) Two roles for guanine nucleotides in stimulus secretion sequence of neutrophils. Nature 319: 504-507
Bittner MA, Holz RW & Neubig RR. (1986) Guanine nucleotide effects on catecholamine secretion from digitonin-permeabilized adrenal chromaffin cells. J. Biol. Chem. 261: 10182-10188
Chaplinski TJ & Niedel JE. (1982) Cyclic nucleotide-induced maturation of human promyelocytic leukemia cells. J. Clin. Invest. 70: 953-964
Cockcroft S & Gomperts BD. (1985) Role of guanine nucleotide

binding proteins in the activation of polyphosphoinositide phosphodiesterase Nature **314**: 534-536

Cockcroft S, Howell TW & Gomperts BD. (1987) Two G-proteins act in series to control stimulus-secretion coupling in mast cells: Use of neomycin to distinguish between G-protein controlling polyphosphoinositide phosphodiesterase and exocytosis. J. Cell Biol. **105**: 2745-2750

Cockcroft S & Stutchfield J. (1988a) G-proteins, the inositol lipid pathway, and secretion. Phil. Trans. R. Soc. Lond. B **320**: 247-265

Cockcroft S & Stutchfield J. (1988b) Effect of pertussis toxin and neomycin on G-protein-regulated polyphosphoinositide phosphodiesterase: A comparison between HL60 membranes and permeabilized HL60 cells. Biochem. J. **256**: 343-350

Geny B, Stutchfield J & Cockcroft S. (1988) Phorbol ester inhibits polyphosphoinositide phosphodiesterase activity stimulated by either Ca^{2+}, fluoride or GTP analogues in HL60 membranes and in permeabilized HL60 cells. Cellular Signalling (in press).

Howell TW & Gomperts BD (1987) Rat mast cells permeabilized with streptolysin O secrete histamine in response to Ca^{2+} at concentrations buffer in the micromolar range. Biochim. Biophys. Acta **927**: 177-183

Stutchfield J & Cockcroft S. (1988a) Guanine nucleotides stimulate polyphosphoinositide phosphodiesterase and exocytotic secretion from HL60 cells permeabilized with streptolysin O. Biochem. J. **250**: 375-382

Stutchfield J & Cockcroft S. (1988b) Effects of phorbol ester on IP_3 production and secretion in permeabilized HL60 cells. Trans. Biochem. Soc. (in press)

Vallar L, Biden TJ & Wollheim CB. (1987) Guanine nucleotides induce Ca^{2+} independent secretion from permeabilized RINm5F cells. J. Biol. Chem. **262**: 5049-5056

HYDROPHOBIC INTERACTIONS IN THE CALCIUM- AND PHOSPHOLIPID DEPENDENT ACTIVATION OF PROTEIN KINASE C

G.T. Snoek.
Centre for Biomembrane and Lipid Enzymology.
University of Utrecht, Padualaan 8, 3584 CH, Utrecht.
The Netherlands.

INTRODUCTION.

The Ca^{++}-activated, phospholipid-dependent protein kinase, protein kinase C, has been implicated as a regulatory element in receptor-mediated signal transduction and as the major cellular receptor for, and mediator of the action of tumourpromoting phorbol esters (Nishizuka, 1984; Leach et al., 1983). During signal transduction induced by ligand-receptor binding, phosphoinositides are hydrolysed in the plasma membrane and diacylglycerol (DAG) is transiently generated (Hokin, 1985; Berridge, 1984; Berridge & Irvine, 1984). DAG and the biologically active phorbol ester, phorbol 12-myristate, 13-acetate (PMA), increase the affinity of protein kinase C for phospholipids and for Ca^{++} ions to the physiological range. As a result the predominantly inactive, cytoplasmic protein kinase C becomes associated with the plasma membrane on incubation with PMA or DAG-generating polypeptides and is fully activated by interaction with the phospholipids in the plasma membrane at physiological Ca^{++} concentrations (Castagna et al.,1982; Kraft & Anderson, 1983, Berridge, 1984; Farrar & Anderson, 1985).

In vitro, protein kinase C can be activated by negatively charged phospholipids such as phosphatidylserine (PS), phosphatidylinositol (PI) and phosphatidic acid (PA), not by neutral phospholipids such as phosphatidylcholine (PC) or phosphatidylethanolamine (PE) (Kaibuchi et al., 1981, Blumberg et al., 1984).

Protein kinase C contains two functionally distinct domains: a hydrophobic, phospholipid-, phorbol ester- and DAG binding domain, and a hydrophylic, catalytic domain. Through Ca^{++}-dependent proteinases, the native enzyme generates a catalytically active phospholipid-independent form of the enzyme (protein kinase M, 50 kDa, Kishimoto et al., 1983; Tapley & Murray, 1984; Girard et al., 1986) and a phorbol ester- or DAG binding 38 kDa fragment (Hoshijma et al., 1986). The interaction of the hydrophobic domain with a lipid bilayer is thus responsible for the phospholipid-dependent activation.

In this paper we describe our studies on the cellular localization and in vitro interactions of protein kinase C with lipid bilayers.

NATO ASI Series, Vol. H29
Receptors, Membrane Transport and Signal Transduction
Edited by A.E. Evangelopoulos et al.
© Springer-Verlag Berlin Heidelberg 1989

Localisation of protein kinase C in control and stimulated HeLa cells.

The recovery of protein kinase C from different cellular compartments with different extraction methods permits comparison of the lipid bilayer-protein kinase C interaction in control and PMA stimulated cells. For this purpose, cytoplasmic and particulate fractions from HeLa cells were prepared, partly purified and assayed for protein kinase C activity as described (Snoek et al., 1986a). In HeLa control cells, when lysed in the absence of Ca^{++} and in the presence of Ca^{++} chelators (EGTA,EDTA), protein kinase C activity was predominantly found in the cytoplasmic fraction and a minor amount in the particulate fraction (Fig. 1, open bars). After stimulation of the cells with PMA (10 min., 50 ng/ml) and lysis in the same buffer, most of the activity had disappeared from the cytoplasm and was found in the particulate fraction (Fig. 1, striped bars). However, when control cells were lysed in the presence of Ca^{++} (1 mM), more than 90% of the protein kinase C activity was found in the particulate fraction (Fig. 1, black bars). When this fraction was incubated with EGTA (1mM), the protein kinase C activity was separated from the particulate fraction (not shown) indicating a Ca^{++}-dependent association of protein kinase C with the particulate fraction. When PMA-stimulated cells were lysed in

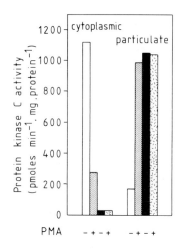

Figure 1. Distribution of protein kinase C activity in HeLa cells. Open and black bars: cells lysed in the absence of Ca^{++}, with Ca^{++} chelators EGTA/EDTA present; cross-hatched and spickled bars: cells lysed in the presence of 1 mM Ca^{++}.

'he presence of 1 mM Ca^{++}, most of the activity was found in the particulate fraction. This activity could not be recovered from the particulate fraction by incubation with EGTA; however the enzyme could be extracted with 1% Nonidet P40 or Triton X100 (Snoek et al., 1988).

These results indicate that in stimulated cells, protein kinase C is attached to the particulate fraction, possibly the plasma membrane, in a Ca^{++}-dependent way. Upon stimulation of the cells with PMA, the interaction between protein kinase C and the particulate fraction is changed and is no longer (only) Ca^{++}-dependent. A possible explanation could be an increase in hydrophobic interactions, i.e., the insertion of the enzyme in the lipid bilayer of the plasmamembrane.

Association of protein kinase C with unilamellar bilayers.

To study, in vitro, the interaction between protein kinase C and lipid bilayers, we have purified protein kinase C from mice brain (Snoek et al. 1986b).

Insertion of proteins into lipid bilayers can be studied by labelling the part of the protein that is in contact with the hydrophobic core of the lipid bilayer. An excellent reagent for this purpose is the foto-activatable [125]Iodo-naphtalene-azide

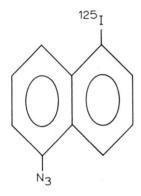

Figure 2. Structure of the photoactivatable labelling reagent ^{125}I-iodonaphtalene-1-azide.

(^{125}I-INA, Fig. 2; Bercovici & Gitler, 1978). INA has been shown to be restricted, by its insolubility in water, to the hydrophobic core of a lipid bilayer. Therein, the reactive nitrene, upon radiation attaches covalently to lipids and to proteins inserted into the lipid bilayer. The ^{125}I-labelled compounds are analysed by SDS-polyacrylamide gelelectrophoresis and autoradiography. We have used this procedure to study the interaction of protein kinase C with large unilamellar lipid vesicles (LUVs) composed of 100% PC and 80% PC/20% PS. In 100% PC LUVs only little labelling of protein kinase C was found, while addition of PS to the LUVs increased the labelling. Addition of PMA or DAG had no effect on protein kinase C labelling in 100% PC vesicles but further increased the labelling of protein kinase C in 80% PC/20% PS vesicles (Fig. 3, Snoek et al., 1986b). Since DAG and PMA also increase the PS-dependent activation of protein kinase C, these results could indicate that DAG and PMA affect the hydrophobic interaction of protein kinase C with the lipid bilayer.

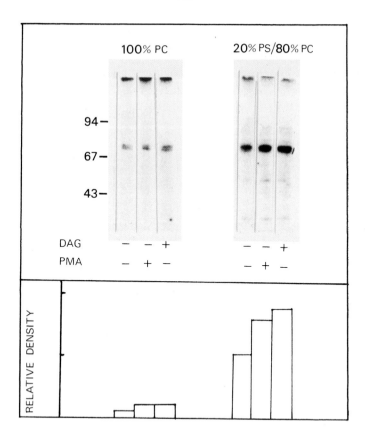

Figure 3. Autoradiogram and scan of SDS-polyacrylamide electrophoregram of protein kinase C incubated with LUVs composed of 100% PC or 80% PC/20% PS. PMA (50 ng/ml) and diolein (3.2 µg/ml) were added as indicated. Methods are described in Snoek et al. 1986b.

The role of hydrophobic interactions in the activation of protein kinase C.

We intended to study the role of hydrophobic interactions in the activation of protein kinase C by changing the hydrophobic core of the lipid bilayer and studying the activation of protein kinase C under these conditions. This is accomplished by using PS preparations with varied fatty acid compositions. The unsaturation index of a phospholipid is defined as the number of unsaturated bonds per one hundred fatty acid molecules. Available were PS isolated from bovine heart (PS-BH), from bovine spinal cord (PS-SC) and from bovine brain (PS-BB) with unsaturation indexes of resp. 112, 90 and 50. We have analysed the activation of protein kinase C by these PS preparations and the effects of DAG on the activation.

Protein kinase C was optimally activated by a PS concentration of 192 µg/ml for all three PS types (Fig. 4A, open symbols). However, the maximal activation of protein kinase C for the three PS types appeared to be dependent on the degree of saturation of PS. The maximal activation is induced by PS-BB, followed by PS-SC and PS-BH; maximal activation is inversily proportional to the unsaturation index of the activating phospholipid.

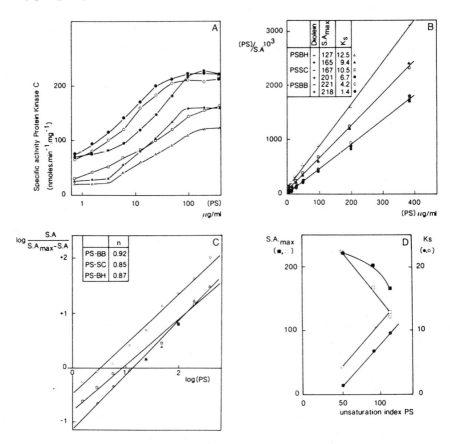

Figure 4. A: PS-dependent activation of protein kinase C in the absence (open symbols) and presence (closed symbols) of diolein.
△,▲PS-BH;□,■ PS-SC;○,● PS-BB.
B: Analysis of results from figure 4A using the Hanes equation (see text); symbols as in figure 4A.
C: Analysis of results from figure 4A using the Hill equation (see text); symbols as in figure 4A.
D: Relation between the unsaturation index of PS and the maximal specific activity (□ ,■) and the K_s (○,●) of protein kinase C in the absence (○,□) and presence (●,■) of diolein. Methods are described in Snoek et al., 1988.

Analysis of the kinetics of the interaction between protein kinase C and PS according to Hanes, i.e. the relationship between [PS] and [PS]/specific activity of protein kinase C, resulted in three linear curves (Fig. 4B, open symbols). In the Hanes equation:

$$[PS]/SA = [PS]/SA_{max} + K_s/SA_{max},$$

where K_s is the dissociation constant of the PS-protein kinase C complex and SA_{max} the maximal specific activity of protein kinase C. With this equation we calculated the K_s and SA_{max} for the reaction between protein kinase C and the three PS types (insert fig. 4B). The Hanes equation yielded the same conclusion as Fig. 4A: maximum PS-dependent activation is inversily proportional to the unsaturation index of the activating PS. On the other hand, the dissociation constant K_s, is proportional to the unsaturaration index (Fig. 4D, open symbols.). To calculate the number of PS molecules reacting with one molecule of protein kinase C, we used the Hill equation:

$$\log[SA/SA_{max}\text{-}SA] = n\log[PS]\text{-}\log K_s.$$

In this equation n is the number of PS molecules reacting with one protein kinase C molecule and K_s is the dissociation constant. By using SA_{max} from the Hanes plots, the n values for all three PS-types were calculated to be approximately 1 (Fig. 4C + insert). These results lead to the conclusion that protein kinase C is differentially activated by PS types that vary in their fatty acid composition, i.e. the hydrophobic part of the molecule. This indicates that hydrophobic interaction between phospholipid and protein kinase C does affect the activation of the enzyme.

Cofactors (activators) of protein kinase C, PMA and DAG are thought to function by increasing the affinity of protein kinase C for Ca^{++} and PS, but the nature of the interaction between the enzyme, phospholipid and the cofactor is not clear.

We analysed the effects of diolein on the PS-dependent activation of protein kinase C under the conditions described above. Optimal activation was induced at a diolein concentration of 3.2 µg/ml (not shown). Addition of diolein to vesicles of PS-BH and PS-SC increased the protein kinase C activation in the concentration range used. However,addition of diolein to PS-BB vesicles hardly affected activation (Fig. 4A, closed symbols). The Hanes plot in all three case was linear (Fig. 4B, closed symbols) and yielded new values for SA_{max} and K_s (insert in Fig. 4B). SA_{max} with PS-BH and PS-SC increased after addition of diolein; the SA_{max} with PS-BB was hardly affected by diolein (as already concluded from fig. 4A). On the otherhand, the dissociation constant K_s of the three PS-protein kinase C complexes were affected in an essential identical way: in each case the absolute value of K_s is decreased. However, the relative effect of diolein on the K_s values is different for the three PS types (for PS-BH, PS-SC and PS-BB resp. a factor of 1.3, 1.6 or 3.0), indicating that the diolein effect on PS-protein kinase C interaction involves an effect on the hydrophobic interaction between the enzyme and the activating phospholipid (Snoek et al. 1988).

Discussion.

Regulation of activation of protein kinase C might be an important feature in the many cellular events where protein kinase C is an intermediate. Both hydrophylic (Ca^{++}, negatively charged phospholipids) and hydrophobic (as described in this paper) interactions are involved. However, little is known yet about the exact nature of the interaction between protein kinase C, PS, Ca^{++} and the cofactors DAG or PMA. Since it has been reported (Sekiguchi et al., 1987;Snoek et al., 1987; Nishizuka, 1988) that different populations and isozymes of protein kinase C demonstrate differences in their affinity for, and responses to Ca^{++}, PS and DAG or PMA, information on isozymes hydrophobic interaction with diversified environmental lipids will help in identifying their specific functions in specific cellular responses.

Acknowledgements. This work was performed at the Hubrecht laboratory , Uppsalalaan 8, 3584 CT Utrecht, The Netherlands in cooperation with A.Feijen, W. van Rotterdam, W.J. Hage and at the Weizmann Institute of Science, Membrane Dept., Rehovot, Israel in cooperation with I. Rosenberg and C. Gitler.
This research was supported by the Netherlands Cancer Foundation (Koningin Wilhelmina Fonds).

References.

Bercovici, T. & Gitler, C. (1978) Biochemistry 17, 1484-1489.
Berridge, M.J. (1984) Biochem. J. 220,345-360.
Berridge, M.J. & Irvine, R.F. (1984) Nature (London) 312,315-321.
Blumberg, P.M., Jaken, S., Konig, B., Sharkey, N.A., Leach, K.L., Jeng, A.Y. & Yek, E. (1984) Biochem. Pharmacol. 33, 933-940.
Castagna, M., Takai, Y., Kaibuchi, K., Sano, K., Kikkawa, U. & Nishizuka, Y. (1982) J. Biol. Chem. 257, 7847-7851.
Farrar, W.L. & Anderson, W.B. (1985) Nature (London) 315,233-235.
Girard, P.G., Mazzei, G.J. & Kuo, J.F. (1986) J. Biol. Chem. 261,370-375.
Hokin, L.E. (1985) Annu. Rev. Biochem. 54,205-235.
Hoshijma, M. Kikuchi,A., Tanimoto, T., Kaibuchi, K. & Takai, Y. (1986) Cancer Res. 46, 3000-3004.
Kaibuchi, K., Takai, Y. & Nishizuka, Y. (1981) J. Biol. Chem. 256, 7146-7149.
Kishimoto, A., Kajikawa, N., Shiota, M. & Nishizuka, Y. (1983) J. Biol. Chem. 258, 1156-1164.
Kraft, A.S. & Anderson, W.B. (1983) Nature (London) 301, 621-623.
Leach, K.L., James, M.L. & Blumberg, P.M. (1983) Proc. Natl. Acad. Sci. U.S.A. 80, 4208-4212.
Nishizuka, Y. (1984) Nature (London) 308, 693-698.
Nishizuka, Y (1988) Nature (London, 334,661-665.
Sekiguchi, K., Tsukuda, M., Ogita, K., Kikkawa, K. & Nishizuka, Y. (1987) Biochem. Biophys. Res. Commun. 145, 797-802.
Snoek, G.T., Mummery, C.L., van den Brink, C.E., van der Saag, P.T. & de Laat S.W. (1986a) Devel. Biol. 115, 282-292.
Snoek, G.T., Rosenberg, I., de Laat, S.W. & Gitler, C. (1986b) Biochim. Biophys. Acta 860, 336-344.
Snoek, G.T., Feijen, A., Mummery, C.L. & de Laat, S.W. (1987) In: Signal transduction and protein phosphorylation. Plenum Press NY.
Snoek, G.T., Feijen, A., Hage, W.J., van Rotterdam, W. & de Laat, S.W. (1988) Biochem. J. 255, 629-637.
Tapley, P.M. & Murray, A.W. (1984) Biochem. Biophys. Res. Commun. 118, 835-841.

ACTIVATION OF TRANSDUCIN BY ALUMINUM OR BERYLLIUM FLUORIDE COMPLEXES

Joëlle BIGAY

Laboratoire de Biophysique Moléculaire et Cellulaire
(Unité Associée 520 du CNRS)
DRF/CENG - BP 85 X, F 38041 - GRENOBLE

INTRODUCTION

Transducin is the GTP binding protein of the phototransduction system. This protein, like other G proteins involved in numerous hormonal or sensorial transduction systems, plays a central role in the enzymatic cascade triggered by the external signal (light), and controlling the concentration of a soluble messenger (cGMP). Similarities between visual and hormonal systems have also been demonstrated to concern the structure of the receptor proteins, and the regulation mechanism of the enzymatic cascades.

That the adenylate cyclase system is activated by fluoride was known from the earliest studies on the enzyme by Sutherland in 1958. Fluoride activation is similar to stimulation in the presence of hormone plus GTP. It was later demonstrated (Pfeuffer 77) that fluoride acts directly on the G protein. Aluminum (or beryllium) was then identified (Sternweis et al 82) as a necessary cofactor for fluoride action. Although the fluoride target and cofactor were identified, no precise mechanism of action had been proposed.

We have studied fluoride effects on the visual system, particularly on its G protein, transducin. Our experimental results, added to the structural similarity between the fluoro-complexes and PO_4^{3-} ion, led us to propose a model in which the fluoride complex interacts with a GDP molecule in the nucleotide site of transducin, thus mimicking the presence of a GTP in this site.

In this report, I will briefly present the characteristics of the phototransduction system and then discuss the alumino- (or beryllo-) fluoride action.

NATO ASI Series, Vol. H29
Receptors, Membrane Transport and Signal Transduction
Edited by A. E. Evangelopoulos et al.
© Springer-Verlag Berlin Heidelberg 1989

THE PHOTOTRANSDUCTION SYSTEM

The photoreceptors cells

In the retina, light is detected by the photoreceptor cells, rods and cones, which transformed this signal into nervous response. As shown in Fig. 1a for a rod cell, the phototransduction machinery of this receptor cells is localized in the outer segment, and is isolated from the metabolic, energetic and genetic machineries which are confined in the inner segment. An important characteristics of the outer segments is to be filled by thousands of flatened membrane vesicles called discs, which bind the proteins of the transduction cascade. The last part of photoreceptors are the synaptics ends that allow transmission of the cell response to the other cells of the retina, and finally to the brain.

More abundant, and easier to isolate than cones, the rod cells have been preferentially studied for a long time, and our work refers only to them.

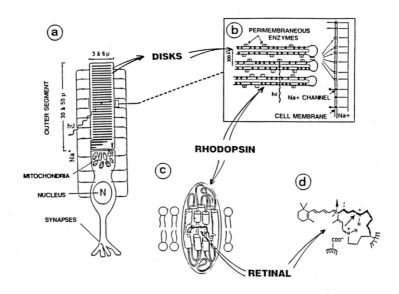

Figure 1: Structure of the photoreceptor cell. a: a rod cell. b: details of the discs environnement. c: model of the rhodopsin molecule. d : isomerization of the 11-cis retinal into all-trans retinal.

The signal

In dark conditions, i.e. in absence of stimulation, a constant flux of Na^+ enters the outer segment via cationic channels located in the plasma membrane, while an equal flux exits from the inner segment. This dark current causes the polarization of the cell membrane to - 40 mV. Upon illumination, the outer segment membrane is transiently hyperpolarized to - 80 mV, due to the closure of the cationic channels.

It is important to note that the kinetics of the process is particularly fast: the rod response to a single photon is maximum in 200 msec and is terminated about 500 msec after the flash.

During the past ten years, numbers of studies have focused on the phototransduction mechanism, and today the complete transduction process from the photon to the closure of the cationic channels is delineated, although all kinetic steps are not yet resolved.

The enzymatic cascade

The first event of the phototransduction process, photoreception, takes place in the disc membrane whereas the final response is on the external membrane (Fig.1b). Between the two membranes, the message propagates via an enzymatic cascade (fig.2) from which five proteins have been caracterized and purified.

Among these proteins, **rhodopsin (Rh)**, the photosensitive pigment, is the most abundant (about 80 % of the total protein content of the outer segment). This transmembraneous protein of 39 kD is folded into seven helix spanning the disc membrane (Fig.1c). The "7 α-helix" structure is now known to be characteristic of a family of hormone receptors interacting with G proteins in membrane transduction systems, such as muscarinic, adrenergic or opiate receptors (Dohlman et al. 87).

Rhodopsin is composed of the protein itself and a chromophore, the retinal. One half of the protein is embedded in the bilayer membrane, while the remaining is divided into two hydrophilic domains on the intradiscal and cytoplasmic faces of the membrane. The retinal molecule is bound in the middle of the last helix and buried in the hydrophobic core .

The first event following illumination of a rod is the photoisomerization of the 11-cis retinal into all-trans retinal (Fig.1d). This isomerization induces a change in the rhodopsin conformation, and two binding sites appear on its cytoplasmic surface, one specific for the interaction with transducin, and the other for the binding of rhodopsin kinase (Fig.2b.step 1).

Transducin is the second important protein of the outer segment. Similarly to other GTP binding proteins, transducin (T) is composed of three distinct peptides (T_α : 39 kD, T_β : 36 kD and T_γ : 8 kD). Transducin is membrane bound at physiological ionic strength, although no anchor exists (like a transmembrane helix or covalently bound lipids), and it can be easily solubilized in absence of detergent, when the disc membranes are put in an hypotonic medium.

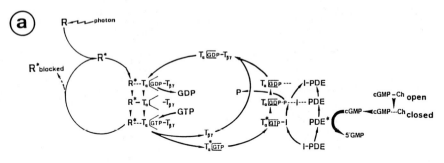

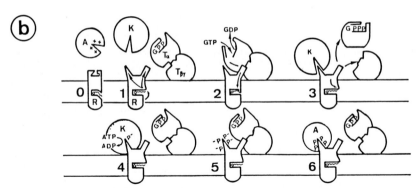

Figure 2: a: The enzymatic cascade. b: Moleculars events controlling rhodopsin and transducin activation.

The guanine nucleotide site is on T_α. In absence of stimulation, a non exchangeable GDP is locked in the nucleotide site, and T_α associates with $T_{\beta\gamma}$ to form a $T_{\alpha.GDP}$-$T_{\beta\gamma}$ complex (Fig.2b.step 2). This complex has a high affinity for photoexcited Rh (Rh*), to which it binds as soon as it is formed. This interaction induces the opening of the nucleotide site of $T\alpha$, and the release of the bound GDP. The Rh*-T complex is transiently emptied of nucleotide, and in this state, a GTP can enter the site and bind very quickly (Fig.2b.step 2), leading to a change in T_α conformation. With a bound GTP, the affinity of T_α for Rh* and for $T_{\beta\gamma}$ are lowered: $T_{\alpha.GTP}$

dissociates rapidly and becomes soluble, while $T_{\beta\gamma}$ remains on the membrane (Fig.2b.step 3). The soluble $T_{\alpha-GTP}$, with a GTP locked in the nucleotide site, is the active form of transducin, that can carry the activation signal to a cGMP specific phosphodiesterase (Fig.2a). The $T\alpha$ subunit also possesses an intrinsic GTPase activity, and is thus able to hydrolyze its bound GTP into GDP. $T_{\alpha.GDP}$ does not activate any more the phosphodiesterase, but reassociates rapidly with $T_{\beta\gamma}$ on the membrane.

The third protein of the cascade is a **cyclic-GMP specific phosphodiesterase** (PDE). As the effector protein, it regulates the intracellular concentration of the soluble messager of the system, cGMP, which directly controls the cationic channels conductivity on the cellular membrane. Like transducin, PDE is peripherally bound to the disc membrane. This protein, composed of four subunits associated in a $PDE_{\alpha\beta\gamma2}$ complex, is maintained in an inactive form by its two $PDE\gamma$ subunits (13 kD), also called inhibitors subunits (Deterre et al. 88).

The interaction of $T_{\alpha.GTP}$ with PDE, more precisely with the PDE subunits, relieves the inhibitory constraint on the active catalytic units, $PDE_{\alpha\beta}$ (84 and 88 kD). Activated PDE is then responsible for the rapid and important hydrolysis of intracellular cGMP.

Under resting condition, i.e. in the dark, a high concentration of cGMP maintains the cationic channels in an open state. At least two cGMP molecules are bound per channel, but they dissociate rapidly when the cGMP level decreases. The channel then closes, and the membrane hyperpolarizes.

Two other important proteins of the outer segment have been purified: rhodopsin kinase (K) and arrestin (A). Both are involved in the rapid turn-off of the signal, and act at the rhodopsin level. **Rhodopsin kinase** is a soluble ATP-dependent kinase of 68 kD, whose activation does not require any cofactor: it is selectively active on Rh*. It can phosphorylate as many as 9 serin or threonin residues on the C-terminal of photoexcited rhodopsin. **Arrestin**, also discovered as "48 kD" or "S-antigen", is a very soluble protein, present at a high concentration in the cytoplasm. Arrestin is a blocker protein that irreversibly binds to the photoactivated and phosphorylated rhodopsin.

Following the interaction of Rh* with T and the release of $T_{\alpha.GTP}$, the liberated Rh* can catalyze successive GDP/GTP exchanges on hundreds transducins, thus providing a high amplification for the cascade. The liberated Rh* can also bind rhodopsin kinase (Fig.2b.step 3), and is thus phosphorylated (Fig.2b.step 4). Phosphorylated Rh* is still able to activate transducins, although with a lower efficiency, depending on the number of phosphates bound (Fig.2b.step 5). This ability will be completely abolished by the binding of arrestin, which recognizes specifically the multiphosphorylated Rh* (Fig.2b.step 6).

ALUMINO- OR BERYLLO-FLUORIDE ACTIVATION OF TRANSDUCIN.

Although fluoride was widely used as an activator of the adenylate cyclase system for about thirty years, its action has been elucidated only recently. Our interest in fluoride began with the increasing knowledge on the G proteins family and the evidences that fluoride acts on G proteins in hormonal systems.

In the phototransduction system, it was demonstrated that PDE activity could be activated by fluoride in the dark, i.e. in absence of stimulation (Sitaramayya et al. 77); this is similar to adenylate cyclase activation by fluoride in absence of hormone. It was then shown that this PDE activation correlates with transducin activation (Stein et al. 85), confirming that in our system fluoride action is also mediated by the G protein.

Aluminum is a required cofactor

It was reported by Sternweis and Gilman (82) that Al^{3+} is a required cofactor for adenylate cyclase activation by F^-. We have therefore verified such a requirement for fluoride activation of transducin, and found similar results. Our observation was that transducin is stoichiometrically activated by micromolar Al^{3+} in presence of millimolar fluoride (Bigay et al. 85). Under such condition the effective species would be an AlF_4^- (Martin 86) or AlF_3,OH^- (Martin 88) complex. It is here worth to notice that trace amounts of Al^{3+} are easily etched from the glassware by NaF or KF solution, which could explain fluoride activation in absence of added Al. Therefore to determine precisely the influence of Al^{3+}, Al^{3+} free solution of fluoride have to be prepared in plastic vessel and preserved from any contact with glass.

Importance of the bound GDP

A striking observation was that the Al/F complex, which activates transducin in the dark, has no effect on T in presence of Rh* unless one added GDP. This important remark led us to investigate carefully the influence of different nucleotides on alumino-fluoride activation of transducin. To determine this activation, a simple test is the solubilization of Tα in the supernatant of disc suspension, quantitized by SDS-PAGE.

Under dark condition, i.e. when a GDP is locked in the nucleotide site, the solubilization of T induced by fluoride is independent of the presence of nucleotide in the incubation medium. But as soon as Rh is illuminated, the interaction of T with Rh* induces the nucleotide site to open, and the lost of the GDP molecule. In this condition, the Al/F action is inhibited unless a GDP, present in the incubation medium, binds in the nucleotide site. We thus demonstrated that the GDP molecule is essential for the Al/F effect. If GDP or its analog GDPαS is present, Al/F activates T as efficiently as in absence of Rh*. But in presence of GDPβS, an other GDP analog known to bind in the nucleotide site of T, Al/F does not solubilize Tα.

The influence of the nucleotide on the Al/F induced activation of T in presence of Rh*, led us to investigate Al/F effects on the purified Tα subunit, which carries the nucleotide site. The activation of pure Tα was monitored by limited trypsin proteolysis pattern. Tα indeed presents three trypsin proteolytic sites when it binds a GDP, but one of them, located near the nucleotide site, is no more accessible when a GTP (or the non-hydrolyzable analog GTPγS) is bound. The difference in the proteolytic patterns allow distinction between the active and the inactive conformation of Tα.

We have thus observed that the Al/F complex induces the trans-conformation of inactive $T_{\alpha.GDP}$ into active $T_{\alpha.GDP-Al/F}$, similar to the active $T_{\alpha.GTP\gamma S}$ species. Al/F induces similar trans-conformation of Tα bearing a GDPαS ($T_{\alpha.GDP\alpha S}$), but it has no effect on $T_{\alpha.GDP\beta S}$. As GDPαS and GDPβS are both analogs of GDP with a sulfur atom in place of an oxygen on the diphosphate chain (substitution on the α-phosphate for GDPαS, and on the β-phosphate for GDPβS), the sensitivity of Al/F to GDPβS and not to GDPαS argues for an interaction of the complex with the β-phosphate of GDP, in the nucleotide site of Tα.

One may also note that the activation of $T_{\alpha.GDP}$ by Al/F complex does not require the Tβγ subunit nor photoexcited rhodopsin, although both are absolutely required for the natural activation of Tα. The artificial activation has also the specific characteristics of being easily reversed either upon dilution of fluoride ions, or in presence of an excess of Tβγ subunits.

AlF$_4^-$ or BeF$_3^-$,H$_2$O are structurally similar to PO$_4^{3-}$

As did Sternweis and Gilman (82), we have tested the influence of Be ion, and we found identical results for all our experiments whether Al or Be were used. According to Be/F equilibrium data, the complex formed under condition of maximal activation of T (i.e. millimolar fluoride), correspond to BeF$_3^-$,H$_2$O (Mesmer et al. 69, Martin 88).

The chemical structure of the Al/F or Be/F complexes can be compared to that of PO$_4^{3-}$ ion: all three complexes are tetrahedral in solution, and the average bond length between the central atom (P, Al or Be) and the peripheral atoms (O or F) is constant ($\approx$ 1,60 to 1,70 Å). Thus with similar shape and size, either an Al/F or a Be/F complex can easily substitute to a PO$_4^3$ in the nucleotide site of Tα.

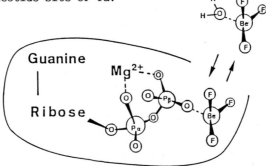

Figure 3: The reversible binding of BeF$_3$,H$_2$O to GDP in the nucleotide binding site of Tα.

A property of Be ions is to form strictly tetrahedral complex. As the activation observed in presence of either Be or Al are absolutely identical, it suggests that in the nucleotide site, both fluoro complexes bind as tetrahedral complexes. From the equilibrium data of Al/F complexes, it is hardly known whether the active complex in solution is AlF$_4^-$ or AlF$_3$OH$^-$ for Al/F, although BeF$_3^-$,H$_2$O appears to be the active Be/F complex. But to bind the β-phosphate of the GDP, and to conserve its tetrahedral structure, the fluoro complex may only be present in the nucleotide site as a trifluoro complex (Fig. 3).

The model

From all these data, we have proposed a model in which an Al/F or Be/F complex reversibly binds to the β-phosphate of the GDP molecule in the nucleotide site of the G protein, at the exact place of the 𝛾-phosphate of a GTP. The presence of this "GTP like" complex in the nucleotide site of the G protein, thus induces activation of the protein (Bigay et al. 87).

This model is in agreement with the abundant literature on the activation of the adenylate cyclase by fluoride, which confirmed that this activation always required in the nucleotide site the presence of a GDP, or an analog, which enables the interaction of the fluoro complex with a free oxygen of its β-phosphate (Fig.4).

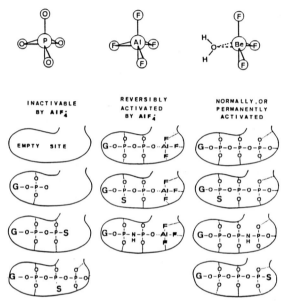

Figure 4: Models for the insertion of Al/F complex into the nucleotide site of Gα (or Tα) depending of the guanine nucleotide bound in the site.

The model of alumino-fluoride action has now been shown to be valid in other systems. For example, it has been demonstrated that Al/F complex blocks the depolymerization of microtubules by forming a GDP-Al/F complex similar to a GDP-Pi complex in the nucleotide site of tubulin (Carlier et al. 88). Similarly, the inactivation of the F1/Fo ATPase by fluoride, seems to be due to the irreversible formation of an ADP-Al/F complex in one of the three nucleotide sites of the enzyme (Lunardi et al. 88).

From these different studies, it is possible to conclude that fluoride action in systems involving proteins interacting with phosphates or nucleoside polyphosphates can probably be explained by our model of alumino-fluoride action.

BIBLIOGRAPHY

Bigay J., Deterre P., Pfister C and Chabre M.: (1985) "Fluoroaluminates activate transducin-GDP by mimiking the ɣ-phosphate of GTP in its binding site". FEBS Lett., **191**, 181-185.

Bigay J., Deterre P., Pfister C and Chabre M.: (1987) "Fluoride complexes of aluminum and beryllium act on G proteins as reversibly bound analogs of the -phosphate of GTP." EMBO J., **6**, 2907-2913.

Carlier M.F., Didry D., Melki R., Chabre M. and Pantaloni D.: (1988) "Stabilization of microtubules by inorganic phosphate and its structural analogs, the fluoride complexes of aluminum and beryllium." Biochemistry, **27**, 3555-3559.

Deterre P., Bigay J., Forquet F., Robert M. and Chabre M.: (1988) "cGMP phosphodiesterase of retinal rods is regulated by two inhibitory subunits." Proc. Natl. Acad. Sci. **85**, 2424-2428.

Dohlman H.G., Caron M.G. and Lefkowitz R.J.: (1987) "A family of receptors coupled to guanine nucleotide regulatory proteins." Biochemistry, **26/10**, 2657-2664.

Lunardi J., Dupuis A., Garin J., Issartel J-P., Michel L., Chabre M. and Vignais P.V.: (1988) "Inhibition of H+-transporting ATPase by formation of a tight nucleoside diphosphate-fluoroaluminate complexe at the catalytic site." Proc. Natl. Acad. Sci. **85**, 8958-8962.

Martin B.R.: (1986) "The chemistry of aluminium as related to biology and medicine". Clinical Chemistry, **32/10**, 1797-1806.

Martin B.R.: (1988) "Ternary hydroxide complexes in neutral solutions of Al3+ and F-", B.B.R.C. **155/3**, 1194-1200.

Mesmer R.E. and Baes C.F.Jr: (1969) "Fluoride complexes of beryllium (II) in aqueous media", Inorganic Chemistry, **8**, 618-626.

Pfeuffer R.: (1977) "GTP binding proteins in membranes and the control of adenylate cyclase activity." J. Biol. Chem., **252**, 7224-7234.

Sitaramayya A., Virmaux N. and Mandel P.: (1977) " On the mechanism of light activation of retinal rod outer segment cyclic GMP phosphodiesterase" Exp. Eye Res. **25**, 163-169.

Stein P.J., Halliday K.R. and Rasenick M.M.: (1985) "Photoreceptor GTP binding protein mediates fluoride activation of PDE.", J. Biol. Chem., **260**, 9081-9084.

Sternweis P.C. and Gilman A.G.: (1982), "Aluminium: a requirement for activation of the regulatory component of adenylate cyclase by fluoride", P.N.A.S., **79**, 4888-4891.

GLUTAMATE RECEPTORS AND GLUTAMATERGIC SYNAPSES

P. Ascher
Laboratoire de Neurobiologie
Ecole Normale Supérieure
46, rue d'Ulm
75230 Paris

L-glutamate is a strong candidate for the role of neurotransmitter at many interneuronal synapses in both vertebrates and invertebrates, as well as at many neuromuscular junctions in invertebrates. It has been clear for many years that glutamate activates a variety of receptors, but the precise characterization of these receptors has lagged behind that of the receptors for other transmitters, such as acetylcholine, noradrenaline or GABA. One of the primary reasons for this delay is that, in vertebrates, glutamate receptors are mostly found in central neurons, which, until recently, have not lent themselves to quantitative pharmacological studies. New techniques (in particular the patch-clamp (Hamill et al., 1981), and the expression of glutamate receptors in oocytes), used in conjunction with newly discovered selective agonists and antagonists, have profoundly renewed the field, and there is now a relatively broad consensus concerning the classification of glutamate receptors. If one considers only the case of vertebrate receptors, the best-known effects of glutamate are fast cationic conductance increases which are probably mediated by receptors directly coupled to the ionic channels ("ionotropic receptors"). However, some responses to glutamate appear to use a different type of receptor that is coupled to a G-protein ("metabotropic receptor") that in turn leads to the production of IP_3 and the release of Ca from intracellular stores. The "ionotropic receptors" can be separated into two main groups : those activated by N-methyl-D-aspartate (NMDA) and those insensitive to this compound ("non-NMDA" receptors). Thus glutamate receptors can be subdivided into three main groups: NMDA (ionotropic), "non-NMDA" ionotropic and metabotropic receptors. I shall first summarize the main properties of these three groups and then attempt to characterize their functions in synaptic transmission.

Most of the data discussed below were obtained either from neurons in primary culture or from brain slices. Because of the accessibility of the neuronal membrane, cultured neurons are the most commonly used preparation for patch-clamp studies. However, in cultures it is difficult to record from identified cells, and very difficult to analyze identified synapses. Neurons in slices are much more readily identified and often retain synaptic connections. Slices from the hippocampus and cerebellum have been most frequently used because of the regularity of their laminar organization. Patch-clamp is difficult to apply to "classical" slices, but techniques have been developed which have allowed the method to be applied to enzymatically treated slices (Gray and Johnston, 1985), to organotypic slices (Llano et al.,

NATO ASI Series, Vol. H29
Receptors, Membrane Transport and Signal Transduction
Edited by A. E. Evangelopoulos et al.
© Springer-Verlag Berlin Heidelberg 1989

1988) as well as to "thin" slices (Konnerth et al., 1988). Finally, glutamate receptors have been expressed by injecting brain mRNA into Xenopus oocytes (e.g. Gundersen et al., 1988; Houamed et al., 1984; Sugiyama et al., 1987; Verdoorn et al., 1987; Kushner et al., 1988; Kleckner and Dingledine, 1988). This type of preparation has been particularly useful in demonstrating the existence of a "metabotropic" glutamate receptor but its robustness also makes it a promising tool for the pharmacological characterization of ionotropic receptors.

1. NMDA receptors

NMDA receptors can be activated by selective agonists (of which NMDA is the most prominent), and blocked by selective competitive antagonists. The best known of these antagonists is D-amino-phosphonovalerate (D-APV) (see Watkins and Olverman, 1987).

Activation of the NMDA receptor by agonists leads to the opening of channels characterized by a main conductance state of 50 pS (in Mg-free solutions, see below). Occasional openings with a lower elementary conductance have been described (Cull-Candy and Usowicz, 1987; Jahr and Stevens, 1987; Ascher et al., 1988a).

The NMDA channel is a cationic channel. In physiological conditions its reversal potential is close to $\emptyset$ mV and most of the current through the channel is carried by the inward flow of Na ions and the outward flow of K ions. From a functional point of view, however, two of the most interesting properties of the NMDA channel are its permeability to Ca and its blockade by Mg.

Permeability to Ca.

The permeability to Ca of the NMDA channel had been suspected very early, but it was only established rigorously by experiments (MacDermott et al., 1986) in which NMDA was found to increase the intracellular Ca concentration in cells voltage clamped near resting potential, i.e. in conditions where Ca entry cannot occur through voltage dependent Ca channels. The Ca permeability of the NMDA channels was confirmed by the observation that current can flow through the NMDA channel from a solution containing only Ca as a cation (see Ascher and Nowak, 1988b). A quantitative evaluation of the Ca permeability of the NMDA channel remains to be done, however, and is made difficult by the uncertainty in the value of the surface potential at the mouths of the NMDA channel (Ascher and Nowak, 1988b).

Magnesium block.

While Ca ions permeate the NMDA channel, Mg ions block it. The blockage is voltage dependent : it increases with hyperpolarization. The effect is best seen if one measures the current produced by NMDA in voltage clamp conditions. In solutions lacking Mg, the current produced by a given concentration of NMDA is an approximately linear function of the transmembrane voltage. In the presence of physiological concentrations of external Mg (1-2 mM) the outward currents are unchanged but the inward currents are reduced. The reduction increases as the potential is made more negative. As a result, the current induced by activation of the NMDA receptor in physiological solutions is very small when the membrane potential is held near its resting level; it increases when the cell is depolarized before decreasing again when the membrane potential approaches 0 mV (the reversal potential). The I-V relation of the NMDA response thus has a characteristic "negative resistance" region (Nowak et al., 1984; Mayer et al., 1984).

The presence of a negative resistance in the I-V relation is classically associated with the possibility of triggering a regenerative change of potential. If an NMDA receptor agonist is applied in the presence of Mg to a cell in which the membrane potential is free to change, the resulting depolarization is likely to be very small if the membrane potential is initially close to the resting level. However, since this small depolarization will reduce the Mg block, it will increase the amount of current flowing through NMDA channels and lead to an additional depolarization. A regenerative cycle will be triggered if it is not opposed by hyperpolarizing conductances such as voltage activated potassium conductances.

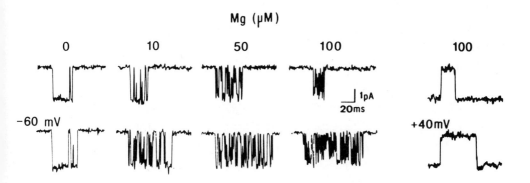

Fig. 1. Outside-out patch. Single-channel currents induced by NMDA (10 μM) in the absence of extracellular Mg^{2+} and in the presence of increasing concentrations of Mg^{2+}. Two examples of long openings are illustrated in each case. The eight records on the left were obtained at -60 mV ; the two records on the right were obtained at +40 mV. In the first column, no Mg^{2+} was present. The concentration of Mg^{2+} was then increased to 10 μM (second column), 50 μM (third column) and 100 μM (fourth and fifth columns). Notice that at -60 mV, increasing the Mg^{2+} concentration increases the flickering. No flickering is detected at +40 mV even in 100 μM-Mg^{2+}. From Ascher and Nowak (1988b).

At the single channel level, Mg produces a characteristic change of the single channel currents recorded in the presence of NMDA agonists (Nowak et al., 1984; Ascher and Nowak, 1988b). While in the absence of Mg the NMDA channel opens on average for a few ms, in the presence of external Mg at hyperpolarized potentials the records show bursts of shorter openings. The duration of these short openings decreases as the Mg concentration is increased (fig. 1). The exact kinetic model accounting for the flickering produced by external Mg block remains to be established. It appears highly probable that the flicker results from rapid entry and exit of Mg ions into and out of the open channel. What is not clear is whether the channel can close when Mg is inside the channel, and if so whether the probability of such closure is different from the probability of a closure when Mg does not occupy the channel (see Ascher, 1988).

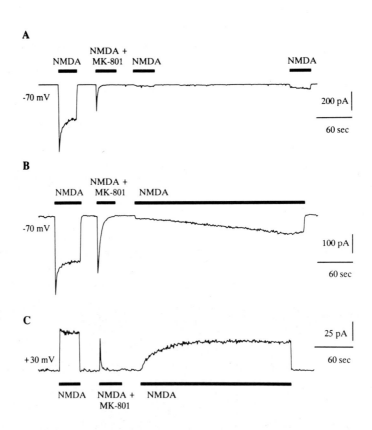

Fig. 2. Recovery from blockade by MK-801 . Whole cell recording from a cultured cortical neuron. A) After blocking the response to 30 μM NMDA + 1 μM glycine with 10 μM MK-801, the recovery of the NMDA-induced current was monitored at -70 mV by applying short pulses of NMDA (+ glycine) at 5 min intervals. Very little recovery was obtained. B) Following blockade by MK-801, the recovery was monitored during a continuous application of NMDA. C) Same experiment as in B, but the recovery was analysed at a holding potential of +30 mV. The recovery at -70 mV had a mean time constant of 90 min, versus a time constant of 4-5 min at +30 mV. Reprinted from Huettner and Bean (1988).

Channel blockers

Three compounds of great pharmacological interest (ketamine, phencyclidine (PCP) and MK-801) have recently been found to block the NMDA channel and their effect is voltage dependent (Honey et al., 1985; Wong et al., 1986 ; MacDonald et al., 1987; Huettner and Bean, 1988). In the case of MK-801, the voltage dependence of the K_D is half to that of the K_D of the Mg block found by Ascher and Nowak (1988b) which, given the difference in charge between MK-801 and Mg, may indicate that both act at the same site. However, the rate of unblock is much smaller than that of Mg, permitting a direct demonstration (fig. 2) of three properties of the MK-801 block :

1) the block starts only after the channel has been opened by an agonist.

2) after a blockade has been established, no recovery occurs if the agonist is removed.

3) the recovery observed in the presence of the agonist is voltage dependent.

These results clearly indicate that MK-801 is an "open-channel blocker", and that it can be trapped inside the channel by channel closure following the removal of the agonist. From a pharmacological point of view, trapping accounts for the long lasting effects of the blocker, while open-channel block explains the "use-dependence" of the block (the block only starts when the agonist is added and the speed of block onset will increase if more agonist is applied). From a biochemical point of view, trappable blockers may provide a way to tag the NMDA receptor-channel complex with a nearly "irreversible" ligand, and help in its purification.

Allosteric modulation of the NMDA receptors

It has recently been discovered that the response to NMDA is greatly enhanced by the addition of glycine (Johnson and Ascher, 1987). Although glycine by itself does not open the NMDA channel, when it is added to NMDA it greatly increases the probability of opening of the channel. The effect of glycine can be observed in outside-out patches, and therefore is not likely to involve second messengers or a G-protein. All the available evidence suggests that the glycine binding site is on the same protein as the NMDA site. The affinity of glycine for its site is quite high (100-300 nM). It has been recently suggested (Kleckner and Dingledine, 1988) that occupancy of the glycine site is an absolute requirement for the opening of the NMDA channel by NMDA agonists. In favor of this hypothesis is the finding that, in oocytes injected with mRNA and responding to NMDA in the presence of glycine, NMDA applied in the complete absence of glycine does not elicit any response. It remains to be established whether the responses to NMDA observed by others under similar conditions (addition of NMDA in the nominal absence of glycine) were due to an undetected contamination by glycine or to differences in sensitivity between the NMDA receptors expressed in oocytes and the NMDA receptors of cultured neurons.

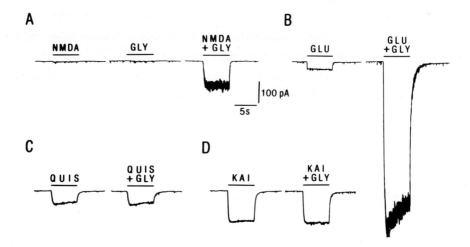

Fig. 3. Effects of 1 μM glycine (Gly) on the currents induced by (a) 10 μM NMDA (b) 10 μM L-glutamate (Glu), (c) 2 μM quisqualate (Quis), and (d) 10 μM kainate (Kai). Whole cell recording. Internal solution, CsCl. Holding potential, - 50 mV. Glycine potentiates the currents produced by NMDA or glutamate, and does not potentiate the currents produced by quisqualate or kainate cells. The quisqualate-and kainate-induced currents show little noise, as expected from the fact that the single channel conductances of the corresponding "non-NMDA" channels are smaller than that of the NMDA channels. In the absence of glycine, the noise of the glutamate-induced current is of similar magnitude to that of the quisqualate-and kainate-induced currents, suggesting that most of the current flows through non-NMDA channels. In the presence of glycine the noise variance increases dramatically, indicating the recruitement of NMDA channels. (From Johnson and Ascher, 1987). Reprinted by permission from Nature. Copyright (c) 1987 Macmillan Magazines Ltd.

A second unsolved question is whether, under physiological conditions, variations of the extracellular glycine concentration can modulate the function of NMDA receptors. Until now, measurements of glycine concentration have only been made in the cerebrospinal fluid. The high levels found (a few μM, Ferraro and Hare, 1985) suggest that the glycine level is close to saturation of the glycine site. The discovery of glycine antagonists like kynurenate, 7-chlorokynurenate and HA-966, the blocking effects of which can be relieved by the addition of glycine (see Kessler et al., 1987; Ascher, Henderson and Johnson, 1988 ; Birch et al., 1988 ; Fletcher and Lodge, 1988 ; Kemp et al., 1988 ; Watson et al., 1988), may make it possible to answer this question in the near future. These compounds have already been used to show that in some cases in which the addition of glycine did not increase the NMDA response under control conditions , it did induce a potentiation if the NMDA response had previously been reduced by addition of the glycine antagonist. This certainly suggests that in control conditions the absence of potentiation was due to the presence of glycine at a saturating concentration.

Recently Zn has been found to block NMDA responses in a non-competitive way. The action of Zn, like that of glycine, is voltage independent. The Zn site appears, however, to be independent of the glycine site. The Zn block could have a functional importance since Zn is present in nerve terminals and is released in some cases from these terminals at concentrations which appear to interfere with NMDA receptor activation (see Peters et al., 1987; Westbrook and Mayer, 1987).

In conclusion, the NMDA receptor bears at least four different binding sites : the NMDA binding site, the glycine binding site, the Zn binding site and the Mg-PCP-MK-801 binding site(s). The solubilization of the receptor has recently been reported (Ambar et al., 1988) as well as the expression of the receptor in oocytes (Verdoorn et al., 1987). With biochemists and molecular biologists closing in on the NMDA receptor, structural data comparable to those obtained for the nicotinic ACh receptor and the GABA-A receptor should soon be available for the NMDA receptor.

2. The ionotropic "non-NMDA" receptors

Ionotropic "non-NMDA" receptors are glutamate receptors which are not activated by NMDA and which mediate rapid conductance changes persisting in "outside-out" patches, in the absence of any intracellular messenger. Although there is some controversy concerning the exact number of such receptors, most authors distinguish two groups : those activated by quisqualate and by AMPA (α-amino-3-hydroxy-5-methylisoxazole-4-propionic acid) and those activated by kainate (see Watkins and Olverman, 1987). The first have often been called "quisqualate receptors" but this may be revised in view of the fact that quisqualate (but not AMPA) is also a powerful activator of the "metabotropic" glutamate receptor. It may thus be more appropriate to characterize the "non-NMDA" ionotropic receptors as the AMPA and kainate receptors.

Both AMPA and kainate activate cationic channels permeable to Na and K, but not to Ca. The quisqualate and AMPA activated channels have main conductance states that are higher than that of kainate activated channels, and smaller than that of NMDA channels. The exact values of the conductance differ slightly according to various authors (Cull-Candy and Usowicz, 1987; Jahr and Stevens, 1987; Ascher et al., 1988b; Cull-Candy et al., 1988). The quisqualate response desensitizes rapidly after addition of high concentrations of quisqualate or L-glutamate, whereas the kainate response does not desensitize (Ishida and Neyton, 1985; Kiskin et al., 1986).

Doubt persists concerning the existence of separate AMPA (quisqualate) and kainate receptors. High concentrations of quisqualate alter kainate responses, and vice-versa (see e.g.

Ishida and Neyton, 1985 ; Kiskin et al., 1986 ; O'Brien and Fischbach, 1986). The most potent antagonist of AMPA, CNQX, is much more effective in displacing AMPA than kainate in binding studies, but blocks with comparable potency quisqualate-and kainate-induced depolarizations (Honoré et al., 1988). These observations, and others, have led some authors to propose that quisqualate and kainate receptors exist in interconvertible forms having a very different affinity for quisqualate and kainate. (Honoré and Drejer, 1988). Others have suggested that there is a single receptor with two sites : an activating one binding both quisqualate and kainate, and a "desensitizing" one binding only quisqualate (Kiskin et al., 1986).

3. The metabotropic receptor

Ponderous and misleading as it is, the term "metabotropic" (Eccles and McGeer, 1979) may be the best adapted to describe the response to glutamate that appears to be mediated by activation of a G-protein, leading to IP3-induced Ca release from intracellular stores. For a recent review, see Sladeczek et al. (1988).

The first observations concerning this receptor were made by Sladeczek et al. (1985) who measured in cultured neurons an increased production of phosphoinositides after the addition of L-glutamate, NMDA, quisqualate or kainate. The most potent agonist was quisqualate. Although these experiments suggested that there was a glutamate receptor coupled to phospholipase C (PLC), the demonstration remained incomplete because of the possibility that PLC activation resulted from an increase in intracellular Ca (see Eberhard and Holz, 1988) secondary to the opening of Ca channels agonist-induced by depolarization. The effect could thus have been a remote consequence of an "ionotropic" response. This ambiguity was not completely relieved in the later experiments of Nicoletti et al. (1986 a,b) who analyzed the effects of extracellular Ca removal. They found that the accumulation of IP_1 produced by glutamate was strikingly reduced (but not abolished) whereas the accumulation of IP_3 produced by glutamate was not changed.

The first unequivocal demonstration that some glutamate agonists could activate phospholipase C in the complete absence of both extracellular Ca and depolarization was given by Sugiyama et al. (1987). These authors discovered that in Xenopus oocytes injected with brain mRNA, quisqualate induced a characteristic "oscillatory" Cl conductance change. Similar changes had previously been reported after expression of various receptors in oocytes and had been found to result from an oscillatory variation of Ca concentration activating a Ca-sensitive Cl current (abundant in native Xenopus oocytes). The situation turned out to be similar. The Cl current disappeared when the oocyte was injected with EGTA. When the response was analyzed at potentials for which the Cl current was hyperpolarizing, no initial depolarizing component was observed. This strongly suggests that, in this case at least, the activation of the PLC is not secondary to Ca entry but rather, through the production of IP₃, releases Ca from intracellular stores.

The receptor characterized by Sugiyama et al. (1987) is activated by quisqualate but not by AMPA, kainate, or NMDA. It is not blocked by any of the classical blockers of either "NMDA" or "non NMDA" responses. Sugiyama et al. (1987) thus justifiably proposed that this was a new receptor type. Glutamate thus resembles transmitters like ACh or GABA, which are known to activate both "ionotropic" (nicotinic, GABA-A) and "metabotropic" (muscarinic, GABA-B) receptors.

The quisqualate metabotropic response has recently been characterized further in cultured neurons by Miller and Murphy (1988) and by Sugiyama et al.(1988). Using fura-2 to detect changes in internal Ca, these authors showed that quisqualate (but not NMDA, AMPA or kainate) produces an increase in intracellular Ca in the absence of extracellular Ca, indicating the release of Ca from an intracellular compartment.

The ionic conductance changes produced by the metabotropic glutamate receptor have not yet been identified. One can expect increases in intracellular Ca to activate - in various combinations - K conductances, Cl conductances and non-selective cationic conductances (Marty, Tan and Trautmann, 1984).

An interaction between ionotropic and metabotropic receptors has been recently reported by Palmer et al. (1988). These authors observed that activation of the ionotropic glutamate receptors inhibits the phosphoinositide metabolism induced by activation of the metabotropic glutamate receptor. The effect was particularly marked for NMDA receptors, and was reduced by reducing the extracellular Ca concentration. This suggests that in certain conditions an increase in intracellular Ca may inhibit the phospholipase C - an hypothesis opposite to that discussed previously to account for the stimulation of phosphoinositide metabolism by NMDA, AMPA or kainate.

4. Other modes of action of L-glutamate

There are a number of observations which cannot easily be put into any of the categories treated above, and which may indicate the existence of additional glutamate receptor types. I shall consider a few examples.

A *presynaptic* Cl conductance increase has been observed in response to the application of glutamate on the synaptic endings of salamander cones (Sarantis et al., 1988 ; but see Tachibana and Kaneko, 1988). The effect was mimicked by kainate. Cl conductance increases induced by glutamate had until then only been reported in invertebrate muscle (see, e.g. Cull-Candy, 1976) and invertebrate neurons (e.g., Kehoe, 1978).

Presynaptic receptors appear to account for the effects of L-APB (2-amino-4-phosphonobutyric acid) at a well characterized synapse of the guinea-pig hippocampus, the synapse between mossy fibers and CA3 pyramidal neurons. At this synapse (one of the few central synapses where excitatory quanta can be resolved), L-APB reduces the amplitude of the synaptic potentials without altering either the amplitude of miniature synaptic potentials or the post-synaptic depolarization produced by L-glutamate (Cotman et al., 1986). The ionic mechanisms involved have not been characterized.

A conductance decrease leading to *hyperpolarization* is produced by L-glutamate in "depolarizing (on-center) bipolar cells" (Shiells et al., 1981; Atwell et al., 1987). The reversal potential is near -15 mV (Shiells et al., 1981) suggesting the closure of cationic channels. The effect is mimicked by both kainate and L-APB (Shiells et al., 1981, Slaughter and Miller, 1981). The interpretation of these experiments is made difficult by the fact that they were done using whole retina, and by the fact that the effect of glutamate was only observed at high concentrations (0.5 - 1 mM).

Nawy and Copenhagen (1987) have suggested that L-glutamate has a second hyperpolarizing effect on depolarizing bipolar cells and that, in addition to the closure of cationic channels, it opens channels with a reversal potential near -40 mV. This second effect could be similar to the Cl conductance increase described by Sarantis et al. (1988).

It should be noted that L-APB has been found to bind to a Cl dependent site which appears to be an uptake site (Pin et al., 1984). L-APB has also been reported to block the stimulation of PI metabolism produced by quisqualate in adult rat hippocampus (Nicoletti et al., 1986a) but not in cerebellar granule cells in culture (Nicoletti et al., 1986b).

5. Physiological functions of the glutamate receptors

Glutamatergic synapses were initially thought to resemble the vertebrate cholinergic nerve-striated muscle junction, and it was thus presumed that glutamatergic synaptic potentials, like end-plate potentials, were fast, phasic depolarizations corresponding to a brief opening of cationic channels, and capable of following relatively high frequencies of stimulation. This picture is probably valid for synaptic potentials involving non-NMDA ionotropic receptors. However, such synaptic potentials seem to occur rarely in isolation. In many cases, glutamatergic synapses combine two, and possibly more of the glutamate receptors described in the preceding section. As a result, they appear to be capable of complex behavior. Four examples of such synapses are discussed below.

a) Sustained activation of NMDA receptors can lead to rhythmic fluctuations in membrane potential .

Many observations have indicated that continued application of NMDA receptor agonists can produce a regular series of bursts of action potentials (see e.g., Herrling et al., 1983). Tetrodotoxin, which eliminates action potentials, reveals an underlying slow potential oscillation. This slow oscillation disappears in the presence of NMDA receptor antagonists, or in Mg free solutions (Wallén and Grillner, 1987).

The most likely explanation of this rhythm (Wallén and Grillner, 1987) assumes that the depolarizing limb of the oscillation is a regenerative depolarization due to the presence of a negative resistance region in the I-V relation of the NMDA response. This would bring the membrane potential to a steady depolarized value were it not for the fact that a repolarization process starts soon after the depolarization. The most likely source of the repolarization is the opening of K channels - the best candidate being Ca dependent K channels which would be activated by the entry of Ca through NMDA activated channels. If repolarization occurs, the NMDA channels will tend to be blocked by Mg, and this may in turn lead to a regenerative repolarization due to the negative resistance. The intracellular Ca concentration will then decrease and K channels subsequently close, allowing redevelopment of the NMDA current.

The rhythmic electrical activity produced by NMDA may be at the origin of a number of biological rhythm, in particular in swimming, walking or respiration. In the best studied case - fictive swimming of the isolated lamprey spinal cord - Grillner and his colleagues have performed a number of tests (alteration of the Mg concentration, pharmacological experiments, etc...) which strongly support the hypothesis that a "tonic" release of glutamate acting on NMDA receptors could be a key factor underlying the rhythmic activation of motoneurons (Grillner et al., 1987).

b) The dual "NMDA - non-NMDA" excitatory synapse

At many presumed glutamatergic synapses (see MacDermott and Dale, 1987 ; Blake et al., 1988) the excitatory synaptic potential consists of two separate components : (1) an early component, with a fast rise and a fast decay, insensitive to Mg and to NMDA antagonists, blocked by CNQX, and corresponding to the activation of the "non-NMDA" cationic channels, and (2) a component, with a slower rise and a slower decay, amplified by Mg free solution, blocked by NMDA antagonists, and corresponding to the activation of NMDA receptors (fig. 4) . This second component is sometimes seen only after subtraction of the first component, or after amplification by Mg free solutions. Why the NMDA component has a slower time course than the non NMDA component is not yet clear. Spectral analysis and single channel analysis of NMDA and non-NMDA responses have not revealed striking differences between the durations of channel opening (Ascher and Nowak, 1988 a-b ; Cull-Candy et al., 1988). Possible explanations involve either intrinsic differences in the rates of opening (e.g. a very low probability of opening of NMDA channels, even at very high agonist concentrations (see Huettner and Bean, 1988) or a peculiar distribution of NMDA receptors on the cell surface, forcing the agonist to diffuse a long distance from its release site to the receptors.

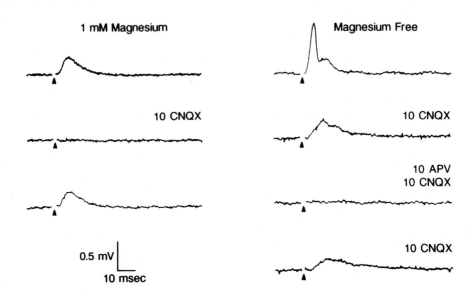

Fig. 4. Differential blockade of the NMDA and non-NMDA components of a dual synaptic potential in rat hippocampal slices. The records correspond to extracellular recordings of an excitatory synaptic potential elicited by low frequency stimulation of Schaffer collateral-commissural fibers. In the presence of 1 mM Mg, the synaptic potential is completely blocked by CNQX, an antagonist of the non-NMDA ionotropic receptors. A magnesium free solution reveals the presence of a second component which is blocked by the NMDA antgagonist APV.The Mg free solution also potentiates the non-NMDA, CNQX-sensitive component by increasing transmitter release. From Blake, Brown and Collingridge (1988).

Whatever the explanation for the slowness of the NMDA component, dual EPSPs have turned out to possess an interesting frequency dependence. At a low frequency of presynaptic stimulation, the non-NMDA component accounts nearly entirely for the transmission. At a high frequency stimulation, the summation of the depolarizations contains an increasingly larger NMDA component, the size of which depends both on the frequency and on the duration of the presynaptic stimulation. As a result, the same synapse can potentially transmit two types of signals : a frequency independent one and a frequency dependent one, each using a separate receptor.

c) Long-term potentiation of synaptic transmission in the hippocampus

Long-term potentiation in the hippocampus was first described in detail in 1973 (see Bliss and Lynch, 1988) in one of the three main types of excitatory synapses of the hippocampus: that between the (presynaptic) fibers of the perforant path and the (postsynaptic) granule cells of the dentate gyrus. Stimulation of the perforant path induces in the granule cell an excitatory synaptic potential (EPSP) which, at low frequency of stimulation (intervals of a few seconds), has a stable amplitude. After a brief tetanic stimulation (e.g. 100 Hz, 1s) of the presynaptic pathway, single shock low frequency stimulation now produces a larger EPSP. The potentiation was shown to last for hours in the anesthetized rabbit, and for days or weeks in animals with chronically implanted electrodes. A similar phenomenon exists in two other types of synapses of the hippocampus (the synapse between Schaffer collaterals/commissural fibers of the CA3 pyramidal cells and the CA1 pyramidal cells, and the synapse between the mossy fibers and the CA3 pyramidal fibers) and in other cerebral structures (see Bliss and Lynch, 1988). The mechanism of LTP is not identical in all cases. The following discussion only concerns the hippocampal synapses (perforant path/granule cells; Schaffer/commissural/CA1) for which the key role of NMDA receptors in the induction of LTP has been well established.

Collingridge et al. first reported in 1983 that NMDA antagonists had little effect on the EPSPs elicited by low frequency stimulation, but blocked the induction of long term potentiation. The presence of an NMDA component in the low frequency EPSPs was at that time unnoticed because, as mentioned in the previous section, this component is usually quite small when the membrane potential is near resting potential and when the extracellular solutions contains physiological concentrations (mM) of Mg. This NMDA component is, however, the key element of the LTP induction, since during repetitive stimulation the cumulative depolarizing effects of the summating non-NMDA and NMDA components relieve the Mg block and allow the expression of the NMDA response.

The report of Collingridge et al. was soon followed by others indicating that LTP induction probably involved a rise in intracellular Ca. Lynch et al. (1983) (see also Malenka et al., 1988) showed, in particular, that intracellular injection of EGTA prevented the development of LTP. At that time the permeability to Ca of NMDA channels was not known, and it was assumed that Ca entry was secondary to the opening by action potentials of voltage dependent Ca channels. However, it is now considered much more likely that the key pathway for Ca entry is the NMDA channel. Unlike Ca entry through voltage dependent channels, Ca entry through NMDA channels is restricted to the postsynaptic region facing the stimulated presynaptic terminal. This restricted localization of Ca entry accounts for the well-known fact that LTP is selective for the activated pathway : in cells receiving multiple excitatory inputs, tetanic stimulation of a given input only potentiates the EPSP corresponding to this input.

Although there is general agreement on the hypothesis that tetanic stimulation acts by triggering a local entry of Ca through NMDA channels, the steps following Ca entry are obscure. Ca is widely assumed to activate an enzyme, but this enzyme has not been identified (see Kennedy, 1988). Furthermore there is a continuing discussion about whether the potentiation of the non-NMDA EPSP is due to a presynaptic effect (increased release of transmitter) and/or to a post-synaptic effect (increased sensitivity of the non-NMDA receptor for the transmitter). The first interpretation (see Bliss and Lynch, 1988) implies a "retrograde" signal through which Ca entry in the post-synaptic target triggers increased transmitter release by the presynaptic neuron. Recently Bliss et al. (1988) and Williams and Bliss (1988) have suggested that the retrograde signal may be arachidonic acid or one of its metabolites. It is known that Ca can stimulate phospholipase A2, and NMDA receptor activation has now been found to produce arachidonic acid (Dumuis et al., 1988 ; Lazarewicz et al., 1988). Since arachidonic acid diffuses rapidly through lipids, it could cross from the post-synaptic to the presynaptic side and since arachidonic acid alters ionic permeabilities in many cells, and in particular neurons (Piomelli et al., 1987), it could be acting presynaptically in response to a post synaptically elicited message.

LTP, defined as a potentiation lasting hours is preceded by an immediate potentiation with fast onset decaying in about 30 minutes (STP) and possessing distinct pharmacological properties (Reymann et al., 1988 a-b ; Kauer et al., 1988). STP can be mimicked by direct application of NMDA (Kauer et al., 1988) ; it is not blocked by sphingosine or polymyxin B (PKC inhibitors) (Kauer et al., 1988; Malinow et al., 1988;Reymann et al., 1988b), but it is blocked by calmidazolium (a calmodulin inhibitor) (Reymann et al., 1988a). In contrast LTP is not mimicked by the application of NMDA alone (Kauer et al., 1988). Its induction is blocked by NMDA antagonists like APV (Collingridge et al., 1983) and by a variety of protein kinases inhibitors such as H-7, K-252band sphingosine (Reymann et al., 1988b; Malinow et al., 1988). The maintenance of LTP is reversibly blocked by H-7 (applied after the tetanus), but not by sphingosine (Malinow et al., 1988) which suggests that a constitutive protein kinase remains active during LTP. Finally, L-APB (see "other modes of action of glutamate") has been reported to block the maintenance, but not the induction of the LTP (Matthies and Reymann, 1989).

d) Long-term depression

A phenomenon which at first glance appears symmetrical to LTP is the long term depression (LTD) of synaptic transmission displayed in the cerebellum by the excitatory synapse between parallel fibers (presynaptic element) and Purkinje cells (post synaptic element). In this case, the trigger for the process is the simultaneous activation of the parallel fibers and of another set of excitatory presynaptic fibers, the climbing fibers. Both the parallel fibers and the climbing fibers are presumed to use glutamate (or a related amino acid) as transmitter. However, a marked difference exists in the way in which they innervate the (post-synaptic) Purkinje neuron. Thousands of parallel fibers terminate on a given Purkinje neuron, each making only a few contacts. In contrast, each Purkinje neuron receives input from a single climbing fiber, which makes numerous synaptic contacts. Activation of the climbing fiber leads to massive activation of the Purkinje neuron. LTD was first observed in vivo (see Ito et al., 1982), and more recently in vitro (Sakurai, 1987) : simultaneous activation of climbing fibers and parallel fibers (e.g. at 4Hz for 25s) produced a depression of the parallel fiber mediated EPSP in the Purkinje cell. The depression was about 30 % on average and lasted for more than 50 min.

In contrast to LTP, LTD does not involve NMDA receptors (which seem absent in Purkinje neurons) and a post synaptic reduction in glutamate sensitivity has been demonstrated. It can be observed if, instead of using the synaptic potential elicited by parallel fiber activation, as a test response, one uses the depolarisation produced by a pulse of glutamate activating the "non-NMDA" ionotropic receptors (Ito et al., 1982).

The two stimulations (climbing fibers and parallel fibers), which are usually paired to elicit LTD, can each be replaced by a more direct stimulation. The climbing fiber stimulation can be replaced by an electrical stimulation of the Purkinje neuron, provided that this stimulation is adjusted so that it triggers a Ca spike in the Purkinje neuron dendrites (Crepel and Krupa, 1988). The parallel fiber stimulation can be replaced by a local application of glutamate, or of quisqualate, on the dendrites of the Purkinje neuron (Kano and Kato, 1987). It thus appears that what is required to produce the long term depression of the glutamate sensitivity is the conjunctive activation of a quisqualate receptor and Ca entry into the post-synaptic Purkinje neuron.

The experiments of Kano and Kato (1987) were done before the characterization of the metabotropic quisqualate receptor. It is thus not clear whether quisqualate as a "conjunctive" stimulus is acting via the ionotropic receptor (which is certainly present, since it is the one which is depressed by LTD) or by a metabotropic receptor. Evidence in favor of an involvement of diacylglycerol in LTD has been produced by Crepel and Krupa (1988) who have reported that stimulation of PKC by phorbol esters induced a selective LTD of glutamate induced responses in Purkinje neurons.

Conclusion

The study of glutamate receptors has progressed extremely rapidly in the last five years, and shows no sign of slowing. As this review is sent to the editor, the latest issue of Nature (Garthwaite et al., 336, 385-388) reports a new twist in the NMDA story. Ca entry through NMDA activated channels appears to produce a unstable compound resembling the "endothelium derived relaxing factor" (EDRF) which could be nitric oxide (NO). This compound, diffusing out of the cell, stimulates cGMP production in neighbouring cells. Thus, after arachidonic acid, another potential "retrograde" factor is uncovered.

A few years ago it appeared possible to separate transmitters into fast and slow, excitatory and inhibitory. L-glutamate and GABA were respectively the prototypes of the "fast excitatory transmitters" and "fast ex-inhibitory transmitters", while the role of slow transmitters was attributed to monoamines and peptides (Iversen, 1984). Even at the time at which this classification was proposed, the data on invertebrate neurons already warned us against such a generalization. In molluscan neurons, for example, each transmitter which had been studied in detail had been found to have both fast and slow, and both excitatory and inhibitory effects (see Ascher and Kehoe, 1975; Kehoe, 1978). But the recent developments show that even in vertebrates such a generalization cannot be maintained, and that glutamate in particular can no longer be defined as fast and excitatory. The metabotropic glutamate receptor clearly triggers a slow response, through an increase in internal Ca that is likely to induce both inhibitory and excitatory conductance changes. And even the ionotropic NMDA receptor now appears capable of eliciting slow responses, and actually, through its involvement in LTP, is clearly involved in long term processes.

Acknowledgements. I thank Drs. J.W. Johnson, J.S. Kehoe and W. Sather for their comments on the manuscript and Ms. R. Bouaziz for typing. Our work is supported by the CNRS (URA-295) and the Université Pierre et Marie Curie.

References

Ambar I, Kloog Y, Sokolovsky M (1988) Solubilization of rat brain phencyclidine receptors in an active binding form that is sensitive to N-methyl-D-aspartate receptor ligands. J Neurochem 51: 133-140.

Ascher P (1988) Magnesium block of the NMDA receptor channel. In: Cavalheiro EA, Lehmann J, Turski L (eds) Frontiers in excitatory amino acid research. Alan R Liss New York, pp 151-157.

Ascher P, Bregestovski P, Nowak L (1988 a) N-methyl-D-aspartate-activated channels of mouse central neurones in magnesium free solutions. J Physiol 399: 207-226.

Ascher P, Henderson G, Johnson JW (1988b) Dual inhibitory actions of kynurenate on the N-methyl-D-aspartate (NMDA) activated responses of cultured mouse cortical neurones. J Physiol 406: 141P.

Ascher P, Kehoe JS (1975) Amine and amino acid receptors in Gastropid neurons. In Iversen LL, Iversen S, Snyder S (eds) Handbook of Psychopharmacology, Plenum, New York 4: 265-310.

Ascher P, Nowak L (1988 a) Quisqualate and kainate activated channels in mouse central neurones in culture. J Physiol 399: 227-245.

Ascher P, Nowak L (1988 b) The role of divalent cations in the N-methyl-D-aspartate responses of mouse central neurones in culture. J Physiol 399: 247-266.

Attwell D, Mobbs P, Tessier-Lavigne M, Wilson M (1987). neurotransmitter-induced currents in retinal bipolar cells of the axoloth, *Ambystoma mexicanum*. J Physiol 387: 125-161.

Birch PJ, Grossman CJ, Hayes AG (1988) Kynurenate and FG 9041 have both competitive and non-competitive antagonist actions at excitatory amino acid receptors. Eur J Pharmacol 151: 313.

Blake JF, Brown MW, Collingridge GL (1988) CNQX blocks acidic amino acid induced depolarizations and synaptic components mediated by non-NMDA receptors in rat hippocampal slices. Neurosci Lett 89: 182-186.

Bliss TVP, Clements MP, Errington ML, Lynch MA, Williams JH (1988) Long-term potentiation is accompanied by an increase in extracellular release of arachidonic acid. Soc Neurosci Abstr 14: 564.

Bliss TVP, Lynch MA (1988) Long-term potentiation of synaptic transmission in the hippocampus: properties and mechanisms. In: Landfield PW, Deadwyler SA (eds) Long-term potentiation: from biophysics to behavior. Alan R Liss, New York pp 3-72.

Collingridge GL, Kehl SJ, McLennan H (1983) Excitatory amino acids in synaptic transmission in the Schaffer collateral-commissural pathway of the rat hippocampus J Physiol 334: 33-46.

Cotman CW, Flatman JA, Ganong AH, Perkins MN (1986) Effects of excitatory amino acids antagonists on evoked and spontaneous excitatory potentials in guinea-pig hippocampus. J Physiol 378: 403-415.

Crepel F, Krupa M (1988) Activation of protein kinase C induces a long-term depression of glutamate sensitivity of cerebellar Purkinje cells. An in vitro study. Brain Res 458: 397-401.

Cull-Candy SG (1976) Two types of extrajunctional L-glutamate receptors in locust muscle fibres. J Physiol 255: 449-464.

Cull-Candy SG, Howe JR, Ogden DC (1988) Noise and single channels activated by excitatory amino acids in rat cerebellar granule neurones. J Physiol 400: 189-222.

Cull-Candy SG, Usowicz MM (1987) Multiple conductance channels activated by excitatory amino acids in cerebellar neurones. Nature 325: 527-528.

Dumuis A, Sebben M, Haynes L, Pin JP, Bockaert J (1988) NMDA receptors activate the arachidonic acid cascade system in striatal neurons. Nature 336: 68-70.

Eberhard DA, Holz RW (1988) Intracellular Ca^{2+} activates phospholipase C. Trends Neurosci 11 : 517-520.

Eccles JC, McGeer PL (1979) Ionotropic and metabotropic neurotransmission. Trends Neurosci. 2: 39-40.

Ferraro TN, Hare TA (1985) Free and conjugated amino acids in human CSF: influence of age and sex. Brain Res 338: 53-60.

Fletcher EJ, Lodge D (1988) Glycine reverses antagonism of N-methyl-D-aspartate (NMDA) by 1-hydroxy-3-aminopyrrolidone-2 (HA-966) but not by D-2-amino-phosphono-valerate (D-AP5) on rat cortical slices. Eur J Pharmacol 15: 161-162.

Gray R, Johnston D (1985) Rectification of single GABA-gated chloride channels in adult hippocampal neurons. J. Neurophysiol 54: 134-142.

Grillner S, Wallén P, Dale N, Brodin L, Buchanan J, and Hill R, (1987) Transmitters membrane properties and network circuity in the control of locomotion in lamprey. Trends Neurosci 10: 34-41.

Gundersen CB, Miledi R, Parker I (1984) Messenger RNA from human brain induces drug and voltage-operated channels in Xenopus oocytes. Nature 308: 421-424.

Hamill OP, Marty A, Neher E, Sakmann B, Sigworth FJ (1981) Improved patch-clamp techniques for high resolution current recording from cells and cell-free membrane patches. Pflügers Archiv 391: 85-100.

Herrling PL, Morris R, Salt TE (1983) Effects of excitatory amino acids and their antagonists on membrane and action potentials of cat caudate neurones. J Physiol 339: 207-222.

Honey CR, Miljkovic Z, MacDonald JF (1985) Ketamine and phencyclidine cause a voltage-dependent block of responses to L-aspartic acid. Neurosci Lett 61: 135-139.

Honoré T, Davies SN, Drejer J, Fletcher EJ, Jacobsen P, Lodge D, Nielsen FE (1988) Quinoxalinediones : potent competitive non-NMDA glutamate receptor antagonists. Science 241: 701-703.

Honoré T, Drejer J (1988) Binding characteristincs of non-NMDA receptors In: Lodge D (ed) Excitatory amino acids in health and disease. Wiley, Chichester New York. pp 91-106.

Houamed KM, Bilbe G, Smart TG, Constanti A, Brown DA, Barnard EA, Richards BM (1984) Expression of functional GABA, glycine and glutamate receptors in Xenopus oocytes injected with rat brain mRNA. Nature 310: 318-321.

Huettner JE, Bean BP (1988) Block of N-methyl-aspartate-activated current by the anticonvulsant MK-801: selective binding to open channels. Proc Natl Acad Sci USA 85: 1307-1311.

Ishida AT, Neyton J (1985) Quisqualate and L-glutamate inhibit retinal horizontal cell responses to kainate. Proc Natl Acad Sci USA 82: 1837-1841.

Ito M, Sakurai M, Tongroach P (1982) Climbing fiber-induced depression of both mossy fiber responsiveness and glutamate sensitivity of cerebellar Purkinje Cells. J Physiol. 324: 113-134.

Iversen LL (1984) Amino acids and peptides fast and slow chemical signals in the nervous system ? Proc Roy Soc Lond, B 221: 245-260.

Jahr CE, Stevens CF (1987) Glutamate activates multiple single channel conductances in hippocampal neurones. Nature 325: 522-525.

Johnson JW, Ascher P (1987) Glycine potentiates the NMDA response in cultured mouse brain neurons. Nature 325: 529-531.

Kano M, Kato M (1987) Quisqualate receptors are specifically involved in cerebellar synaptic plasticity. Nature 325: 276-279.

Kauer JA, Malenka RC, Nicoll RA (1988) NMDA application potentiates synaptic transmission in the hippocampus. Nature 334: 250-252.

Kehoe J (1978) Tranformation by concanavalin A of the response for molluscan neurones to L-glutamate. Nature 274: 866-869.

Kemp JA, Foster AC, Leeson PD, Priestley T, Tridgett R, Iversen LL, Woodruff GN (1988) 7-Chlorokynurenic acid is a selective antagonist at the glycine modulatory site of the N-methyl-D-aspartate receptor complex. Proc Natl Acad Sci USA, 85: 6547-6550.

Kennedy MB (1988) Synaptic memory molecules. Nature 335: 770-772.

Kessler M, Baudry M, Terramani T, Lynch G (1987) Complex interactions between a glycine binding site and NMDA receptors. Soc Neurosci Abstr 13: 760.

Kiskin NI, Krishtal OA, Tsyndrenko AY (1986) Excitatory amino acid receptors in hippocampal neurons : kainate fails to desensitize them. Neurosci Lett 63: 225-230.

Kleckner NW, Dingledine R (1988) Requirement for glycine in activation of NMDA-receptors expressed in *Xenopus* oocytes. Science 241: 835-837.

Konnerth A, Takahashi T, Edwards F, Sakmann B (1988) Single channel and synaptic currents recorded in neurons of mammalian brain and spinal cord slices. Soc Neurosc Abstr 14: 1046.

Kushner L, Lerma J, Zukin RS, Bennett MVL (1988) Coexpression of N-methyl-D-aspartate and phencyclidine receptor in *Xenopus* oocytes injected with rat brain mRNA. Proc Natl Acad Sci USA 85: 3250-3254.

Lazarewicz JW, Wroblewski JT, Palmer ME, Costa E (1988) Activation of N-methyl-D-aspartate-sensitive glutamate receptors stimulates arachidonic acid release in primary cultures of cerebellar granule cells. Neuropharmacology 27: 765-769.

Llano I, Marty A, Johnson JW, Ascher P, Gähwiler BH (1988) Patch-clamp recording of amino-acid-activated responses in "organotypic" slice cultures. Proc Natl Acad Sci USA 85: 3221-3225.

Lohmann SM, Walter U, Miller PE, Greengard P, de Camilli P (1981) Immunohistochemical localization of cyclic GMP-dependent protein kinase in mammalian brain. Proc Natl Acad Sci USA 78: 653-657.

Lynch G, Larson J, Kelso S, Barrionuevo G, Schottler F (1983) Intracellular injections of EGTA block induction of hippocampal long-term potentiation. Nature 305: 719-721.

MacDermott AB, Dale N (1987) Receptors ion channels and synaptic potentials underlying the integrative actions of excitatory amino acids. Trends Neurosci 10: 280-284.

MacDermott AB, Mayer ML, Westbrook GL, Smith SJ, Barker JL (1986) NMDA-receptor activation increases cytoplasmic calcium concentration in cultured spinal cord neurons. Nature 321: 519-522.

MacDonald JF, Miljkovic Z, Pennefather P (1987) Use dependent block of excitatory amino acid currents in cultured neurons by ketamine. J Neurophysiol 58: 251-266.

Malenka RC, Kauer JA, Zucker RS, Nicoll RA (1988) Postsynaptic calcium is sufficient for potentiation of hippocampal synaptic transmission. Science 242: 81-84.

Malinow R, Madison DV, Tsien RW (1988) Persistent protein kinase activity underlying long-term potentiation. Nature 335: 820-824.

Marty A, Tan YP, Trautmann A (1984) Three types of calcium-dependent channel in rat lacrimal glands. J Physiol 357: 293-325.

Matthies H, Reymann KG (1989) 2-amino-4-phosphonobutyrate (APB) selectively eliminates late phases of long-term potentiation (LTP) in rat hippocampal CA1 cells *in vitro*. J Physiol (in press).

Mayer M, Westbrook GL, Guthrie PB (1984) Voltage-dependent block by Mg^{2+} of NMDA responses in spinal cord neurones. Nature 309: 261-263.

Mayer ML, Westbrook GL (1987) The physiology of excitatory amino acids in the vertebrate central nervous system. Progr Neurobiol 28: 197-276.

Miller RJ, Murphy SN (1988) A unique glutamate receptor regulates intracellular calcium mobilization in hippocampal neurons. Soc. Neurosci Abstr. 14: 40-2.

Nawy S, Copenhagen DR (1987) Multiple classes of glutamate receptor on depolarizing bipolar cells in retina. Nature 325: 56-58.

Nicoletti F, Meek JL, Iadarola MJ, Chuang DM, Roth BL, Costa E (1988a) Coupling of inositol phospholipid metabolism with excitatory amino acid recognition sites in rat hippocampus. J Neurochem 46: 40-46.

Nicoletti F, Wroblewski JT, Novelli A, Alho H, Guidotti A, Costa E (1988b) The activation of inositol phospholipid metabolism as a signal-transducing system for excitatory amino acids in primary cultures of cerebellar granule cells. J Neurosci 6: 1905-1911.

Nowak L, Bregestovski P, Ascher P, Herbet A, Prochiantz A (1984) Magnesium gates glutamate-activated channels in mouse central neurones. Nature 307: 462-465.

O'Brien R, Fischbach GD (1986) Characterization of excitatory amino acid receptors expressed by embryonic chick motoneurons *in vitro*. J Neurosci 6: 3275-3283.

Palmer E, Monaghan DT, Cotman CW (1988) Glutamate receptors and phosphoinositide metabolism stimulation via quisqualate receptors is inhibited by N-methyl-D-aspartate receptor activation. Molec Brain Res 4: 161-165.

Peters S, Koh J, Choi DW (1987) Zinc selectively blocks the action of N-methyl-D-aspartate on cortical neurons. Science 236: 589-593.

Pin JP, Bockaert J, Récasens M (1984) The Ca^{2+}/Cl^--dependent L-[^{3}H] glutamate binding a new receptor or a particular transport process ? FEBS Lett 175: 31-36.

Piomelli D, Volterra A, Dale N, Siegelbaum SA, Kandel ER, Schwartz JH, Belardetti F (1987) Lipoxygenase metabolites of arachidonic acid as second messengers for presynaptic inhibition of *Aplysia* sensory cells. Nature 328: 38-43.

Reymann KG, Brödemann R, Kase H, Matthies H (1988a) Inhibitors of calmodulin and protein kinase C block different phases of hippocampal long-term potentiation.Brain Res 461: 388-392.

Reymann KG, Frey U, Jork R, Matthies H (1988b) Polymyxin B, an inhibitor of protein kinase C, prevents the maintenance of synaptic long-term potentiation in hippocampal CA1 neurons. Brain Res 440: 305-314.

Sakurai M (1987) Synaptic modification of parallel fibre-Purkinje cell transmission in *in vitro* guinea pig cerebellar slices. J Physiol 399: 463-480.

Sarantis M, Everett K, Attwell D (1988) A presynaptic action of glutamate at the cone output synapse. Nature 332: 451-453.

Shiells RA, Falk G, Naghshineh S (1981) Action of glutamate and aspartate analogs on rod horizontal and bipolar cells. Nature 294: 592-594.

Sladeczek F, Pin JP, Récasens M, Bockaert J, Weiss S (1985) Glutamate stimulates inositol phosphate formation in striatal neurones. Nature 317: 717-719.

Sladeczek F, Récasens M, Bockaert J (1988) A new mechanism for glutamate receptor action: phosphoinositide hydrolysis. Trends Neurosci 11: 545-548.

Slaughter MM, Miller RF (1981) 2-Amino-4-Phosphonobutyric acid: a new pharmacological tool for retina research. Science 211: 182-185.

Sugiyama H, Ito I, Hirono C (1987) A new type of glutamate receptor linked to inositol phospholipid metabolism. Nature 325: 531-533.

Sugiyama H, Ito I, Okada D, Hirono C, Ohmori H, Shigemoto T, Furuya S (1988) Functional and pharmacological properties of glutamate receptors linked to inositol phospholipid metabolism. In Frontiers in Excitatory Amino Acid Research, ed. Cavalhevio EA, Lehmann J, Turski L Alan R Liss, New York. pp 21-28.

Tachibana M, Kaneko A (1988) L-glutamate induced depolarization in solitary photoreceptors: a process that may contribute to the interaction between photoreceptors *in situ*. proc Natl Acad Sci 85: 5315-5319.

Verdoorn TA, Kleckner NW, Dingledine R (1987) Rat brain N-methyl-D-aspartate receptors expressed in *Xenopus* oocytes. Science 238: 1114-1116.

Wallén P, Grillner S (1987) N-methyl-D-aspartate receptor-induced, inherent oscillatory activity in neurons active during fictive locomotion in the lamprey. J Neurosci 7: 2745-2755.

Watkins JC, Olverman HJ (1987) Agonists and antagonists for excitatory amino acid receptors. Trends Neurosci 10: 265-272.

Watson GB, Hood WF, Monahan JB, Lanthorn TH (1988) Kynurenate antagonizes actions of N-methyl-D-aspartate through a glycine sensitive receptor. Neurosci Res Comm 2: 169-174.

Westbrook GL, Mayer ML (1987) Micromolar concentrations of Zn^{2+} antagonize NMDA and GABA responses of hippocampal neurons. Nature 328: 640-643.

Williams JH, Bliss TVP (1988) Induction but not maintenance of calcium-induced long-term potentiation in dentate gyrus and area CA1 of the hippocampal slice is blocked by nordihydroguaiaretic acid. Neurosci Lett 88: 81-85.

Wong EHF, Kemp JA, Priestley T, Knight AR, Woodruff GN, Iversen LL (1986) The novel anticonvulsant MK-801 is a potent N-methyl-D-aspartate antagonist. Proc Natl Acad Sci USA 83: 7104-7108.

MECHANISMS OF GLUTAMATE EXOCYTOSIS FROM ISOLATED NERVE TERMINALS

David Nicholls, Anne Barrie, Harvey McMahon, Gareth Tibbs and
Rosemary Wilkinson
Department of Biochemistry
University of Dundee
Dundee DD1 4HN
Scotland, U.K.

INTRODUCTION

The isolated nerve terminal, or synaptosome, is the simplest
system in which the presynaptic events of neurotransmission,
from the initial plasma membrane depolarization to the final
release of neurotransmitter, can be followed. Synaptosomes are
produced during mild homogenization of brain tissue, when the
axon is torn off from the terminal, and the membrane rapidly
reseals. When incubated in physiological media, synaptosomes
use their high glycolytic activity to maintain overall ATP/ADP
ratios in excess of 5; generate plasma membrane potentials
(δu_p) in the region of -60mV (Blaustein and Goldring, 1975; Scott
and Nicholls, 1980), and lower cytosolic free Ca^{2+} concentrations
($[Ca^{2+}]_c$) to 0.1-0.3µM (Richards et al., 1984; Hansford and
Castro, 1985; Nachshen, 1985a; Ashley, 1986; Adam-Vizi and
Ashley, 1987). The intra-synaptosomal mitochondria play an
essential role in all functional terminals, since inhibition
of respiration leads to a fall in ATP/ADP ratio sufficient to
largely inhibit the highly energy-dependent process of transmitter release (Kauppinen et al., 1988), despite a 10-fold
increase in glycolysis as the Pasteur effect attempts to compen-
sate for the loss of oxidative phosphorylation (Kauppinen and
Nicholls, 1986). The in situ mitochondria maintain a membrane
potential (δum) of about 150mV (Scott and Nicholls, 1980), and
can utilize pyruvate either from glycolysis, or added directly
to the synaptosomal preparation (Kauppinen and Nicholls, 1986).

L-Glutamate is the major excitatory neurotransmitter in
the mammalian brain (Fonnum, 1984). Glutamatergic synapses may
be associated with processes of learning and memory (Collingridge
and Bliss, 1987), and also in a number of important pathological
conditions, including ischaemic brain damage (Rothman and Olney,

NATO ASI Series, Vol. H29
Receptors, Membrane Transport and Signal Transduction
Edited by A. E. Evangelopoulos et al.
© Springer-Verlag Berlin Heidelberg 1989

1987). Together with L-aspartate and the major inhibitory transmitter gamma-aminobutyrate (GABA), L-glutamate is present at
very high concentration in preparations of synaptosomes from
the cerebral cortex. The guinea-pig preparation used in this
laboratory contains an initial 29nmol glutamate/mg protein,
equivalent to 9mM, without taking into account any preferential
accumulation into the glutamate-releasing sub-population of
terminals, which may only account for 15% of the synaptosomes
(Beart, 1976). This high concentration facilitates investigation
of the transmitter, but also raises the question how such an
ubiquitous amino acid can possess a specific transmitter role.
At the post-synaptic membrane, specificity can be assured by
glutamate-selective receptors, and distinct classes of such
receptors can be categorized by pharmacological means (Foster
and Fagg, 1984). At the presynaptic terminal, there is a requirement for a specific release mechanism, which can be triggered
by plasma membrane depolarization, and also for a means to
prevent or reverse any non-evoked leakage of amino acid from
the terminal. The highly active, Na^+-coupled, acidic amino
acid transporter (Kanner, 1983) can maintain the extra-synapto-
somal glutamate concentration below 1μM: in this report we
shall discuss recent experiments from our laboratory pertinent
to the nature of the release mechanism.

Transmitter glutamate is released from a non-cytosolic, osmot-ically resistant pool

There is considerable debate as to whether transmitter
glutamate is released from synaptic vesicle stores by direct
exocytosis, or from the cytoplasm by the operation of a hypothetical, Ca^{2+}-gated channel (Nicholls, 1988). Glutamate accumulating synaptic vesicles have been extensively characterized
(Naito and Ueda, 1985), and in addition the exocytotic mode is
consistent with the sensitivity of glutamate release to botulinum
neurotoxin type B (Sanchez-Prieto et al., 1987a); a high energy
dependency (Sanchez-Prieto et al., 1987b; Sanchez-Prieto and
Gonzales, 1988; Kauppinen et al., 1988); and activation by
phorbol esters (Diaz-Garcia et al., 1988), all of which are

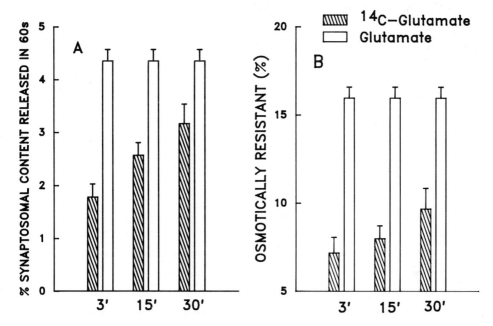

Fig. 1. Exogenous ^{14}C-L-glutamate only slowly exchanges into the pool of glutamate which is released in a Ca^{2+}-dependent manner by 60s exposure to 30mM KCl (A), and also exchanges slowly into the pool of glutamate which is resistant to hypo-osmotic lysis of synaptosomes (B), consistent with the trans-mitter pool being located in slowly exchanging, osmotically resistant synaptic vesicles.

characteristic of authenticated exocytotic processes. However more persuasive proof would come from fractionation studies. Fig. 1A shows that some 4.5% of the total endogenous glutamate is released from synaptosomes in a Ca^{2+}-dependent manner within 60s of depolarization in the presence of Ca^{2+}. However, despite the fact that exogenous labelled glutamate equilibrates through the acidic amino acid carrier with the cytosol within 3min (data not shown), the incorporation into the Ca^{2+}-dependent pool is much slower, indicating a kinetic barrier separating the Ca^{2+}-dependent pool from the cytoplasm. Fig. 1B shows that some 15% of the endogenous glutamate is resistant to osmotic lysis, consistent with a location inside small, highly curved vesicles. The slow exchange of exogenous label into this pool would support the idea that the slowly exchanging Ca^{2+}-dependent glutamate pool was located within this osmotically resistant pool, i.e. with a direct exocytotic model.

The Ca^{2+}-dependent glutamate pool does not respond to wide fluctuations in the glutamate/aspartate ratio

Although less well characterized than glutamate, L-aspartate is also a candidate neurotransmitter. However, unlike glutamate, aspartate has been reported not to be accumulated into synaptic vesicles from cerebral cortex (Naito and Ueda, 1985). Consistent with this, we find no Ca^{2+}-dependent release of aspartate from preparations which show substantial Ca^{2+}-dependent glutamate release (Fig. 2B). It has been argued that a low Ca^{2+}-dependent aspartate release is simply a reflection of a relatively low concentration of this amino acid in the relevant terminals (Szerb, 1988); however, a 60min preincubation with either 100µM L-aspartate or a similar incubation with 100µM L-glutamate, to

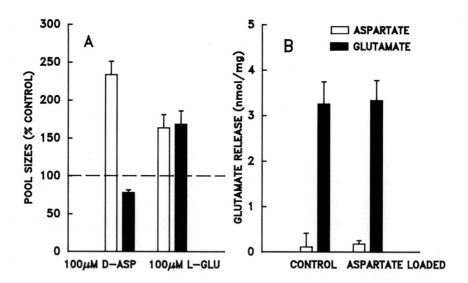

Fig. 2. The total intra-synaptosomal pool sizes of glutamate and aspartate can be widely manipulated by preincubating synaptosomes for 60min with either 100µM aspartate or 100µM glutamate (A); however even the aspartate loaded synaptosomes show no Ca^{2+}-dependent release of the amino acid. Results are for D-aspartate loading, similar results were obtained with L-aspartate.

cause a large perturbation in the intra-synaptosomal amino acid pools (Fig. 2A), is completely without effect on the extent of Ca^{2+}-dependent glutamate release, and fails to produce a Ca^{2+}-dependent release of aspartate (Fig. 2B). Thus the Ca^{2+}-dependent glutamate pool of nerve terminals is very resistant to large changes in cytosolic amino acid concentration or composition.

Glutamate excoytosis is highly energy-dependent

Glutamate exocytosis has an energy requirement which can only fully be satisfied by aerobic glycolysis. Despite the high capacity for anaerobic glycolysis in the preparation, inhibition of respiration by rotenone causes a decrease in gross ATP/ADP ratios from 5-6 to 1.7 within 6min (Kauppinen et al., 1988), and this in turn causes a substantial inhibition of the extent of Ca^{2+}-dependent glutamate release (Fig. 3).

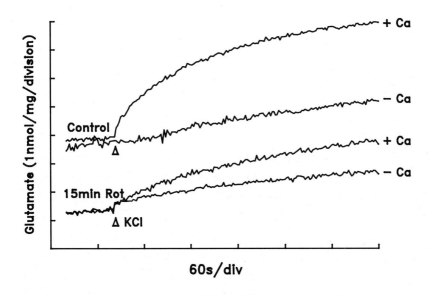

Fig. 3. Ca^{2+}-dependent release of glutamate from synaptosomes monitored continuously by fluorometry (Nicholls et al., 1987). Addition of 30mM KCl to control synaptosomes gives an extensive Ca^{2+}-dependent release which is largely inhibted in synaptosomes with respiration inhibited by 15 preincubation with rotenone.

The more extreme energy depletion caused by rotenone plus iodoacetate (to additionally block glycolysis) totally inhibits glutamate exocytosis within 60s. At the same time the Ca^{2+}-independent release of cytosolic glutamate is considerably accelerated (data not shown), as the Na^+ electrochemical gradient declines, allowing the glutamate carrier to reverse.

The discrimination between transient and continuous Ca^{2+} channels in the synaptosomal membrane

Electrophysiological studies on neuronal cell bodies have shown the presence of three classes of voltage-activated Ca^{2+}-channels which have been termed L (long-lasting), T (transient) and N (neuronal) (Fox et al., 1987). Despite this nomenclature, it is not clear whether the Ca^{2+}-channels in the presynaptic nerve terminal, which is generally too small for electrophysiological investigation, conform to this categorization. There is a biphasic kinetic of $^{45}Ca^{2+}$ entry into synaptosomes following depolarization of synaptosomes by elevated KCl (Drapeau and Blaustein, 1983; Nachshen, 1985b) a very rapid entry decreasing after <1s to a continuous influx which remains above the polarized control. The changes in $[Ca^{2+}]_c$ during KCl depolarization reflect this, a transient spike being followed by a continuous elevation (Fig. 4A).

4-aminopyridine (4AP) depolarizes synaptosomes by inhibiting K^+-channels. Fig. 4B shows that 4AP causes the continuous Ca^{2+} elevation but little or no transient rise. When, however, 30mM KCl is added after 4AP (Fig. 4C) an extensive spike in $[Ca^{2+}]_c$ is seen, presumably caused by a transient Ca^{2+} influx causes which then declines as cytosolic Ca^{2+} is sequestered, either by intra-synaptosomal organelles or by net extrusion across the plasma membrane. Thus it appears that 4AP is able to fully activate a continuous Ca^{2+} entry, but not the transient channel, which is only activated with subsequent KCl depolarization.

Unless the voltage-gated Na^+-channel is modified by alkaloids such as veratridine, which stabilizes the activated conformation, there is no single potential at which the

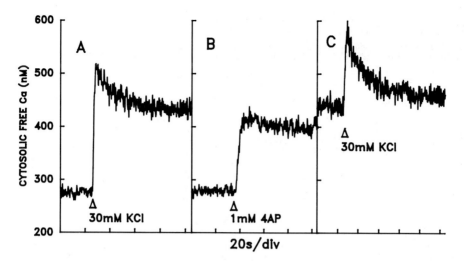

Fig. 4. Fura-2 traces of $[Ca^{2+}]_c$ following depolarization by 30mM KCl, 4-AP, or 30mM KCl 5min after 4AP. Data are means of 3-4 traces sampled at 0.2s.

Na^+-channel is continuously conductive, and therefore likely to contribute significantly to the steady-state ion fluxes in long-term synaptosomal experiments. Therefore Na^+-channel inhibitors such as tetrodotoxin (TTx) should be without effect. However, the 4AP-evoked increase in $[Ca^{2+}]_c$ and the resulting glutamate release are almost entirely sensitive to TTx (Fig. 5). Even the basal glutamate release from polarized synaptosomes is slightly inhibited, although that evoked by 30mM KCl is not affected by the toxin (traces c and d).

Unless the gating properties of the voltage-gated Na^+-channel are modified in the presence of 4AP, the TTX sensitivity suggests that the membrane potentials of the small (1μm) diameter synaptosomes could be undergoing large statistical fluctuations when the K^+-conductance is inhibited by 4AP, sufficient for voltage-gated Na^+-channels to fire repetitively, generating action potentials and amplifying the depolarization. The depolarization would also activate voltage-gated Ca^{2+}-channels and

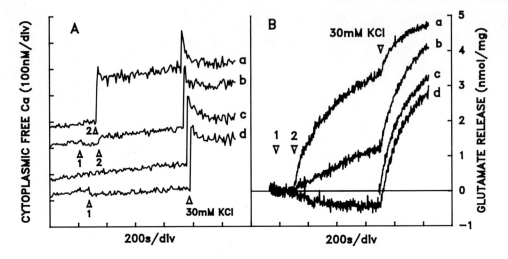

Fig. 5. Parallel fura-2 and glutamate release determinations to show the inhibitory action of TTx. (1), addition of 1µM TTx (traces b,d); (2) addition of 1mM 4AP (traces a,b). 30mM KCl was added where indicated to all experiments. The glutamate releases in B are expressed relative to that in the absence of additions.

elevate $[Ca^{2+}]_c$. This amplification would be lost in the presence of TTx. In the population of synaptosomes, these action potentials would be randomly distributed, and so would appear as a uniform depolarization when determined for example by tetraphenyl phosphonium (TPP$^+$) cation distribution.

The statistical fluctuations, and hence the amplification of depolarization by Na$^+$-channels, would not occur when depolarization was induced by elevated [K$^+$], since the high conductance of the membrane, and consequent low noise fluctuation, would prevent the potential from fluctuating sufficiently to reverse the inactivation of the Na$^+$-channel. It is at first site possible that both the 4AP-activated continuous $[Ca^{2+}]_c$ elevation and the KCl-evoked transient spike could be caused by the same class of transient Ca^{2+}-channel. In the presence of 4AP, repetitive action potentials could lead to repeated firing of transient Ca^{2+}-channels, which, since they would not be synchronized between synaptosomes, would appear as a continuous, non-inactivating, elevation in $[Ca^{2+}]_c$. Subsequent addition of high KCl would cause a synchronous activation of all Ca^{2+}-channels, leading to the observed spike. However, two factors argue against this: firstly, the same plateau in $[Ca^{2+}]_c$ is seen in the presence of high KCl after the inital spike. Since this is TTx insensitive, showing that potential fluctuations are insufficient for repetitive firing of Na^+-channels, it is unlikely that a similar repetitive firing of transient Ca^{2+}-channels would be possible. Secondly, the addition of KCl subsequent to 4AP should prevent further firing of 4AP-activated Ca^{2+}-channels, with a consequent decrease of $[Ca^{2+}]_c$ to levels characteristic of polarized synaptosomes, whereas the plateau of elevated $[Ca^{2+}]_c$ is still retained.

Two independent modes of voltage-activated Ca^{2+}-entry are therefore indicated, one of which shows no inactivation, and one of which is transient. Since the transient Ca^{2+} entry is not evoked by 4AP, this implies that it requires a more intense depolarization. The ability to separately evoked the continuous and transient Ca^{2+} entry allows their respective roles in transmitter release to be investigated. Both transient and continuous Ca^{2+} channels are coupled to glutamate exocytosis (Fig. 5), determined by assaying the glutamate released continuously, by including glutamate dehydrogenase and $NADP^+$ in the incubation (Nicholls and Sihra, 1986; Nicholls et al., 1987). With 30mM KCl the Ca^{2+}-dependent release of glutamate is rapid, and amounts to 2-3nmol. min^{-1}. mg^{-1}. 1mM 4AP also induces Ca^{2+}-dependent glutamate release, although at a somewhat lower rate. Exocytosis induced by 4AP is incomplete over the time course of the experiment, allowing the "spike" in $[Ca^{2+}]_c$ by subsequent KCl to evoke a rapid exocytosis of residual releasable vesicles.

Presynaptic modulation of glutamate release

Autoreceptors on nerve terminals modulate the release of
a neurotransmitter by sensing the concentration of that
transmitter in the extra-cellular medium and regulating its
further release. They are normally inhibitory and serve to
decrease release; however glutamate is implicated in theories
of learning and memory which involve synaptic facilitation
(Collingridge and Bliss, 1987), an increased transmitter release
and/or post-synaptic receptor sensitivity as a result of synaptic
activity. Inhibitory presynaptic autoreceptors would antagonize
this action, whereas excitatory autoreceptors would promote
facilitation.

Three classes of excitatory post-synaptic glutamate receptor
have been described, selective respectively for N-methyl-D-
aspartate (NMDA), kainate and quisqualate (Foster and Fagg,

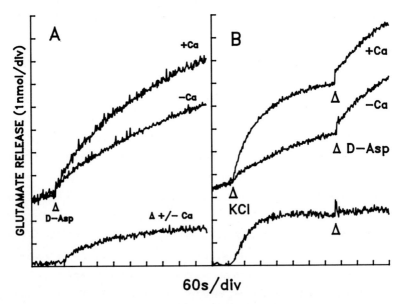

Fig. 6. Ca^{2+}-dependent release of glutamate evoked by the uptake-
induced depolarization on adding 100µM D-aspartate. Note that
prior depletion of the pool by KCl (B) removes the Ca^{2+}-depen-
dency. Bottom traces are differences showing the net Ca^{2+}-
dependent release. Data from McMahon et al., 1988.

1984). NMDA is without presynaptic effect, while kainate and
quisqualate in the 100μM concentration range respectively inhibit
the plasma membrane Na^+-coupled acidic amino acid carrier
(responsible for the uptake of glutamate and aspartate into
the terminal) and induce a Ca^{2+}-independent release of glutamate
from the terminal (McMahon et al., 1988). Despite this lack of
conventional presynaptic autoreceptors, both glutamate and D-
aspartate cause a reversible depolarization of the plasma
membrane, monitored by a TPP^+-selective electrode (McMahon et
al., 1988). This depolarization correlates with the net uptake
of the amino acids into the synaptosomes and can be prevented
by inhibitors of the acidic amino acid carrier. The electrogenic
uptake into the glutamate secreting nerve terminals causes the
depolarization observed with the TPP^+ electrode, and can be
sufficient to cause Ca^{2+} entry and to induce the Ca^{2+}-dependent

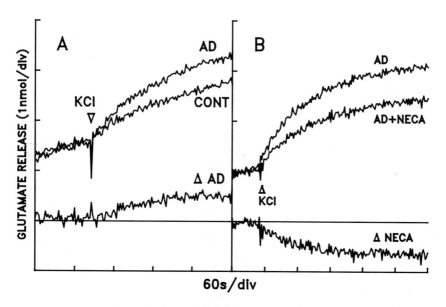

Fig. 7. Adenosine deaminase removes endogenous inhibitory adenosine from presynaptic receptors, and potentiates the Ca^{2+}-
dependent release of glutamate evoked by sub-optimal (10mM)
KCl. In (B) this stimulation is reversed by the addition of
100μM NECA, and adenosine receptor agonist. Bottom traces are
differences.

release of transmitter glutamate from the terminals (Fig. 6). This novel form of positive feed-back could have significance in the short-term potentiation of glutamate release.

A more conventional, inhibitory, presynaptic modulation of glutamate release is caused by adenosine. When synaptosomes are depolarized by sub-optimal KCl the extent of glutamate release is highly sensitive to positive or negative modulation. Removal of endogenous adenosine by adenosine deaminase potentiates release (Fig. 7) and this is reversed by the adenosine agonist 5'-N-ethylcarboxyamidoadenosine (NECA), suggesting that this negative modulation is mediated by the A2 sub-class of adenosine receptors. Protein kinase C involvement in the regulation of glutamate exocytosis is indicated by the stereo-specific action of 4β- (but not 4α-) phorbol didecanoate on the release of glutamate evoked by sub-optimal KCl (Fig. 8).

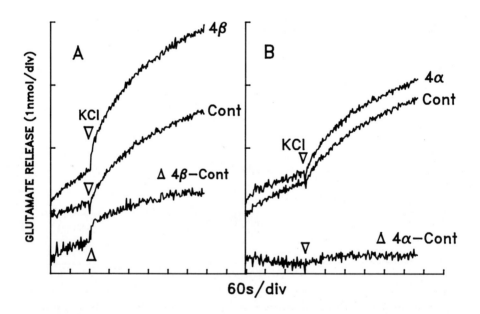

Fig. 8. Phorbol esters cause a stereo-specific stimulation of glutamate release. 1µM 4β- (A) or 4α- (B) phorbol didecanoate was added 3min prior to depolarization of synaptosomes by sub-optimal (10mM) KCl. The bottom trace is the difference +/- ester.

Conclusions

The isolated nerve terminal provides an interface between
cellular biochemistry and neurochemistry, allowing the techniques
of the former to be applied to some of the challenging problems
in the latter discipline. The wide variety of biochemical
methods, some of which have been detailed in this paper, which
can be assembled to investigate synaptosomal transmitter release
are providing increasingly precise information on the mechanism
and regulation of this process.

Acknowledgements

Research in the authors' laboratory is funded by the
Wellcome Trust, the Medical Research Council and Merck, Sharp
and Dohme Research Laboratories.

References

Adam-Vizi V, Ashley RH (1987) Relation of acetylcholine release
to Ca^{2+} uptake and intra-terminal Ca^{2+} concentration in
guinea-pig cortex synaptosomes. J. Neurochem. 49:1013-1021

Ashley RH (1986) External calcium intrasynaptosomal free calcium
and neurotransmitter release. Biochim. Biophys. Acta
854:207-212

Beart PM (1976) The autoradiographic localization of L-[³H]
glutamate in synaptosomal preparations. Brain Research
103:350-355

Blaustein MP, Goldring JM (1975) Membrane potentials in pinched-
off presynaptic nerve terminals monitored with a fluorescent probe. J. Physiol. (London) 247:589-615

Collingridge GL, Bliss TVP (1987) NMDA receptors - their role
in long-term potentiation. Trends Neurosci. 10:288-293

Diaz-Guerra MJM, Sanchez-Prieto J, Bosca L, Pocock J, Barrie A,
Nicholls DG (1988) Phorbol ester translocation of protein
kinase C in guinea-pig synaptosomes and the potentiation
of calcium-dependent glutamate release. Biochim. Biophys.
Acta (in press)

Drapeau P, Blaustein MP (1983) Initial release of ³H-dopamine
from rat striatal synaptosomes: correlation with calcium
entry. J. Neurosci. 3:703-713

Fonnum F (1984) Glutamate as a neurotransmitter. J. Neurochem. 42:1-11

Foster AC, Fagg GE (1984) Acidic amino acid binding sites in mammalian neuronal membranes: their characteristics and relationship to synaptic receptors. Brain Res. Rev. 7: 103-164

Fox AP, Nowycky MC, Tsien RW (1987) Kinetic and pharmacological properties distinguish three types of calcium channels in chick sensory neurones. J. Physiol. (Lond.) 394:149-172

Hansford RG, Castro F (1985) Role of Ca^{2+} in pyruvate dehydrogenase interconversion in brain mitochondria and synaptosomes. Biochem. J. 227:129-136

Kauppinen RA, Nicholls DG (1986a) Synaptosomal bioenergetics the role of glycolysis pyruvate oxidation and responses to hypoglycaemia. Eur. J. Biochem. 158:159-165

Kauppinen RA, McMahon H, Nicholls DG (1988) Ca^{2+}-dependent and Ca^{2+}-independent glutamate release energy status and cytosolic free Ca^{2+} concentration in isolated nerve terminals following in vitro hypoglycaemia and anoxia. Neuroscience (in press)

McMahon HT, Barrie AP, Lowe M, Nicholls DG (1988) Glutamate release from guinea-pig synaptosomes: extra-cellular glutamate may exert a positive feedback by reuptake-induced depolarization rather than receptor activation. J. Neurochem. (in press)

Nachshen DA (1985a) Regulation of cytosolic calcium concentrations in presynaptic nerve endings isolated from rat brain. J. Physiol. (London) 363:87-101.

Nachshen DA (1985b) The early time-course of potassium stimulated calcium uptake in presynaptic nerve terminals isolated from rat brain. J. Physiol. (London) 361:251-268

Naito S, Ueda T (1985) Characterization of glutamate uptake into synaptic vesicles. J. Neurochem. 44:99-109

Nicholls DG (1988) The release of glutamate aspartate and GABA from isolated nerve terminals. J. Neurochem. (in press)

Nicholls DG, Sihra TS (1986) Synaptosomes possess an exocytotic pool of glutamate. Nature 321:772-773

Nicholls DG, Sihra TS, Sanchez-Prieto J (1987) Calcium dependent and independent release of glutamate from synaptosomes monitored by continuous fluorometry. J. Neurochem. 49: 50-57

Richards CD, Metcalfe J, Heskith TR (1984) Changes in free-calcium levels and pH in synaptosomes during transmitter release. Biochim. Biophys. Acta 803:215-220.

Rothman SM, Olney JW (1987) Excitotoxicity and the NMDA receptor. Trends Neuro. Sci. 10:299-302

Sanchez-Prieto J, Gonzales MP (1988) Anoxia induces a large Ca^{2+}-independent release of glutamate in isolated nerve terminals. J. Neurochem. (in press)

Sanchez-Prieto J, Sihra TS, Evans D, Ashton A, Dolly JO, Nicholls DG (1987a) Botulinum toxin A blocks glutamate exocytosis from guinea pig cerebral cortical synaptosomes. Eur. J. Biochem. 165:675-681

Sanchez-Prieto J, Sihra TS, Nicholls DG (1987b) Characterization of the exocytotic release of glutamate from guinea pig cerebral cortical synaptosomes. J. Neurochem. 49:58-64

Scott ID, Nicholls DG (1980) Energy transduction in intact synaptosomes: influence of plasma-membrane depolarization on the respiration and membrane potential of internal mitochondria determined in situ. Biochem. J. 186:21-33

Szerb JC (1988) Changes in the relative amounts of aspartate and glutamate released and retained in hippocampal slices during depolarization. J. Neurochem. 50:219-224

CHARACTERISTICS OF THE EPIDERMAL GROWTH FACTOR RECEPTOR

J. Boonstra[1], L.H.K. Defize[2]. P.M.P. van Bergen en Henegouwen[1], S.W. de Laat[2] and A.J. Verkleij[1].

[1] Department of Molecular Cell Biology, University of Utrecht, Padualaan 8, 3584 CH Utrecht, the Netherlands;

[2] Hubrecht Laboratory, Netherlands Institute for Developmental Biology, Uppsalalaan 8, 3584 CT Utrecht, the Netherlands.

INTRODUCTION

Polypeptide growth factors have been recognized as important deter-minants in the regulation of cellular proliferation and differentiation (Carpenter and Cohen, 1979; de Laat *et al.*, 1983; Schlessinger *et al.*, 1983). As such these factors are believed to play a crucial role in embryonal development on the one hand and in carcinogenesis on the other (Gospodarowicz, 1981; de Laat *et al.*, 1983; Sporn and Roberts, 1985; Thornburn *et al.*, 1985; Burgess, 1986; Heldin *et al.*, 1987). This notion is strongly supported by the findings that products of certain oncogenes have significant homology to either growth factors or to their receptors (Doolittle et al., 1983; Downward et al., 1984; Heldin and Westermark, 1984). The wide interest in growth factors and their mechanisms of action has resulted in detailed knowledge on the molecular properties of both factors and their receptors, while in addition many details are known about the effects of growth factors in their target cells (Carpenter and Cohen, 1979; de Laat *et al.*, 1983; Heldin and Westermark, 1984; Gospodarowicz, 1985; Moolenaar, 1986; Schlessinger, 1986; 1988; Carpenter, 1983, 1987). Nevertheless, the precise nature of the molecular mechanism of action of growth factors is still unknown.

NATO ASI Series, Vol. H29
Receptors, Membrane Transport and Signal Transduction
Edited by A. E. Evangelopoulos et al.
© Springer-Verlag Berlin Heidelberg 1989

One of the most intensively studied growth factors so-far is epidermal growth factor (EGF). EGF, a polypeptide of 53 amino acid residues, evokes its effects in the target cells by binding to a specific plasma membrane receptor, a transmembrane glycoprotein of 170,000 daltons (Carpenter and Cohen, 1979).

In this contribution we will not review the molecular characteristics of EGF and EGF-receptor in detail, since recent review articles concerning this subject are available (Carpenter and Zendegui, 1986; Carpenter, 1987; Schlessinger, 1986, 1988), but instead describe recent results which may provide a better insight into the molecular mechanisms of activation of the EGF-receptor by EGF. In this context we will consider four parameters of interest, i.e. EGF binding properties, receptor oligomerization, receptor association to the cytoskeleton and receptor kinase activation and phosphorylation.

For a better understanding we will first briefly summarize the most important properties of the EGF-receptor. The protein part of the mature receptor is a single polypeptide chain of 1186 amino acid residues, as deduced from the nucleotide sequence of cDNA clones (Ullrich *et al.*, 1984). The receptor can be divided into three domains, i.e. the extracellular, N-terminal domain of 621 amino acid residues containing the EGF binding site, the intramembraneous domain of 23 amino acids, and the intracellular, C-terminal domain of 542 amino acids. The extracellular domain is characterized by a high amount of cysteine residues, localized in two major domains of each 160 amino acid residues long (Ullrich *et al.*, 1984). The EGF-binding site is localized between these two cysteine rich domains (Lax *et al.*, 1988). In addition, the extracellular domain contains approximately 11 N-linked oligosaccharide chains (Soderquist and Carpenter, 1986).

The amino acids comprising the intramembraneous stretch have a strong hydrophobic nature, typical for the transmembrane segments of many intrinsic membrane proteins.

Finally, the cytoplasmic domain of the receptor contains protein tyrosine kinase activity. Binding of EGF to the receptor induces activation of this kinase (Ushiro and Cohen, 1980), leading to the phosphorylation of the receptor itself and other cellular proteins. In intact cells the phosphorylation of the receptor occurs mainly on tyrosine residue 1173 (Downward *et al.*, 1984). A number of additional tyrosine residues are phosphorylated when EGF is added to solubilized membranes or purified receptor preparations. The lysine residue at position 721 functions most likely as part of the ATP binding site (Russo *et al.*, 1985; Honegger *et al.*, 1987). In addition to phosphotyrosine, the EGF receptor also contains phosphorylated serine and threonine residues, these latter being the substrates of other cellular kinases, such as protein kinase C (Hunter and Cooper, 1985). The cytoplasmic domain of the receptor exhibits considerable sequence homology with other tyrosine kinases (Russo *et al.*, 1985; Hunter and Cooper, 1985), in particular with the protein product of the erb B oncogene from avian erythroblastosis virus (Downward *et al.*, 1984; Ullrich *et al.*, 1984). A linear representation of the EGF receptor structure is shown in Fig. 1.

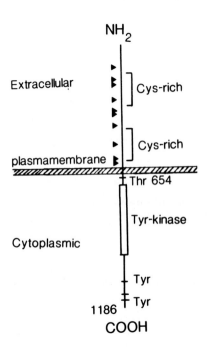

Fig. 1:   Schematic linear representation of the EGF-receptor structure. The horizontal hatched bar represents the plasma membrane. The numbers indicate individual amino acids according to the numbering system proposed by Ullrich *et al.* (1984). For further details see text.

Binding characteristics

The kinetics of EGF binding to its receptor are of a complicated nature, although the factor binds to the receptor in a 1:1 stoichiometry (Weber *et al.*, 1984). Usually, receptor binding characteristics are established by determination of the relationship between specific ligand binding and ligand concentration under steady state conditions, followed by analysis according to the method of Scatchard. This analysis yields the apparent dissociation constant K_D, and the maximal binding capacity of the cells from which the total receptor number can be calculated. Curvilinearity of the Scatchard graph is usually interpreted as indicative for the presence of different classes of binding sites with different K_D's. EGF binding characteristics in a wide variety of cell lines, analyzed by the Scatchard method, reveals such a curvilinear relationship, and therefore indicates the presence of at least two classes of EGF binding sites, i.e. a high-affinity, low capacity class and a low-affinity, high capacity class (Boonstra *et al.*, 1985a,b; Wiegant *et al.*, 1986; King and Cuatrecasas, 1982; Honegger *et al.*, 1987; Livneh *et al.*, 1987). However, conclusions about the presence of two classes of EGF binding sites based upon Scatchard analysis, should be considered with caution as discussed in detail previously (Boonstra *et al.*, 1985a; Carpenter, 1987). But other lines of evidence favour the presence of high and low affinity binding sites in EGF-receptor containing cells. Firstly, the tumor promoter phorbol myristate acetate (PMA) has been demonstrated to specifically influence high-affinity EGF-binding in a variety of cell lines (Boonstra *et al.*, 1985a; Shoyab *et al.*, 1979; Magun *et al.*, 1980; Lockyer *et al.*, 1983; Hunter and Cooper, 1985; King and Cuatrecasas, 1982; Lee and Weinstein, 1978). Secondly, treatment of cells with other growth factors such as platelet derived growth factor (PDGF) or bombesin, each binding to its own distinct receptor, results in loss of high-affinity binding (Olashaw *et al.*, 1986; Zachary *et al.*, 1986; Davis and Czech, 1985a), while an increase in high-affinity EGF receptors is observed after exposure of cells to gluco-corticoids (Fanger *et al.*, 1984) or cyclic AMP (Boonstra *et al.*, 1987). Thirdly, dissociation of EGF from its receptor was demonstrated to follow biphasic characteristics (Van Bergen en Henegouwen *et al.*, 1988). The most compelling evidence in favour of the existence of two classes of binding sites is obtained by studies using antibodies directed against the EGF-receptor (Gregoriou and Rees, 1984; Defize *et al.*, 1986, 1989a). In particular the monoclonal anti-EGF receptor antibody 2E9 (Defize *et al.*, 1986) has been shown to exhibit unique properties in this respect. The 2E9 antibody is directed against the extracellular protein core of the receptor of a variety of human cells (Defize *et al.*, 1988, 1989a). The antibody binds to the EGF receptor in a 1:1 ratio and has no direct effect on a

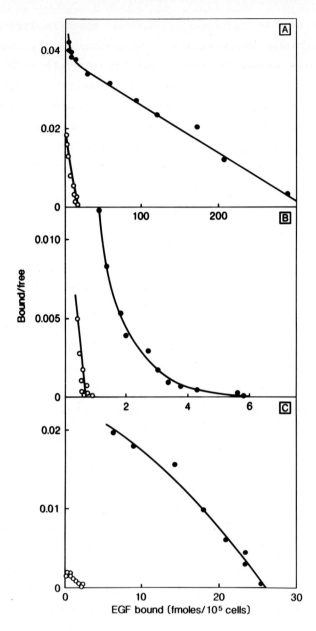

Fig.2: Effect of 2E9 on EGF binding.
Specific EGF binding was determined by Scatchard analysis as
described in detail previously (Boonstra *et al.*, 1985a). The
effect of the monoclonal anti EGF-receptor antibody 2E9 was
measured by a preincubation of the cells in the presence of 300
nM 2E9 for 3 hours at room temperature, followed by EGF binding
as described in detail elsewhere (Defize *et al.*, 1989a). Untreated (•-•) and 2E9 treated (o-o) preparations. A: A431 cells;
B: Hela cells; C: Transfected NIH/3T3 cells. Data taken from
Defize *et al.*, 1989a.

variety of rapid responses evoked by EGF in intact cells (Defize *et al.*, 1989a,b,c). Most importantly however, the antibody is demonstrated to block specifically EGF-binding to the low-affinity receptor population in a variety of cell lines, such as A431 and Hela cells (Fig. 2A,B) and a variety of transformed human keratinocyte cell lines (Boonstra and Ponec, 1989), while leaving EGF binding to the high-affinity population undisturbed (Defize *et al.*, 1988; 1989a). Furthermore, in 3T3 cells transfected with a mutated EGF receptor that exhibited no high-affinity EGF binding site, a complete inhibition of EGF binding by 2E9 was observed (Fig. 2C). These results clearly demonstrate the existence of two classes of EGF binding sites in EGF-responsive cells.

The unique properties of 2E9 described above, enabled us to study the role of the high-affinity class in the EGF-induced responses. Thus, low-affinity EGF-binding was excluded by addition of saturating amounts of 2E9, subsequently followed by measurements of EGF effects (this occurring via high affinity binding only) on the generation of inositol phosphates, the rise in intracellular pH, the release of Ca^{2+} from intracellular stores, the phosphorylation of EGF-receptor on Thr-654, the induction of c-fos, the induction of morphological changes and the growth inhibition of A431 cells. It was demonstrated that these early and late cellular responses to EGF can occur exclusively via binding of EGF to the high-affinity receptor (Defize *et al.*, 1989c).

Receptor oligomerisation

Considering the single transmembrane domain of the EGF-receptor and the rapid activation by EGF of the EGF-receptor protein tyrosine kinase activity (see below), an important question concerns the mechanism underlying the signal transfer between the extracellular EGF-binding domain and the intracellular kinase domain. A variety of early experiments have indicated receptor-receptor interactions as essential in the EGF-induced signal transduction pathway.

Using fluorescently labeled EGF, it was demonstrated that homogenously distributed receptors aggregate into patches on the cell surface at 37°C, these patches being immobile (Schlessinger *et al.*, 1978). On ultrastructural level, an aggregation of EGF-receptors was demonstrated using ferritin-conjugated EGF (Haigler *et al.*, 1979). Rotational diffusion measurements demonstrated a clear reduction of the lateral mobility of EGF-recep-

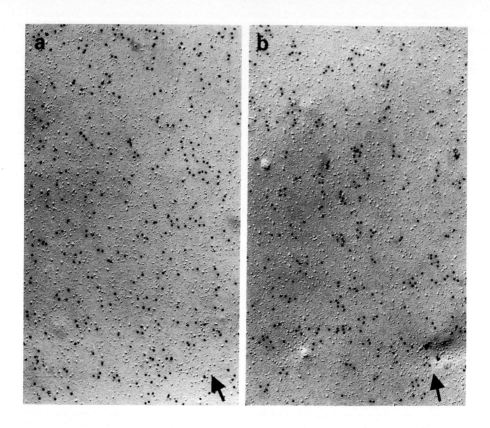

Fig. 3:   Cell-surface distribution of EGF-receptors in A431 cells as visualized by label-fracturing. A431 cells were immunogold labeled
using 2E9 as the primary antibody and processed for label-
fracture as described in detail elsewhere (van Belzen *et al.*,
1988). A: untreated cells; B: cells treated for 5 minutes with 50
ng/ml EGF at 37°C.

tors after binding of EGF at 37°C in intact cells and membrane preparations
(Zidovetski *et al.*, 1981, 1986), indicating the formation of clusters of
receptors comprising 10-50 receptors. The possible importance of micro-
clustering of EGF-receptors for the biological response to EGF was initially demonstrated by the use of monoclonal antireceptor antibodies
(Schreiber *et al.*, 1983). Monovalent Fab fragments of the antibody failed
to induce receptor clustering and DNA synthesis. Cross-linking the cell-
bound Fab fragments with secondary antibodies resulted in restoration of
clustering and DNA-synthesis (Schreiber *et al.*, 1983).

Recently, an electronmicroscopical method has been developed, the so-
called label-fracture method (Pinto da Silva and Kan, 1984), that allows a
irect visualization and quantitative analysis of EGF-induced receptor

clustering (Boonstra *et al.*, 1985c; van Belzen *et al.*, 1988). Thus, it was demonstrated in A431 cells that in the absence of EGF the EGF receptors located on the cell surface were not randomly distributed (Fig. 3A), as deduced from Poisson variance analysis. Following treatment of the A431 cells with EGF, receptor clustering increased rapidly (Fig. 3B), reaching a maximum within 10 minutes (Fig. 4). Maximal clustering was maintained for 1 hour, after which the lateral distribution of receptors returned to the control situation within another hour (Van Belzen *et al.*, 1988). Direct evidence for the EGF induced formation of receptor dimers in intact cells was obtained by the use of chemical cross-linking reagents followed by polyacrylamide gel electrophoresis (Fanger *et al.*, 1986). These studies

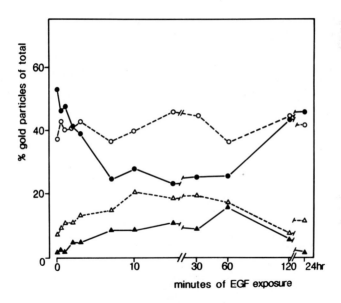

Fig. 4:   EGF-induced EGF receptor clustering.
          A431 cells were treated for with 50 ng/ml EGF for various period
          of time at 37°C as described in Van Belzen *et al.* (1988) and
          cluster sizes were determined from label fracture images.
          ●-●: 1 particle, o-o: 2-3 particles; Δ-Δ: 4-5 particles;
          ▲-▲: 6-10 particles. Taken from Van Belzen *et al.*, 1988.

revealed predominantly monomeric receptors in control cells and a rapid formation of dimeric receptors after EGF binding (Fanger *et al.*, 1986; Cochet *et al.*, 1988).

Similar results were obtained using similar methods on purified EGF-receptors (Yarden and Schlessinger, 1987b). In these preparations the dimerization was shown to be fully reversible and involved saturable, non covalent interactions that were stable at neutral pH and in non-ionic detergents (Yarden and Schlessinger, 1987b). Finally, also two sizes of solubilized receptors were demonstrated using sedimentation analysis (Boni-Schetzler *et al.*, 1987) leading to the general consensus that EGF induces receptor-receptor interactions.

Relationship between oligomerisation and affinity

An important issue in this context concerns the nature of the monomeric and oligomeric forms of the receptor with respect to EGF binding affinity. Using purified receptor preparations, some evidence has been obtained that the dimeric receptors have a higher affinity than the monomeric receptors (Yarden and Schlessinger, 1987a,b). On the other hand, using dexamethasone or glucocorticoids the number of high-affinity sites was demonstrated to increase, without significant changes in the amount of receptor dimers (Fanger *et al.*, 1986). However, the most straight forward evidence against the high-affinity/dimer concept was obtained using the monoclonal antibody 2E9. Exposure of A431 cells to saturating amounts of 2E9 prevents EGF binding to low-affinity binding sites. After such a preincubation the cells

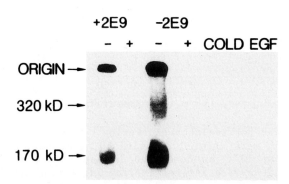

Fig. 5:   Cross-linking of ^{125}I-labeled EGF to untreated (two rightmost lanes) or 2E9 pretreated (two left-most lanes) A431 cells. Cells were incubated in the absence or presence of 2E9 for one hour at 37°C, followed by a 2 hr incubation with ^{125}I-EGF at 4°C. Cross-linking was performed using EDAC as described previously (Defize *et al.*, 1989a). To some cells, an excess of cold EGF was added This as indicated by a+. The cross-linked complexes were separated on a 7-9% linear polyacrylamide gel. Autoradiography was with Kodak XAR-5 film as described (Defize *et al.*, 1989a). Taken from Defize *et al.*, 1989a.

were incubated with ^{125}I-EGF followed by chemical cross-linking. Poly-acrylamide gel electrophoresis and autoradiography revealed the presence of monomers and dimers in control cells and cells treated with 2E9 (Fig. 5). Importantly, quantitation of the autoradiograms demonstrated no differences in the relative amount of dimers in control and 2E9-treated cells (Defize *et al.*, 1989a). It was, therefore, concluded that high-affinity receptors comprise both monomers and dimers.

Association of the EGF-receptor to the cytoskeleton

Numerous membrane proteins have been demonstrated to be associated to the cytoskeletal network of a variety of cells, including receptors for nerve growth factor (NGF) (Schechter and Bothwell, 1981; Vale and Shooter, 1983b), N-formylated peptide (Jesaitis *et al.*, 1984; Painter *et al.*, 1987), cyclic AMP (Ludérus and van Driel, 1988), acetylcholine (Bloch, 1986), fibronectin (Tamkun *et al.*, 1986) and others (Braun *et al.*, 1982; Wheeler *et al.*, 1985). The EGF-receptor appears to be no exception. A variety of

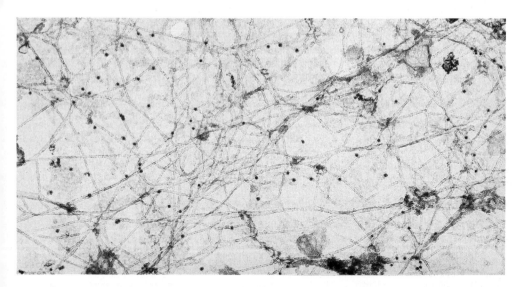

Fig. 6:   Structural association of EGF receptors to the cytoskeleton. A431 cells were processed for lysis squirting as described in detail previously (Boonstra *et al.*, 1989). The isolated ventral membranes and associated cytoskeletons were fixed and consequutively labeled with the rabbit antibody 2-81-7, directed against the cytoplasmic domain of the EGF-receptor, and protein A/gold conjugate and processed as described previously (Boonstra *et al.*, 1989). The average diameter of the gold particles was 10 nm. Arrowheads indicate clearly visible association of the EGF receptor with cytoskeletal elements. Magnification: 100,000x. Taken from Boonstra *et al.*, 1989.

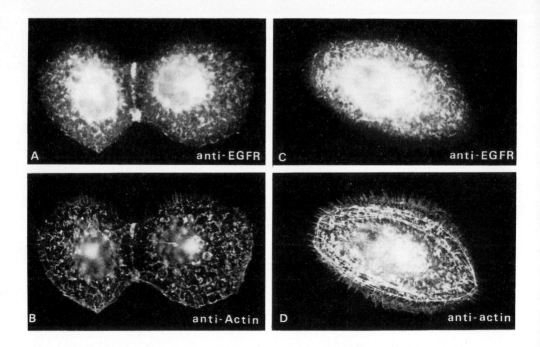

Fig. 7:   Co-localization of EGF-receptors and actin filaments.
           A431 cells were processed for immunofluorescence using 2E9 and
           anti actin antibodies as described by Defize *et al.* (1986).
           Note the strong co-localization of EGF-receptors and actin (A,B)
           but the lack of co-localization between EGF receptors and stress-
           fibers (C,D).

electronmicroscopical methods in combination with immunogold labeling
suggested strongly a structural association between the EGF-receptor and
the cytoskeleton of A431 cells (Fig. 6) (Wiegant *et al.*, 1986; Boonstra *et
al.*, 1989), in line with the observation that the EGF-receptor kinase is
associated to the cytoskeleton of A431 cells (Landreth *et al.*, 1985).

An interesting aspect of the interaction between the EGF-receptor
and the cytoskeleton involves the nature of the cytoskeletal component. It
has been suggested, as judged by the diameter of the elements as observed
in the electron microscope, that actin is involved in the association
(Wiegant *et al.*, 1986). This suggestion has been supported by a
co-localization of EGF-receptors and actin filaments by immunofluorescence
microscopy (Fig. 7). It is tempting to suggest that actin filaments are
involved in receptor clustering, but support for this suggestion is lacking
so-far.

Relationship between cytoskeleton association, affinity and oligomerization

Determination of the binding characteristics of the cytoskeleton-associated receptors of A431 cells by Scatchard analysis, revealed that these
receptors are predominantly of the high-affinity class (Fig. 8) (Wiegant *et
al.* 1986; Van Bergen en Henegouwen *et al.*, 1988). Similarly, a different
cytoskeleton association of the two affinity classes of EGF receptors (Vale
and Shooter, 1983a) and of NGF receptors (Schechter and Bothwell, 1981;
Vale *et al.*, 1985) has been reported in pheochromocytoma cells.

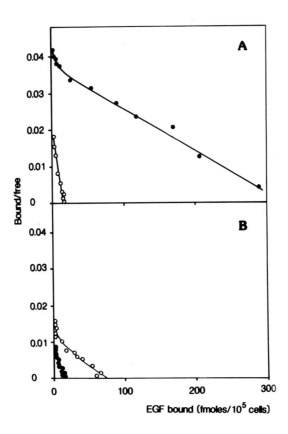

Fig. 8:   Association of EGF receptors to the A431 cytoskeleton.
           EGF binding was determined by Scatchard analysis as described in
           detail previously (Boonstra *et al.*, 1985a). Cytoskeletons were
           isolated as described in detail previously (Van Bergen en
           Henegouwen *et al.*, 1988) using 0.5% Triton X-100 for 10 minutes
           at 4°C. The effect of 2E9 was measured as described in Legend of
           Fig. 2.A: Untreated cells (●-●) and cells preincubated with 40
           ng/ml 2E9 (o-o) as described previously (Defize *et al.*, 1989a).
           B: Cells extracted with Triton X-100 followed by EGF binding
           (●-●) and cells exposed to EGF followed by Triton X-100 extrac-
           tion (o-o). Taken from Van Bergen en Henegouwen *et al.*, 1988.

Of particular interest is the observation that incubation of A431 cells in the presence of EGF for 2 hours at 4°C resulted in a significant increase in the number of EGF-receptors associated to the cytoskeleton (Van Bergen en Henegouwen *et al.*, 1988). Scatchard analysis revealed that these newly cytoskeleton-associated receptors are of the low-affinity subclass (Fig. 8). Moreover, incubation of intact A431 cells with a saturating amount of ^{125}I-labeled 2E9 (300 nM), followed by EGF for various periods of time also revealed an increase in 2E9-labeled (low-affinity) receptors associated to the cytoskeleton (Defize *et al.*, 1989b). Since 2E9 prevents binding of EGF to low-affinity receptors, this provides evidence that binding of EGF to the high-affinity receptors alone is sufficient to cause low-affinity receptors to become associated to the cytoskeleton. The most simple explanation of this phenomenon is an EGF-induced (via high-affinity receptors) dimerization of low and high-affinity receptors.

These results strongly suggest that the cytoskeleton is involved in determination of receptor affinity and also in receptor oligomerization in intact cells, but further proof is needed to establish this in more detail.

Receptor kinase activity and receptor phosphorylation

An EGF-activated protein kinase activity was first detected in membranes of A431 cells (Carpenter *et al.*, 1978). In these membranes, EGF stimulated the phosphorylation of added proteins such as casein, histones, antibodies to pp 60^{v-src} and a variety of peptides (Cohen *et al.*, 1980; Pike *et al.*, 1982; Hunter and Cooper, 1985) as well as a number of endogenous proteins, including the receptor itself (King *et al.*, 1980). Furthermore, the EGF-stimulated kinase was demonstrated to phosphorylate proteins specifically on tyrosine residues (Ushiro and Cohen, 1980; Hunter and Cooper, 1981). This finding was exciting because of the fact that tyrosine phosphorylation was known to be linked to the mitogenic action of pp60^{src} (Hunter and Sefton, 1980). In addition to autophosphorylation, the EGF-receptor is phosphorylated by other kinases as well, mainly on serine and threonine residues. In particular threonine 654 has been demonstrated to represent a substrate of the Ca^{2+} and phospholipid dependent protein kinase C (Davis and Czech, 1985b; Hunter *et al.*, 1984; Lin *et al.*, 1986). The protein kinase C activation, and hence EGF-receptor phosphorylation can be induced by phorbol esters but also by growth factors such as EGF itself, PDGF or bombesin.

However, recently it has been shown that threonine 654 of the EGF-receptor can be phosphorylated upon PDGF stimulation of protein kinase C deficient fibroblasts, indicating that other kinases may use threonine 654 as their substrate as well (Davis and Czech, 1987). An interesting feature of the phosphorylation of threonine 654 by protein kinase C, activated by phorbol ester or other growth factors, concerns the abolishment of high-affinity EGF binding (Shoyab *et al.*, 1979; Boonstra *et al.*, 1985a; Iwashita and Fox, 1984; Lee and Weinstein, 1978; Fearn and King, 1985; King and Cuatrecasas, 1982). This process has been called "receptor transmodulation" (Schlessinger, 1986) and suggested that Thr-654 phosphorylation is directly involved in receptor-affinity. In order to resolve the molecular mechanism of EGF-receptor transmodulation, various cell lines were generated expressing mutations in the EGF-receptor in which Thr654 was substituted for alanine (Lin *et al.*, 1986) or tyrosine (Livneh *et al.*, 1987, 1988). The tyr-654 mutant, creating a potential phosphate acceptor, exhibited both low and high-affinity EGF binding sites and was not phosphorylated by protein kinase C. However, addition of phorbol esters still abolished high-affinity binding, probably by phosphorylation on serine and threonine residues other than Thr-654. These data showed that Thr-654 is not directly involved in the regulation of receptor-affinity, but acts most likely in receptor internalization and EGF-induced mitogenesis (Lin *et al.*, 1986; Livneh *et al.* 1987, 1988).

a) Relationship between receptor kinase activation and receptor affinity.

Information on this relationship in intact cells comes largely from a variety of EGF-receptor mutants. Thus mutated EGF-receptors transfected into NIH/3T3 fibroblasts demonstrated that a four amino acid insertion at residue 708 abolished the protein-tyrosine kinase activity *in vitro* and *in vivo*. Nevertheless, the cells still expressed high- and low-affinity binding sites (Prywes *et al.*, 1986; Livneh *et al.*, 1987).

Furthermore, another mutant in which Lys721, a key residue in the ATP binding site, was replaced with an alanine residue, displayed also high- and low affinity EGF-binding but no tyrosine kinase activity (Honegger *et al.*, 1987). These results indicate that there is no simple mutual relationship between tyrosine-activity and high or low EGF-binding. On the other hand a deletion of 63 amino acids from the C-terminal end of the receptor, thereby removing two autophosphorylation sites, abolished the high-affinity state of the receptor (Livneh *et al.*, 1986), indicating a possible involvement of the phosphorylation state of the receptor in receptor affinity.

Using antibody 2E9, it was demonstrated that exclusive EGF binding to the high-affinity binding site elicits phosphorylation of 60% of the total receptor population, thus including EGF inaccessible, low affinity sites (Defize *et al.*, 1989c). This indicates that high-affinity sites are able to cross-phosphorylate or activate low affinity sites, may be subsequently leading to activation of the low-affinity kinase.

Relationship between tyrosine kinase activity and oligomerisation

The relationship between oligomerisation and receptor tyrosine kinase activity has been studied primarily in purified receptor preparations.

Evidence has been obtained that the dimeric receptor is more active as a kinase than the monomeric form (Yarden and Schlessinger, 1987a,b; Boni-Schnetzler and Pilch, 1987), suggesting that receptor oligomerization leads to activation of the tyrosine kinase. Furthermore, in the presence of detergents, bivalent antibodies were able to activate protein tyrosine kinase in intact cells and membrane preparations indicating that antibody mediated induction of receptor oligomerization results in activation of EGF-receptor protein tyrosine kinase (Yarden and Schlessinger, 1987a; Defize *et al.*, 1989c).

On the other hand, freshly prepared receptor preparations were shown to contain mainly monomeric receptors that displayed a high basal tyrosine kinase activity, while aging of the receptor resulted in dimerization and decreased basal tyrosine kinase activity. In these latter preparations the ability of EGF to stimulate phosphorylation was strongly increased (Biswas *et al.*, 1985). From these data it was concluded that receptor oligomerisation leads to inactivation of protein tyrosine kinase activity.

Finally, using sphingosine it was demonstrated in intact cells that tyrosine kinase activation may also be achieved in the absence of receptor oligomerization (Northwood and Davis, 1988). Similar results were obtained using a low ionic strength medium (Koland and Cerione, 1988). These latter data demonstrate that activation of receptor tyrosine kinase does not strictly require receptor oligomerisation.

Mechanism of action

The data presented thus far can be used to construct a model for the activation of the EGF-receptor protein tyrosine kinase by EGF. Two such models have been proposed, based largely upon the data summarized above, i.e. the intramolecular and intermolecular activation models (Gill et al., 1987; Yarden and Schlessinger, 1987a, 1987b).

According to the intramolecular activation model, EGF binding to the receptor leads to a conformational change in the receptor, which is in one way or another transferred to the intracellular domain. This in turn leads to autophosphorylation of the C-terminal part. Due to the autophosphorylation the C-terminus folds back thus exposing the kinase to other intracellular substrates (Bertics et al., 1985). Evidence in favor of the intramolecular activation has been obtained from a careful study on the kinetics of autophosphorylation (Weber et al., 1984; Bertics et al., 1985) and by using specific conditions or activators (Koland and Cerione, 1988; Northwood and Davis, 1988).

According the intermolecular activation model, EGF-binding leads to an interaction between two receptor molecules (receptor oligomerisation) and this interaction leads to activation of the receptor kinase (Schlessinger 1986, 1988). The driving force of receptor oligomerization is the affinity of the receptor for EGF which is supposed to be higher in the oligomeric state. Although receptor activation is dependent upon receptor-receptor interaction, also in this model phosphorylation of the receptor is supposed to be an intramolecular event. The model is supported by the observation that several bivalent EGF-receptor reagents are able to activate EGF-receptor tyrosine kinase, while immobilization of the receptor on a solid substrate prevents receptor activation. (Yarden and Schlessinger, 1987a, 1987b). Furthermore, cross-linking of EGF receptors increased their affinity towards EGF. Finally, the kinetics of EGF-receptor autophosphorylation had a parabolic dependence on the concentration of EGF-receptor, suggesting involvement of receptor-receptor interactions (Schlessinger and Yarden, 1987a,b). However, both models face serious arguments against their operation, and antibody 2E9 may be helpful to discriminate between the models.

In a series of recent experiments, it has been demonstrated that the antibody by itself has no stimulatory effect on EGF-receptor associated protein tyrosine kinase activity in intact cells. Furthermore, a complete inaccessiblity of low affinity receptor to EGF (in the presence of 2E9)

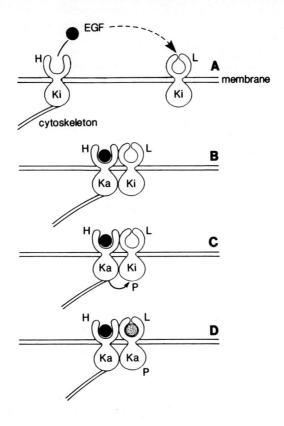

Fig. 9:   Model of EGF-receptor activation in intact cells.

A: High-affinity (H) and low-affinity (L) receptors are monomeric and inactive as a kinase (K_i). The H is associated to the cytoskeleton (C) causing a structural difference cellular domain of the receptor, leading to the high-affinity.

B: Binding of EGF, predominantly to H, causes activation of the tyrosine kinase of the high-affinity receptor. The H and L receptors may form dimers.

C: Upon dimerization, the H receptors cross-phosphorylates the L-receptor.

D: Phosphorylation of the L-receptors results in activation of its kinase activity, either in the presence or absence of bound EGF.

only slightly affects the ability of EGF to activate receptor protein tyrosine kinase, demonstrating that the high-affinity receptors are the major class through which EGF stimulates protein tyrosine kinase activity in intact cells. More importantly, it is shown that EGF binding to high-affinity receptors only, results in phosphorylation of low-affinity receptors, demonstrating the occurrence of cross-phosphorylation or activation (Defize *et al*., 1989c), in line with the intermolecular activation model.

Recently, evidence has been obtained that the EGF-receptor can indeed cross-phosphorylate other plasma membrane receptor-like structures (Stern and Kamps 1988; King *et al.*, 1988).

Taking all data presented into consideration, we propose the following model for receptor activation in intact cells. (Fig. 9)

A. High (H) and low (L) affinity receptors are present in a monomeric form and are inactive as a kinase. Structural differences between the extra-cellular domains determine the affinity. These structural differences may be caused by interactions with particular cellular components, such as for example the cytoskeleton.

B. Binding of EGF causes a conformational change in the high-affinity population leading to facilitated interaction with low affinity receptors, and thus to association to the cytoskeleton of low-affinity receptors. Furthermore, it is assumed that binding of EGF to the high-affinity site is sufficient to activate the kinase activity of this receptor population.

C. Inside the receptor dimers, the activated kinase of the high-affinity site cross-phosphorylates the low-affinity site.

D. The phosphorylation of the low-affinity site may result in activation of the kinase-activity, either in the presence or absence of EGF binding to the low-affinity site. Following phosphorylation, the complexes may dissociate again.

This model combines several features of the above mentioned models and is based primarily upon results obtained in intact cells. Therefore we feel that, although the model is highly speculative, it may serve as a working hypothesis to unravel the molecular mechanism of EGF-receptor kinase activation in more detail in the future.

ACKNOWLEDGEMENTS

We are grateful to our collegues at the Department of Molecular Cell Biology, University of Utrecht and at the Hubrecht laboratory, Netherlands Institute of Developmental Biology, for their contributions reported in this article and their stimulating discussions.

REFERENCES

Bertics PJ, Weber W, Cochet C and Gill GN (1985) Regulation of the epidermal growth factor receptor by phosphorylation. J Cell Biochem 29:195-208

Biswas R, Basu M, Sen-Majumdar A and Das M (1985) Intrapeptide autophosphorylation of the epidermal growth factor receptor: regulation of kinase catalytic function by receptor dimerization. Biochem 24:3795-3802

Bloch RJ (1986) Actin at receptor-rich domains of isolated acetylcholine receptor clusters. J Cell Biol 102:1447-1458

Böni-Schnetzler M and Pilch PF (1987) Mechanism of epidermal growth factor receptor autophosphorylation and high-affinity binding. Proc Natl Acad. Sci USA 84:7832-7836

Boonstra J and Ponec M (1988) High-affinity EGF_2 receptors present in transformed keratinocytes cultured under low Ca^{2+}, proliferating conditions, submitted

Boonstra J, Mummery CL, van der Saag PT and de Laat, SW (1985a) Two receptor classes for epidermal growth factor on pheochromocytoma cells, distinguishable by temperature, lectins and tumor promoters. J Cell Physiol 123:347-352.

Boonstra J, de Laat SW and Ponec M (1985b) Epidermal growth factor receptor expression related to differentiation capacity in normal and transformed keratinocytes. Exp Cell Res 161:421-433

Boonstra J, van Belzen N, van Maurik P, Hage WJ, Blok FJ, Wiegant FAC and Verkleij AJ (1985c) Immunocytochemical demonstrations of cytoplasmic and cell-surface EGF-receptors in A431 cells using cryo-ultramicrotomy, surface replication, freeze etching and label fracture. J Microsc Oxford 140:119-129

Boonstra J, Mummery CL, Feyen A, de Hoog WJ, van der Saag PT and de Laat SW (1987) Epidermal growth factor expression during morphological differentiation of pheochromocytoma cells, induced by nerve growth factor or dibutyryl cyclic AMP. J Cell Physiol 131:409-417

Boonstra J, van Belzen N, van Bergen en Henegouwen, PMP, Hage WJ, van Maurik P, Wiegant FAC and Verkleij AJ (1989) The epidermal growth factor receptor. In: Immunogold labeling (eds.) AJ Verkleij and JLM Leunissen, CRC Press, in press.

Braun J, Hochman P and Unanue E (1982) Ligand-induced association of surface immunoglobulin with the detergent-insoluble cytoskeletal matrix of the B-lymphocyte. J Immunol 128:1198-1204

Burgess AW (1986) Growth factors, receptors and cancer. Bio Essays 5:15-18

Carpenter G (1983) The biochemistry and physiology of the receptor-kinase for EGF. Mol Cell Endocrinol 31:1-19

Carpenter G (1987) Receptors for epidermal growth factor and other polypeptide mitogens. Ann Rev Biochem 56:881-914

Carpenter G and Cohen S (1979) Epidermal growth factor. Ann. Rev. Biochem. 48:193-216

Carpenter G and Zendegui JG (1986) Epidermal growth factor, its receptor, and related proteins. Exp Cell Res 164:1-10

Carpenter G, King L and Cohen S (1978) Epidermal growth factor stimulates phosphorylation in membrane preparations in vitro. Nature 276:409-410

Cochet C, Kashles O, Chambaz EM, Borrello I, King CR and Schlessinger J (1988) Demonstration of epidermal growth factor receptor dimerization in living cells using a chemical covalent cross-linking agent. J Biol Chem. 263:3290-3295

Cohen S, Carpenter G and King L (1980) Epidermal growth factor-receptor-protein kinase interactions. Co-purification of receptor and epidermal growth factor-enhanced phosphorylation activity. J Biol Chem 255:4834-4842

Davis RJ and Czech MP (1985a) Platelet-derived growth factor mimics phorbol diester action on epidermal growth factor receptor phosphorylation at threonine-654. Proc Natl Acad Sci USA 82:4080-4084

Davis RJ and Czech MP (1985) Tumor-promoting phorbol esters cause the phosphorylation of epidermal growth factor receptors in normal human fibroblasts at threonine 654. Proc Natl Acad Sci USA 82:1974-1978

Davis RJ and Czech MP (1987) Stimulation of epidermal growth factor receptor threonine 654 phosphorylation by platelet-derived growth factor in protein kinase C-deficient human fibroblasts. J Biol Chem 262:6832-6841

Defize LHK, Moolenaar WH, van der Saag PT and De Laat SW (1986) Dissociation of cellular responses to epidermal growth factor using anti-receptor monoclonal antibodies. EMBO J 5:1187-1192

Defize LHK, Mummery CL, Moolenaar WH and de Laat SW (1987) Antireceptor antibodies in the study of EGF-receptor interaction. Cell Differentiat 20:87-102

Defize LHK, Arndt-Jovin DJ, Jovin TM, Boonstra J, Meisenhelder J, Hunter T, de Hey HT and de Laat SW (1988) A431 cell variants lacking the blood group A antigen display increased high-affinity EGF-receptor number, protein tyrosine kinase activity and receptor turnover. J Cell Biol 107: 939-949.

Defize LHK, Boonstra J and de Laat SW (1989a) A monoclonal antibody allows specific occupation of high affinity epidermal growth factor receptors with EGF: structural aspects of high affinity EGF-binding in intact cells, submitted

Defize LHK, Boonstra J, Meisenhelder J, Hunter T, van Bergen en Henegouwen PMP and De Laat SW (1989b) High-affinity epidermal growth factor receptors (EGFR) play a major role in the EGF-induced protein-tyrosine kinase activity in intact A431 cells, submitted

Defize LHK, Meisenhelder J, Kruyer W, Tertoolen LGJ, Tilly BC, Hunter T, Boonstra J, Moolenaar WH and De Laat SW (1989c) Early signal transduction by epidermal growth factor occurs through binding to a subclass of high affinity receptors, submitted.

De Laat SW, Boonstra J, Moolenaar WH, Mummery CL, Van der Saag PT and Van Zoelen EJJ (1983) The plasma membrane as the primary target for the action of growth factors and tumour promotors in development. In: Development in Mammals Vol 5, (ed): MH Johnson, Elsevier Science Publ, pp 33-106

Doolittle RF, Hunkapiller MW, Hood LE, Devare SG, Robbins KC, Aaronson SA and Antionades HN (1983) Simian sarcoma virus onc gene, v-sis is derived from the gene (or genes) encoding a platelet-derived growth factor. Science 221:275-277

Downward J, Yarden Y, Mayes E, Scrace G, Totty N, Stockwell P, Ullrich A, Schlessinger J and Waterfield MD (1984) Close similarity of epidermal growth factor receptor and V-erb-B oncogene protein sequences. Nature 307:521-527

Downward J, Parker P and Waterfield MD (1984b) Autophosphorylation sites on the epidermal growth factor receptor. Nature 311:483-485

Fanger BO, Viceps-Madore D and Cidlowski JA (1984) Regulation of high- and low-affinity epidermal growth factor receptor by glucocorticoids. Arch Biochem Biophys 235:141-149

Fanger BO, Austin KS, Earp HS and Cidlowski JA (1986) Cross-linking of epidermal growth factor receptors in intact cells: detection of initial stages of receptor clustering and determination of molecular weight of high-affinity receptors. Biochem 25:6414-6420

Fearn JC and King AC (1985) EGF receptor affinity is regulated by intracellular calcium and protein kinase C. Cell 40:991-1000

Gill GN, Santon JB and Bertics PJ (1987) J Cell Physiol suppl 5:35-41

Gospodarowics D (1981) Epidermal and nerve growth factors in mammalian development. Ann Rev Physiol 43:251-263

Gospodarowics D (1985) Fibroblast growth factor In: Growth and Maturation Factors, Vol III (ed) G Guroff, New York, Wiley Press pp

Gregorion M and Rees AR (1984) Properties of a monoclonal antibody to epidermal growth factor receptor with implications for the mechanism of action of EGF. EMBO J 3:929-937

Haigler HT, McKanna JA and Cohen S (1979) Direct visualization of the binding and internalization of a ferritin conjugate of epidermal growth factor in human carcinoma cells A431. J Cell Biol 81:382-395

Heldin C-H and Westermark B (1984) Growth factors: mechanism of action and relation to oncogenes. Cell 37:9-20

Heldin C-H, Betsholtz C, Claesson-Welsh L and Westermark B (1987) Subversion of growth regulatory pathways in malignant transformation. Biochim Biophys Acta 907:219-244

Honegger AM, Dull TJ, Felder S, Van Obberghen E, Bellot F, Szapary D, Schmidt A, Ullrich A and Schlessinger J (1987) Point mutation at the ATP binding site of EGF receptor abolishes protein tyrosine kinase activity and alters cellular routing. Cell 51:199-209

Hunter T and Sefton BM (1980) Transforming gene product of Rous sarcoma virus phosphorylates tyrosine. Proc Natl Acad Sci USA 77:1311-1315

Hunter T and Cooper JA (1981) Epidermal growth factor induces rapid tyrosine phosphorylation of proteins in A431 human tumor cells. Cell 24:741-752

Hunter T and Cooper JA (1985) Protein-tyrosine kinases. Ann Rev Biochim 54:897-930

Hunter T, Ling N and Cooper NA (1984) Protein kinase C phosphorylation of the EGF-receptor at a threonine residue close to the cytoplasmic face of the plasma membrane. Nature 311:480-483

Iwashita S and Fox CF (1984) Epidermal growth factor and potent phorbol tumor promoters induce epidermal growth factor receptor phosphorylation in a similar but distinctively different manner in human epidermoid carcinoma A431 cells. J Biol Chem 259:2559-2567

Jesaitis AJ, Naemura JR, Sklar LA, Cochrane CG and Painter RG (1984) Rapid modulation of N-formyl chemotactic peptide receptors on the surface of human granulocytes: formation of high-affinity ligand-receptor complexes in transient association with cytoskeleton. J Cell Biol 98:1378-1387

King AC and Cuatrecasas P (1982) Resolution of high and low affinity epidermal growth factor receptors: inhibition of high affinity component by low temperature, cycloheximide, and phorbol ester. J Biol Chem 257:3053-3060

King LE, Carpenter G and Cohen S (1980) Characterization by electrophoresis of epidermal growth factor stimulated phosphorylation using A431 membranes. Biochemistry 19:1524-1528

King CR, Borrello I, Bellot F, Comoglio P and Schlessinger J (1988) EGF binding to its receptor triggers a rapid tyrosine phosphorylation of the erb B-2 protein in the mammary tumor cell line SK-BR-3. EMBO J 7:1647-1651

Koland JG and Cerione RA (1988) Growth factor control of epidermal growth factor receptor kinase activity via an intramolecular mechanism. J Biol Chem 263:2230-2237

Landreth GE, Williams LK and Rieser GD (1985) Association of the epidermal growth factor receptor kinase with the detergent-insoluble cytoskeleton of A431 cells. J Cell Biol 101:1341-1350

Lax I, Burgess WH, Bellot F, Ullrich A, Schlessinger J and Givol D (1988) Localization of a major receptor-binding domain for epidermal growth factor by affinity labeling. Mol Cell Biol 8:1831-1834

Lee L-S and Weinstein IB (1978) Tumor promoting phorbol esters inhibit binding of epidermal growth factor to cellular receptors. Science 202:313-315

Lin C-R, Chen C-W, Lazar CS, Carpenter DC, Gill GN, Evans RN and Rosenfeld MG (1986) Protein kinase C phosphorylation at Thr654 of the unoccupied EGF-receptor and EGF binding regulate functional receptor loss by independent mechanisms. Cell 44:839-846

Livneh E, Prywes R, Kashles O, Reiss N, Sasson I, Mory Y, Ullrich A and Schlessinger J (1986) Reconstitution of human epidermal growth factor receptors and its deletion mutants in cultured hamster cells. J Biol Chem. 261:12490-12497

Livneh E, Reiss N, Berent E, Ullrich A and Schlessinger J (1987) An insertional mutant of epidermal growth factor receptor allows dissection of diverse receptor function. EMBO J 6:2669-2676

Livneh E, Dull TJ, Berent E, Prywes R, Ullrich A and Schlessinger J (1988) Release of a phorbol ester-induced mitogenic block by mutation at Thr-654 of the epidermal growth factor receptor. Mol Cell Biol 8:2302-2308

Lockyer JM, Bowden GT and Magun BE (1983) The effect of fluocinolone acetonide and 12-O-tetradecanoyl-phorbol- -acetate on the binding and biological activity of epidermal growth factor in rat fibroblasts. Carcinogen 4:653-658

Ludérus MEE and van Driel R (1988) Interaction between the chemotactic cAMP receptor and a detergent-insoluble membrane residue of Dictyostelium discoideum. Modulation by guanine nucleotides. J Biol Chem 263:8326-8331

Magun BE, Matusian LM and Bowden GT (1980) Epidermal growth factor. Ability of tumor promoter to alter its degradation, receptor affinity and receptor number. J Biol Chem 255:6373-6381

Moolenaar WH (1986) Effects of growth factors on intracellular pH regulation. Ann Rev Physiol 48:363-376

Northwood IC and Davis RJ (1988) Activation of the epidermal growth factor receptor tyrosine protein kinase in the absence of receptor oligomerization. J Biol Chem 263:7450-7453

Olashaw NE, O'Keefe EJ and Pledger WJ (1986) Platelet-derived growth factor modulates epidermal growth factor receptors by a mechanism distinct from that of phorbol esters. Proc Natl Acad Sci USA 83:3834-3838

Painter RG, Zahler-Bentz, K and Dukes RE (1987) Regulation of the affinity state of the N-formylated peptide receptor of neutrophils: role of guanine nucleotide-binding proteins and the cytoskeleton. J Cell Biol 105:2959-2971

Pike LJ, Marquardt H, Todaro GJ, Gallis aB, Casnellie JE, Bornstein P and Krebs EG (1982) Transforming growth factor and epidermal growth factor stimulate the phosphorylation of a synthetic, tyrosine-containing peptide in a similar manner. J Biol Chem 257:14628-14631

Pinto da Silva P and Kan FWK (1984) Label-fracture: a method for high resolution labeling of cell surfaces. J Cell Biol 99:1156-1161

Prywes R, Livneh E, Ullrich A and Schlessinger J (1986) Mutations in the cytoplasmic domain of EGF receptor affect EGF binding and receptor internalization. EMBO J 5:2179-2190

Russo MW, Lukas TJ, Cohen S and Staros JV (1985) Identification of residues in the nucleotide binding site of the epidermal growth factor receptor/-kinase. J Biol Chem 260:5205-5208

Schlechter AL and Bothwell MA (1981) Nerve growth factor receptors on PC12 cells: Evidence for two receptor classes with differing cytoskeletal association. Cell 27:867-874

Schlessinger J (1986) Allosteric regulation of the epidermal growth factor receptor kinase. J Cell Biol 103:2067-2072

Schlessinger J (1988) The epidermal growth factor receptor as a multifunctional allosteric protein. Biochem 27:3119-3123

Schlessinger J, Schechter Y, Willingham MC and Pastan I (1978) Direct visualization of binding aggregation, and internalization of insulin and epidermal growth factor on living fibroblastic cells. Proc Natl Acad Sci USA 75:2659-2663

Schlessinger J, Schreiber AB, Levi A, Lax I, Libermann T and Yarden Y (1983) Regulation of cell proliferation by epidermal growth factor. CRC Crit Rev Biochem 14:93-111

Schreiber AB, Libermann T, Lax J, Yarden Y and Schlessinger J (1983) Biological role of epidermal growth-factor-receptor clustering investigation with monoclonal anit-receptor antibodies. J Biol Chem 258:846-853

Shoyab M, De Laroo JE and Todaro GJ (1979) Biologically active phorbol esters specifically alter affinity of epidermal growth factor membrane receptors. Nature 279:387-391

Soderquist AM and Carpenter G (1986) Biosynthesis and metabolic degradation of receptors for epidermal growth factor. J Membr Biol 90:97-105

Sporn MB and Roberts AB (1985) Autocrine growth factors and cancer. Nature 313:745-747

Stern DF and Kamps MP (1988) EGF-stimulated tyrosine phosphorylation of p185^{neu}: a potential model for receptor interactions. EMBO J 7:995-1001

Tamkun JW, DeSimone DW, Fonda D, Patel RS, Buck C, Horwitz AF and Hynes RO (1986) Structure of integrin, a glycoprotein involved in the transmembrane linkage between fibronectin and actin. Cell 46:271-282

Thorburn GD, Young IR, Dolling M, Walker DW, Browne CA and Carmichael GG (1985) Growth factors in fetal development In: Growth and Maturation Factors, vol III (ed G Guroff) pp 175-201 New York, Wiley Press

Ullrich A, Coussens L, Hayflick JS, Dull TJ, Gray A, Tam AW, Lee J, Yarden Y, Libermann TA, Schlessinger J, Downward J, Mayes ELV, Whittle N, Waterfield MD and Seeburg PH (1984) Human epidermal growth factor receptor cDNA sequence and aberrant expression of the amplified gene in A431 epidermoid carcinoma cells. Nature 309:418-425

Ushiro H and Cohen S (1980) Identification of phosphotyrosine as a product of epidermal growth factor-activated protein kinase in A431 cell membranes. J Biol Chem 255:8363-8365

Vale RD and Shooter EM (1983a) Epidermal Growth Factor receptors on PC12 cells: Alteration of binding properties by lectins. J Cell Biochem 22:99-109

Vale RD and Shooter EM (1983b) Conversion of nerve growth factor-receptor complexes to a slowly dissociating, Triton X-100 insoluble state by anti nerve growth factor antibodies. Biochem 22:5022-5028

Vale RD, Ignatius MJ and Shooter EM (1985) Association of nerve growth factor receptors with the Triton X-100 cytoskeleton of PC-12 cells. J Neurosci 5:2762-2770

Van Belzen N, Rijken PJ, Hage WJ, De Laat SW, Verkleij AJ and Boonstra J (1988) Direct visualization and quantitative analysis of epidermal growth-factor induced receptor clustering. J Cell Physiol 134:413-420

Van Bergen en Henegouwen PMP, de Kroon J, Van Damme H, Verkleij AJ and Boonstra J (1988) Ligand induced association of epidermal growth factor receptor to the cytoskeleton of A431 cells. J Cell Biochem submitted

Weber W, Bertics PJ and Gill GN (1984) Immunoaffinity purification of the epidermal growth factor receptor. Stoichiometry of binding and kinetics of self-phosphorylaiton. J Biol Chem 259:14631-14636

Wheeler ME, Gerrard JM and Carroll RC (1985) Reciprocal transmembranous receptor-cytoskeleton interactions in Con canavalin A-activated platelt. J Cell Biol 101:993-1000

Wiegant FAC, Blok FJ, Defize LHK, Linnemans WAM, Verkleij AJ and Boonstra J (1986) Epidermal growth factor receptors associated to cytoskeletal elements of epidermoid carcinoma (A431) cells. J Cell Biol 103:87-94

Yarden Y and Schlessinger J (1987a) Self-phosphorylation of EGF-receptor: evidence for a model of intermolecular allosteric activation. Biochemistry 26:1434-1442

Yarden Y and Schlessinger J (1987b) EGF induces rapid, reversible aggregation of the purified EGF-receptor. Biochemistry 26:1443-1451

Zachary I, Sinnett-Smith JW and Rozengurt E (1986) Early events elicited by bombesin and structurally related peptides in quiescent Swiss 3T3 cells. I. Activation of protein kinase C and inhibition of epidermal growth factor binding. J Cell Biol 102:2211-2222

Zidovetski R, Yarden Y, Schlessinger J and Jovin TM (1981) Rotational diffusion of epidermal growth factor complexed to cell surface receptors reflects rapid microaggregation and endocytosis of occupied receptors. Proc Natl Acad Sci USA 78:6981-6985

Zidovetski R, Yarden Y, Schlessinger J and Jovin TM (1986) Microaggregation of hormone-occupied epidermal growth factor receptors on plasma membrane preparations. EMBO J 5:247-250

THREE-DIMENSIONAL STRUCTURAL MODELS FOR EGF AND INSULIN RECEPTOR INTERACTIONS AND SIGNAL TRANSDUCTION

T. Blundell, N. McDonald, J. Murray-Rust, A. McLeod, S. Wood
Laboratory of Molecular Biology
Department of Crystallography
Birkbeck College
Malet Street
London WC1E 7HX

INTRODUCTION

Several hormones and growth factors are known to activate their cellular signalling through a receptor tyrosine kinase (RTK). These receptors have similar molecular structures: a large cystine-rich, extracellular ligand binding domain, a single hydrophobic transmembrane helix and a cytoplasmic tyrosine kinase. They can be divided into three subclasses. The first of these includes the EGF and TGFα receptor and the neu (also called HER-2 or C-erb-B) proto-oncogene; these are closely related to the viral oncogene v-erb-B. The second class includes the receptors for insulin and IGF1; these are synthesised as a single peptide chain, processed into α-and β-chains and exist in the native state as covalently bound dimers. The third subclass includes receptors for PDGF and CSF-1 and related oncogene products v-fms and c-kit. The primary structures and functions of these three subclasses have been reviewed by Carpenter (1987) and by Yarden and Ullrich (1988). The ligands for all these receptors are cystine rich polypeptides of sufficient size to assume a globular structure. Considerable information is available on the conformations of insulin and EGF from X-ray diffraction and nuclear magnetic resonance experiments, and so some speculation can now be made on receptor interactions. Here, we review evidence for the three-dimensional structures of the ligands - hormones and growth factors - and the receptor tyrosine kinases for the EGF and insulin subclasses. We consider both experimental evidence and conclusions based on knowledge-based modelling procedures. We then discuss the implications for hormone receptor interactions and for signal transduction leading to activation of the cytoplasmic tyrosine kinase.

NATO ASI Series, Vol. H29
Receptors, Membrane Transport and Signal Transduction
Edited by A. E. Evangelopoulos et al.
© Springer-Verlag Berlin Heidelberg 1989

HORMONE AND GROWTH FACTOR STRUCTURES

The three-dimensional structure of insulin was first defined by X-ray analysis in 1969 and subsequently molecular structures of insulins from several species - human, porcine and hagfish - have been refined as hexamers, dimers and monomers in native or chemically modified forms (for reviews see Blundell and Wood, 1982; Dodson et al., 1983; Baker et al., 1988). In insulin the A and B chains form a compact globular structure with a hydrophobic core stabilized by two interchain and one intrachain disulphides. Recent data suggest that at least five residues may be cleaved from either NH_2- or COOH- termini of the B-chain without loss of the "insulin fold"; however, the complete A-chain is probably required for retention of native conformation. It is also probable that the B-chain termini have unstructured conformations in the monomer which is the active form; in particular, the region B25-B30 is free to move in the monomer to expose the underlying conserved region involving, for example, A3 Val. The receptor appears to interact closely with a variety of hydrophobic residues such as A3 Val, A19 Tyr, B12 Val, B24 Phe and B25 Phe, and more weakly with polar residues such as A1 Gly, A4 Glu and A21 Asn. Surprisingly several invariant residues, for example A21 Asn, originally thought to be important for receptor binding, have recently been varied by site-directed mutagenesis and found to be not essential for activity (Markussen et al., 1988). Nevertheless, receptor interactions involve regions of insulin that are distant in the sequence and are brought together in the tertiary structure.

The single chain insulin-like growth factors (IGF 1 and 2), which have a short connecting or C-peptide and an extension at the COOH-terminus of the A-chain (D-peptide) can also assume the insulin fold and have part of insulin's receptor binding region retained, consistent with their ability to bind the insulin receptor, albeit with reduced affinity (Blundell and Humbel, 1980). Insulin-like molecules also exist in molluscs, MIP (Smit et al., 1988) and in silkworms, PTTH or bombyxin (Nagasawa et al.,1986); their sequences are also consistent with an insulin-fold, but the receptor binding regions are quite different (Jhoti et al., 1987).

Two-dimensional nmr experiments have shown that EGF also has a globular structure, although the NH_2-terminal region may be flexible. The tertiary structure is organised as two domains, residues 1-32 and residues 33-53. In the first domain, the major feature is a turn (15-19) followed by a β-hairpin with antiparallel β-sheet structure (19-23 and 28-32). There is a small hydrophobic core contributed by residues Pro 7, Tyr 10, Tyr 13, His 22 and Tyr 29. The second domain adopts a double loop structure with a hydrophobic core from Val 34, Ile 35, Tyr 37, Trp 49 and Trp 50. There appears to be good agreement between three separate nmr studies (Montelione et al., 1987; Cooke et al., 1987 and Kohda et al., 1988) for the main chain in the two domains, but less certainty about the relationships between the domains. EGF has recently been crystallised in a form suitable for X-ray diffraction by Higuchi et al. (1988), but at the time of writing the structure is not determined. Modifications and truncations of mouse EGF have implicated the second domain in receptor binding. Removal of the last five residues for mouse EGF results in an equipotent analogue; however, the removal of Leu 47, a conserved residue in the EGF family, gives an inactive analogue with no binding affinity. This suggests that this residue may play an important role in EGF receptor recognition or in maintaining a biologically active conformation.

EGF-like domains occur in many different molecules and all structures are consistent with an EGF-fold, although there is variation in the lengths of the loops between the second and third cystines (A), the third and fourth (B) and the fifth and sixth (C). Appella et al. (1988) have suggested that EGF-related molecules may be classified as three types. Type 1 includes EGF, TGFα, uPA and tPA, and has conserved loops A and C and a sequence variable loop B of 10-11 amino acids. The coagulation factors VII_2, IX, X, protein S and protein C form type 2 with a shorter loop B. The third type includes the LDL receptor, factor VII_1 and EGF precursor domains; these have the first disulphide in a different position. Many of these molecules involve cell surface interactions and may bind receptors in an analogous way, but with different specificity, to EGF.

In summary, it appears that both the insulin and EGF families are characterised by small, globular, disulphide-linked proteins in which a characteristic tertiary structure or fold is adopted. Evolution has explored sequence variation and insertions and deletions of loops consistent with the tertiary structure. Tertiary structures of each are "soft" and can be easily disturbed, with NH_2- and COOH- termini often in less-structured conformations. In each case, receptor interactions appear to involve extensive surface regions of these globular ligands, and so tertiary structure is thought to be important for receptor activation.

MODELLING OF RECEPTOR STRUCTURE

Information concerning the three-dimensional structures will come
from three major sources: modification by chemical techniques or directed
mutagenesis; biophysical techniques such as X-ray diffraction, nuclear
magnetic resonance or circular dichroism; and knowledge-based modelling
procedures. The chemical and molecular genetic methods are most
straightforward to carry out, but need molecular models for their proper
design and interpretation. The structure elucidation is, therefore, likely
to be a cyclic process of modelling and testing by experiment leading to
reappraisal of the model.

Because the structures probably contain a single transmembrane helix
linking large intracellular and extracellular domains, it is less likely
that they will be crystallised using techniques that have proved successful
for intrinsic membrane proteins (see for example the chapter of J.
Rosenbusch in this volume). We are adopting the approach of defining the
structures of the domains separately. As a consequence of the high
resistance to proteolysis of the native receptor, this must be achieved by
engineering genes for separate expression of the extracellular and
cytoplasmic domains, which have now been expressed in milligram quantities
for the EGF receptor (Greenfield and Waterfield, 1988; Schlessinger, 1988
unpublished results) and the insulin receptor (Ellis, 1988) in baculovirus
systems. The extracellular domains have proved to be stable and soluble in
aqueous solution, and have a single class of ligand binding sites with an
affinity similar to that of the native receptor. These are now being used
in crystallisation studies, hopefully for X-ray diffraction. Circular
dichroism of the EGF receptor extracellular domain has already demonstrated
the presence of both α-helix and β-sheet, and conformational changes
consequent to EGF binding.

Modelling procedures are limited by the paucity of information about
the three-dimensional structure of any membrane protein. Assumptions
concerning transmembrane regions are based mainly on the need to satisfy
hydrogen bonding for the NH functions of the peptides. Thus a single
transmembrane strand must be helical, either an α-helix with 3.6_{13} structure
or a three-fold symmetrical helix with 3_{10} structure. β-structures can be
formed only in a β-barrel involving several transmembrane strands. The EGF
receptor has one strongly hydrophobic region (residues 622 to 644) which if

α-helical would be 23 x 1.5Å i.e. 34.5Å long, roughly sufficient to span the hydrophobic region of the membrane that is generally assumed to be ~30Å in width.

Such a transmembrane helix would have a strongly basic "stop sign" of seven arginines, one lysine and one histidine on the cytoplasmic side juxtamembrane region. This region also includes Thr 654, which is phosphorylated by protein kinase C and is important for regulation of the high affinity EGF sites. It is probable, therefore, that this region has specific structure and interactions that are disturbed by the presence of a negatively charged phosphate which may reorganise the contiguous positive charges of arginines and other basic residues.

Although it is unlikely that the cytoplasmic side of the putative α-helix is ever pulled into the membrane, the extracellular region might be as it contains only one charged group (Lys 618). Mutations within the transmembrane region, at positions 627 and 628 had no effect on receptor function except that placing Glu or Gln at residue 627 leads to constitutive activity in neu (Bargmann et al. 1986, 1988). Also, construction of functional chimaeric receptors (Riedel et al. 1986, 1987) show an apparent insensitivity to the sequence of the transmembrane stretch. Both these results are surprising considering possible signalling mechanisms through the cell membrane, and necessitate further studies of this region for clarification.

The intracellular domains include a region homologous with other protein kinases which is assumed to have a bilobal structure with a deep active site cleft characteristic of kinases for which the structure is known. In the EGF receptor there is a conserved pattern of glycine residues (GlyXGlyXXGly) close to the juxtamembrane region at residues 615-700, and this is likely to bind the ribose and phosphate of ATP. Lys 721 is essential for kinase activity and probably binds the β-phosphate of ATP. Both of these groups are predicted to be at one end of a nucleotide binding fold of β-strands and α-helices characteristic of kinases and dehydrogenases (Sternberg and Taylor, 1984). It is also probable that the highly conserved sequence Asp Phe Gly (831 to 833 in the EGF receptor) lies at the end of this nucleotide binding fold adjacent to the conserved glycine sequence and in a position to interact with the adenine moiety of ATP. Although the nucleotide binding domains of kinases are conservatively varied, little is known of the so-called "catalytic domains" that tend to vary and to contain the specificity sites for the substrate, in this case tyrosine.

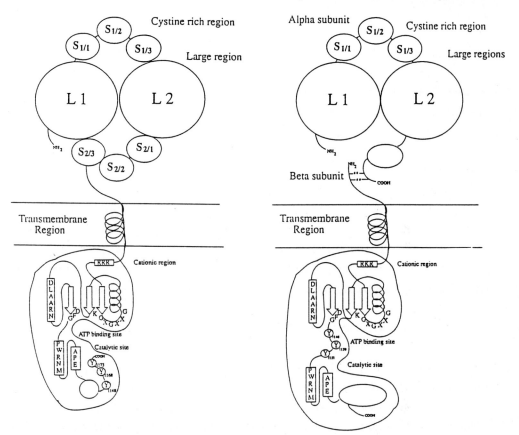

Figure 1. Simplified representation of the EGF, and insulin/IGF 1 receptors showing the large (L) and small (S) subdomains of the extracellular domains in a symmetrical arrangement about a two-fold axis. The intracellular tyrosine kinase domain shows the βαβ nucleotide binding motif and residues thought to be involved in ATP binding. Major autophosphorylation sites are indicated by Y, and sequences specific to all tryosine kinases are boxed.

Few site specific mutagenesis experiments have been attempted for the extracellular domains. However, a comparative analysis of the sequences available for the EGF and insulin subclasses demonstrates some intriguing features (Bajaj et al. 1987). The EGF family comprises four regions (L_1, S_1, L_2, S_2) for which sequence homology indicates gene duplication of an ancestral LS sequence. The S region is cystine rich and appears to comprise a motif of 8 half-cystines repeated two or three times. Nothing is known of whether there is disulphide pairing or free sulphydryls that might be metal bound. In any case, these cystine rich domains are likely to be closely cross-linked and relatively rigid. In the insulin receptor subclass only one cystine rich region occurs.

The other duplicated regions, L_1 and L_2, occur in all the EGF and insulin family members (Bajaj et al., 1987). These regions have few cystines, but have a characteristic sequence pattern indicating five α-β motifs each with a conserved glycine. The loops joining these conserved motifs are hypervariable. It is an attractive hypothesis to consider these regions as ligand binding with the two L domains close together, related by a diad axis and forming an extended recognition site. There exists a superficial analogy with T-cell acceptors if this hypothesis is correct. Recent directed mutagenesis (I Lax, J Schlessinger et al., unpublished results) and photoactive labelled hormone binding studies (D Brandenburg, unpublished results) implicate these domains in ligand binding.

Figure 1 shows a possible arrangement of the whole receptor based on these rather speculative comments on the three-dimensional structures of the domains.

MODELS FOR RECEPTOR BINDING AND SIGNAL TRANSDUCTION

There are two possible models for receptor binding that should be easily distinguished. The less probable one involves the cystine rich domains, perhaps with disulphide exchange with the hormones and growth factors. These ligands all have conserved cystines that appear to be essential for activity, although this may simply reflect their roles in maintenance of three-dimensional structure. More probably, the larger, non-cystine rich domains, L_1 and L_2, are ligand binding, as discussed in the previous section. The interactions will be largely hydrophobic, at least in insulin receptor binding, and will certainly lead to conformational change in both hormone and receptor. This conformational change is necessary for allosteric effects which must be important for signal transduction.

There are several possibilities for signal transduction from the extracellular ligand binding domain to activate the kinase on the cytoplasmic side. The most simple of these is a "bell-toll" or "chain-flush" mechanism, whereby the ligand binding induced conformational change in the receptor results in a pull of the helix perpendicular to the membrane plane (Staros et al., 1985). This may then lead to conformational changes in the kinase leading to activation. This model might be viable for a monomer, but the lack of sequence conservation of the transmembrane region still producing activation makes the mechanism less attractive. It has also been suggested that a rotation of the helix may be important in signal transduction, but this is difficult to understand given that the receptor is free to rotate in the membrane.

A second model involves activity transmitted through a dimer (Schlessinger, 1988). This has many attractive features, not the least that most allosteric activity in proteins is mediated by symmetrical oligomers that serve to enhance the signal and increase the sensitivity of the system. One mechanism for kinase activation would be a purely associative one in which ligand-induced conformational changes in the extracellular domain lead to increased dimer formation; this in turn would lead to a closer specific association of the kinase leading to a bimolecular phosphorylation and activation. However, a dimer would also be advantageous in transmitting other relative movements of the extracellular domains. For example, relative movement of subunits would be an effective mode of information transfer to the kinase within a dimer. The association may not be necessary for the insulin receptor where covalent dimers already exist.

ACKNOWLEDGEMENTS

We are grateful for useful discussions with Dr. M. Bajaj, Dr. M. Waterfield, Dr. C. Greenfield, Dr. J. Schlessinger, Dr. L. Ellis.

REFERENCES

Appella E, Weber IT, Blasi F (1988) Structure and function of epidermal growth factor-like regions in proteins. FEBS Lett 231:1-4

Bajaj M, Waterfield MD, Sclessinger J, Taylor WR, Blundell TL (1987) On the tertiary structures of the extracellular domains of the EGF and insulin receptors. Biochem Biophys Acta 916:220-226

Baker EN, Blundell TL, Cutfield JF et al. (1988) The structure of the 2 Zn pig insulin crystals at 1.5A resolution. Phil Trans Roy Soc Lond (Biol) 319:369-456

Bargmann CI, Weinberg RA (1988) Oncogenic activation of the Neu-encoded receptor protein by point mutation and deletion. EMBO J 7:2043-2052

Bargmann CI, Hung M-C, Weinberg RA (1986) The Neu oncogene encodes an epidermal growth factor receptor-related protein. Cell 45:649-657

Blundell TL, Wood SP (1982) The conformation, flexibility and dynamics of polypeptide hormones. Ann Rev Biochem 51:123-154

Blundell TL, Humbel R (1980) Hormone families: pancreatic hormones and homologous growth factors. Nature 287:781-787

Carpenter G (1987) Receptors for epidermal growth factor and other polypeptide mitogens. Ann Rev Biochem 56:881-914

Cooke RM, Wilkinson AJ, Baron M, Pastore A, Tappin MJ, Campbell ID, Gregory H, Sheard B (1987) The solution structure of human epidermal growth factor. Nature 327:339-341

Dodson GG, Dodson EJ, Hubbard RJ, Reynolds CD (1983) Insulin's structural variations and their relation to activity. Biopolymers 22:281-289

Ellis L (1988) The insulin receptor: a hormone activated transmembrane tyrosine kinase comprised of two largely idependently folded soluble domains. In: Marshall GR (ed) Peptides 302-307

Higuchi Y, Morimoto Y, Horinaka A, Yasuoka N (1988) Crystallization and preliminary X-ray studies on human epidermal growth factor. J Biochem 103:905-6

Jhoti H, McLeod AN, Blundell TL, Ishizaki H, Nagasawa H, Suzuki A (1987) Prothoracicotropic hormone has an insulin-like tertiary structure. FEBS Lett 219:419-425

Kohda D, Go N, Hayashi K, Inagaki F (1988) Tertiary structure of mouse epidermal growth factor determined by two-dimensional NMR. J Biochem 103:741-3

Markussen J, Diers I, Hougaard P, Langkjaer L, Norris K, Snel L, Sørensen AR, Sørensen E, Voigt HO (1988) Soluble prolonged-acting insulin derivatives III Degree of protraction, crystallizability and chemical stability of insulins substituted in positions A21, B13, B23, B27 and B30. Prot Eng 2:157-166

Montelione GT, Wüthrich K, Nice EC, Burgess AW, Scheraga HA (1987) The solution structure of murine epidermal growth factor. Proc Natl Acad Sci USA 84:5226-5230

Nagasawa H, Suzuki A, Ishizaki H (1986) Amino acid sequence of a prothoracicotropic hormone of the silkworm Bombyx mori. Proc Natl Acad Sci USA 83:5840-5843

Reidel H, Dull TJ, Schlessinger J, Ullrich A (1986) A chimaeric receptor allows insulin to stimulate tyrosine kinase activity of epidermal growth factor receptor. Nature 324:68-70

Reidel H, Schlessinger J, Ullrich A (1987) A chimaeric ligand-binding v-erbB/EGF receptor retains transforming potential. Science 236:197-200

Schlessinger J (1988) The epidermal growth factor as a multifunctional allosteric protein. Biochem 27:3119-3123

Smit AB, Vreugdenhil E, Ebberick RHM, Geraerts WPM, Klootwijk J, Joosse J
    (1988) Growth-controlling molluscan neurons produce the precursor of an
    insulin-related peptide. Nature 331:535-538
Staros JV, Cohen S, Russo MW (1985)
    In: Cohen P, Maunslay MD (eds) Molecular mechanisms of transmembrane
 signalling. Elsevier New York, 253-278
Sternberg MJE, Taylor WR (1984) Modelling of the ATP-binding site of
    oncogene products, the epidermal growth factor receptor and related
    proteins. FEBS Lett 175:387-391
Yarden Y, Ullrich A (1988) Receptor tyrosine kinases. Ann Rev Biochem
    57:443-478

POTENTIATION OF NEUROTRANSMITTER RELEASE COINCIDES WITH
POTENTIATION OF PHOSPHATIDYL INOSITOL TURNOVER - A POSSIBLE IN
VITRO MODEL FOR LONG TERM POTENTIATION (LTP)

D. Atlas*, S. Diamant and L. Schwartz
Department of Biological Chemistry
Hebrew University of Jerusalem

*The Otto Lowei Center for Neurobiology

INTRODUCTION

High frequency stimulation induces changes in the gain of
synaptic transmission expressed as long term potentiation (LTP)
(Alger and Taylor, 1976; Bliss and Lomo, 1973; Bliss and Lynch,
1987; Lisman, 1987; Miller and Kennedy, 1987). These changes
have been proposed as models possibly associated with memory
and learning processes (Hebb, 1949; Akers et al., 1986).
Protein phosphorylation, especially those mediated by protein
kinase C (PKC) were suggested as the biochemical link behind
the phenomenon of LTP (Akers et al., 1986; Malenka et al.,
1986; Lynch et al., 1988). Recently neurotransmitter release
was shown to be associated with the action of inositol
trisphosphate (IP_3) and diacylglycerol (DAG) producing
agonists, and the involvement of PLC and PKC demonstrated with
the respective inhibitors (Diamant et al., 1988). In this
report, we show strong potentiation of neurotransmitter release
by muscarinic agonists under depolarizing conditions in rat
cortical slices. Under the same conditions, namely, the
simultaneous presence of receptor-mediated phosphatidyl
inositol turnover and depolarizing agents, such as K^+,
veratridine or oubain, we observe a strong potentiation of
inositol phosphate $([^3H]IP)$ formation. Previously, it was
demonstrated that muscarinic agonists in the presence of
elevated levels of K^+ induce larger amounts of inositol
phosphates (Eva and Costa, 1986). Furthermore, a synergism of

NATO ASI Series, Vol. H29
Receptors, Membrane Transport and Signal Transduction
Edited by A. E. Evangelopoulos et al.
© Springer-Verlag Berlin Heidelberg 1989

ACh and elevated K$^+$ was observed in IP$_4$ formation (Baird and Nahorski, 1986). We also show that agonists which are not involved in the activation of phosphatidyl inositol turnover such as nicotine, do not mediate potentiation of either of [^{3}H]noradrenaline ([^{3}H]NE) release or of [^{3}H]inositol phosphate accumulation. Thus potentiation of neurotransmitter release is tightly correlated with potentiation of IP-accumulation. Dissection of the stimulation period into two consecutive intervals, reveals desensitization of the CCh-induced release in the second period of stimulation. On the other hand, potentiation of release by CCh in the presence of K$^+$ persists in both stimulation periods, irrespective of the CCh-desensitization. These results point to the onset of activation of release initiated by CCh and K$^+$ together, which continues regardless of the CCh-desensitized second period. It is thus proposed that membrane depolarzation in combination with receptor-activated hydrolysis of phosphatidylinositol-4,5-bisphosphate (PIP$_2$) induces a change in synaptic transmission which persists in the absence of CCh, and hence can be regarded as a possible in vitro model for LTP.

Methods

Preparation of cortical slices for [^{3}H]Inositol Phosphate formation

Cross-chopped slices (350 μm x 350 μm) of rat cerebral cortex were prepared using McIlwain tissue chopper, preincubated in 5 ml of oxygenated Krebs-Henseleit buffer (118 mM NaCl, 4.7 mM KCl, 1.2 mM MgSO$_4$, 1.3 mM CaCl$_2$, 25 mM NaHCO$_3$, 1.0 mM NaH$_2$PO$_4$ and 11.1 mM Glucose) for 60 minutes at 37° C with buffer exchange every 10 minutes. The slices were then incubated in the presence of [^{3}H]inositol (15 - 20 μCi/ml) (10 - 30 mCi/mmole) for 90 minutes at 37° C, washed by 2 x 5 ml of Krebs-Henseleit buffer containing 10 mM LiCl and distributed into assay tubes.

Measurement of [³H]Inositol Phosphate Formation

The assay was initiated by addition of 50 μl of packed slices into 0.5 ml buffer containing 10 mM LiCl and the appropritate ligands. The assay was terminated afer 30 minutes (if not otherwise indicated) by addition of 1.5 ml of chloroform/methanol (1:2, by volume). Phases were separated by addition of 450 μl of chloroform and 450 μl of water followed by brief centrifugation. In brain preparations, in the presence of LiCl, the major inositol metabolite is IP, with minor production of IP_2 and IP_3. Therefore, a total water-soluble inositol phosphate fraction was separated using AG - 1 x 8 anion - exchange resin, formate form, and quantitated as previously described (Berridge et al., 1982; Fisher et al., 1981).

The incorporation of [³H]inositol into phospholipid was determined by counting 100 μl aliquots of the lower organic phase.

Preparation of cortical slices and determination of [³H]NE efflux

The assay of [³H]NE release was carried out essentially as described (Diamant et al., 1988). Male albino rats (150-200 g) were sacrificed and their brains rapidly removed. Slices of cerebral cortex were prepared using a McIlwain tissue chopper (350μm) and were incubated for 15 min under oxygenated Krebs-Henseleit buffer (95% O_2 - 5% CO_2) containing the following (mM): 118 NaCl; 4.7 KCl; 1.2 $MgSO_4$; 25 $NaHCO_3$; 1.0 NaH_2PO_4; 0.004 Na_2EDTA and 11.1 Glucose with [³H]NE (19.7 Ci/mmol, final concentration 0.25 μM) at 37°C. Termination of the loading period was followed by 3x3 ml washes for 10 min.

The slices were then distributed into 24 incubation wells
containing 1ml of the oxygenated buffer with additional 1.3 mM
$CaCl_2$. (For more details see Minc-Golomb et al., 1987; Diamant
et al., 1988). The slices were washed for additional 3 x 10 min
consecutive intervals in the incubation wells. At t=46 min the
spontaneous release was monitored for 3 x 2 min intervals
followed by agonist stimulation at t=52 min for a 2 min
duration, if not indicated otherwise. Control runs of slices
under the same experimental conditions without stimulation show
a steady basal efflux of 0.3±0.02% fractional release. The
decline of the CCh induced release back to basal level was
rapid and was observed already at the end of the second period
after stimulation (t = 56 min). At t=60 min the remaining
$[^3H]NE$ was extracted from the tissue by 1ml of 0.1 N HCl for 16
h. Radioactivity was determined using liquid scintillation
spectroscopy with 50% efficiency.

Analysis of the release data

Tritium efflux during 2 min period was expressed as the
percentage of the tritium content of the slices at the begining
of the examined period. The tritium content in the slice was
calculated by adding the amounts of the tritium released during
that period, the following periods and in the content at the
end of the experiment.
Net evoked release was calculated from collected tritium during
two intervals (2 x 2 min) after substracting the basal tritium
efflux in the preceding period (x 2) assuming a constant
spontaneous release. Each point represents an average of 4-12
determinations obtained in two or three separate experiements
as indicated. Results are calculated as the average of all the
experiments ± S.E.M.

RESULTS AND DISCUSSION

Carbachol (CCh) stimulates phosphatidylinositol turnover· in [^{3}H]inositol labled cortical slices via muscarinic receptors (review by Fisher and Agranoff, 1987) and K$^+$ induces IP-accumulation most likely indirectly as a non-receptor mediated effect.

Figure 1A shows the time course of [^{3}H]IP accumulation in rat cortical slices in response to 0.8 mM CCh or 25 mM K$^+$ and in response to both of them added together. CCh-induced incorporation of [^{3}H]inositol into IP was significantly enchanced in the presence of depolarizing K$^+$ concentrations and the potentiation persisted almost linearly up to 60 min (Fig. 1A). At 30 min the degree of potentiation was 5.8 fold above the sum at [^{3}H]IP accumulated by K$^+$ and CCh separately. K$^+$ by itself at concentration (5 - 60 mM) induced a small gradual change in [^{3}H]IP accumulation (Fig. 1B) which could be attributed to a non-receptor mediated activation of PLC. Upon addition of 0.8 mM CCh, a significant synergism of [^{3}H]IP production was observed at K$^+$ concentrations above 5 mM KCl. The K$^+$ dose dependent curves in the presence and in the absence of CCh is different in both the maximal rate of the reaction and most probably in the apparent 50% saturation concentration. A very significant increase in [^{3}H]IP accumulation over 30 min (6-fold) was observed (Fig. 1B).

Figure 1A Time course for accumulation of [^{3}H]inositol phosphate.

Cerebral cortical slices prelabeled with [^{3}H]inositol for 90 min, washed, then incubated in the presence of 10 µM LiCl with no agent added (0), 30 mM KCl (▲), 1 mM CCh (■) and in a combination of CCh and K$^+$ (●) for the time periods as indicated. [^{3}H]Inositol phosphate was separated and analysed. Values are means of three representative experiments.

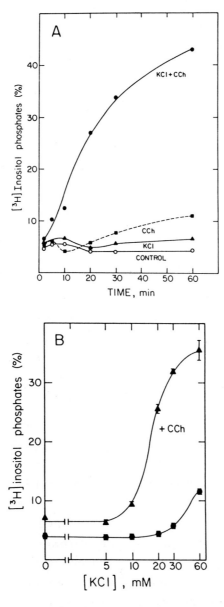

<u>Figure 1B</u> Dose response curve for stimulation of phosphatidyl inositol hydrolysis by K^+. Cerebral cortical slices were preincubated with [^{3}H]inositol in standard incubation medium for 90 min at 37° C. KCl was added at the indicated concentrations in the absence (●) and in the presence of 1 mM CCh (▲). After 30 min, [^{3}H]IP was separated and analyzed. Values represent mean ±SEM of three independent experiments.

The increase in IP accumulation might result from an increase in the number of PLC molecules coupled to muscarinic receptors under depolarizing conditions. Recent results have demonstrated a shift from high to low affinity sites of the muscarinic receptors in synaptoneurosoms depolarized by 50 mM KCl, or by batrachotoxin (Cohen-Armon et al., 1987). Thus an increase in receptor population coupled to PLC might initiate further activation of the enzyme and as a consequence, an increased PIP_2 hydrolysis. A more active conformation of PLC induced by depolarization, and therefore a better rate of catalysis can also explain amplification of IP production. The difference between the two dose response curves is sigmoidal with an inflection point at 15 mM K^+.

Total $[^3H]IP$ production was measured as a function of CCh concentration (0.008 - 3 mM). The CCh dose response curve displays an apparent half maximal effective concentration of 2 x 10^{-4} M which is not changed if receptor activation is carried out in the presence of 30 mM KCl or 2 µM veratridine. On the other hand, a significant increase in V_{max} was observed in the presence of K^+ (~6 fold) or veratridine (~2.5 fold) (Fig. 2). Quinuclidinyl benzylate (QNB) a specific muscarinic antagonist inhibited the potentiation induced by CCh in the presence of depolarizing concentrations of K^+ (Figure 2D), indicating the muscarinic receptor involvement in the potentiation process.

A recent report from our laboratory has shown that receptor mediated activation of PLC, mediate neurotransmitter release (Diamant et al., 1988). Hence, we attempted to examine whether the potentiation of IP-production induced by muscarinic agonists in high K^+ or veratridine, would also be reflected in potentiation of neurotransmitter release.

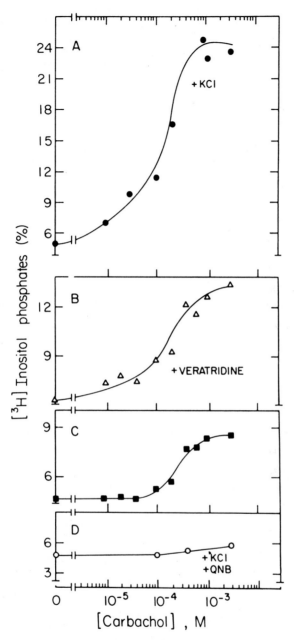

Figure 2 Dose response curve for stimulation of phosphatidyl inositol turnover. Cerebral slices preloaded with [3H]inositol were activated by CCh at the indicated concentrations. (A) in the presence of 30 mM KCl, (B), in the presence of 2 μM veratridine (C) alone (D) in the presence of 30 mM KCl and 10 μM QNB. The reaction was terminated after 30 min and [3H]inositol phosphate was separated and analysed. Values are means of representative experiments (n = 3).

Cortical slices prelabeled with $[^3H]NE$ were stimulated at t=52 min for 2 min either by CCh (1 mM), K^+ (25 mM) or by 1 mM CCh and 25 mM KCl added together. Stimulation by K^+ and CCh combined induces an enhanced fractional release which is more than additive. CCh induced net fractional release of 0.74 ± 0.06%, K^+ alone induced 3.8 ± 0.3% and the two combined induced 7.2 ± 0.3% which is almost two fold above additivity (Figure 3). The degree of potentiation produced by K^+ and CCh combined is 1.7 fold above the K^+-induced $[^3H]NE$ release and 2.5 - 3 fold above the CCh induced release. A longer stimulation period (4 min) shows no change in the CCh-induced fractional release as compared to 2 min (Table 1). The potentiated extent of release induced by CCh and K^+ combined, however, increases synergistically. This indicates that although CCh does not induce further $[^3H]NE$ release in the additional 2 min stimulation interval, its presence is reflected in the continued synergy of release, alluding to a possible retention of a previous signal.

Table 1 CCh- K^+ and CCh + K^+ - induced $[^3H]NE$ release in two consecutive stimulation intervals

	net fractional release (%)	
	stimulation time	
Ligand	2 min	4 min
CCh (1 mM)	0.74 ± 0.06	0.88 ± 0.08
KCl (25 mM)	3.8 ± 0.3	6.34 ± 0.3
KCl + CCh (25 mM)	7.2 ± 0.3	16.36 ± 0.6

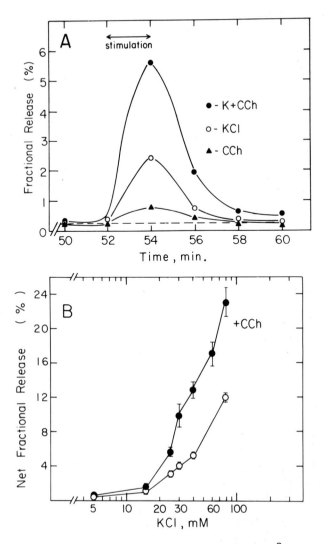

Figure 3 (A) K^+-, CCh- and K^+ + CCh induced [^{3}H]NE release.
Rat cortical slices preloaded with [^{3}H]NE superfused with
physiological buffer. At t=46 min Ca^{2+} (1.3 mM) was added and
spontaneous release of [^{3}H]NE during 2 min intervals (x3) was
monitored. Stimulation was initiated at t = 52 min by 1 mM CCh
($\triangle$), 25 mM KCl (O) and 1 mM CCh in the presence of 25 mM KCl
(o) for a duration of 2 min. The data is presented as a % of
total [^{3}H]NE incorporated into the vesicles. Each point
represents the mean of duplicate determinations (n = 12).
Spontaneous release was 0.25 ± 0.05% as depicted by the dashed
line (B); Dose response of K^+-induced [^{3}H]NE release in the
presence and in the absence of CCh. Cortical slices preloaded
with [^{3}H]NE were superfused with oxygenated buffer 95% O_2/5%
CO_2 physiological buffer without Ca^{2+} at 37° C. At t=46 min,
1.3 mM Ca^{2+} was added and spontaneous release was monitored (2

min x 3). Stimulation was initiated at t = 52 min for 2 min intervals in the absence of CCh (o) and in the presence of CCh at K^+ concentrations as indicated.

As shown in Figure 3, potentiation of release using 1 mM CCh is prominent at 15 mM KCl, up to 60 mM KCl. No potentiation of release was observed at 5 mM KCl, which represents the normal K^+ concentration of the buffer system.

Various cholinergic muscarinic agonists were examined for their ability to induce [3H]IP accumulation in the absence and in the presence of 30 mM KCl (Figure 4A). The order of potency of the various agonists as inducers of IP-accumulation was conserved under depolarizing conditions as compared to stimulation in normal buffer. The maximal degree of potentiation was 5 - 7 fold in the presence of 30 mM KCL and the order potency was ACh > CCh > oxo-M > arecoline >> pilocarpine. Similar results were observed by muscarinic agonists in the presence of 2 μM veratridine (data not shown). Thus, incorporation of [3H]inositol into IP by muscarinic agonists was significantly enhanced by KCl and veratridine. In view of these results we have examined whether potentiation of IP-accumulation by the various agonists is reflected also as an enhancement of neurotransmitter release.

As shown in Figure 4, the combination of muscarinic agonists with 25 mM KCl, similarly to their mutual effect on IP accumulation, induces potentiation of [3H]NE release. Oxo-M, ACh, CCh and arecoline show a 2-fold increase above the K^+ induced release. The extent of potentiation can be evaluated by subtracting the release induced by K^+ only (shown in shaded area) as compared to agonist induced release (black area) (Figure 4). Pilocarpine and bethanechol display a marginal effect on basal outflow of [3H]NE which is not potentiated by K^+. Similar results using the above agonists were obtained if the membrane was depolarized by 2 μM veratridine which at this concentration induces a Ca^{2+} dependent [3H]NE release (data not shown).

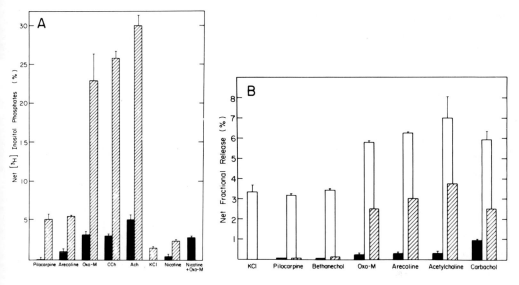

Figure 4   (A) Stimulation of phosphatidyl inositol turnover by various muscarinic agonists in the presence and in the absence of KCl. Cerebral cortical slices preloaded with [^{3}H]inositol for 90 min, were washed and incubated in the presence of 10 mM LiCl with 1 mM of various muscarinic agonists or 1 mM nicotine, in the absence (black columns) and in the presence of 30 mM KCl (striated columns). The values represent an average of 3 experiments and the bars are the SEM. (B) Potentiation of [^{3}H]NE induced release by various muscarinic agonists in the presence and in the absence of KCl. Cortical slices preloaded with [^{3}H]NE (0.25 μM, 30 min) superfused with oxygenated (95% O_2/5% CO_2) physiological buffer were stimulated at t = 52 min for a two min interval by KCl (25 mM) alone, by muscarinic agonists (1 mM each, black columns) and by the various muscarinic agonists (1 mM) in the presence of KCl (25 mM, white columns). The shaded column represents the potentiated [^{3}H]NE fractional release after substracting the net fractional release induced by K$^+$ (25 mM) alone. Each column represents net fractional release (%) induced during 2 min stimulation period. The results are the mean ±SEM of 6 - 8 independent determinations.

The effects of nicotine and K$^+$ added together were additive in both induction of [^{3}H]IP accumulation (Fig. 4A) and [^{3}H]NE release (data not shown). IP-accumulation induced by oxo-M, a

specific muscarinic agonist, was not potentiated in the
presence of nicotine (Figure 4A). Thus, it seems that a change
in $[Ca^{2+}]_i$ induced by nicotine, is not sufficient to enhance
IP-accumulation induced by muscarinic agonists. Further support
for this result was observed by the use of calcium ionophore
A-23187, which potentiates only slightly (~2 fold) the
CCh-induced phosphatidyl inositol turnover (Table 2). Thus a
change in $[Ca^{2+}]_i$ is apparently not the only effect evoked by

Table 2 Ca^{2+} ionophore A-23187 effect on $[^3H]IP$ accumulation

Ligand		net $[^3H]IP$ (%)
-	A-23187 (10 μM)	1.95 ± 0.8
KCl (30 mM)	-	1.71 ± 0.47
KCl and	A-23187	1.21 ± 0.54
CCh (0.8 mM)	-	3.19 ± 0.48
CCh and	A-23187	7.29 ± 0.85*
CCh + KCl	-	30.95± 3.39
CCh+KCl and	A-23187	40.23± 1.68*

*n = 4 - 8.

K^+ and is not sufficient to explain its ability, together with
muscarinic agonists to initiate a sustained state of
depolarization.
Based on these results, our proposed scheme for synergism
observed is:

Scheme 1

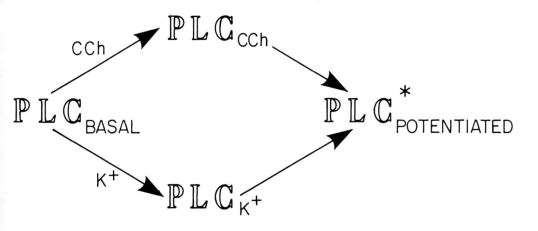

where KCl induces PLC activation (PLC_K) and CCh induces PLC activation (PLC_{CCh}), and both of them combined induce "super" activated enzyme (PLC^*_{K+CCh}).

It is tempting to speculate that enhanced activation of PLC results in generation of elevated levels of second messengers and in turn lead to enhancement of neurotransmitter release. It cannot be excluded that potentiation of release by K^+ and CCh might result from the closure of M-channels (Adams et al., 1982; Higashida and Brown, 1986), acting in concert with slowing of Ca^{2+} currents (Clapp et al., 1987), both processes shown to be activated by agonists mediating PI hydrolysis (Adams et al., 1982; Higashida and Brown, 1986). Since membrane depolarization in vivo is accompanied by neurotransmitter release, it is not impossible to envision a sequential action of depolarization and receptor activation followed by enhanced neurotransmitter release. Recent studies have shown that the hyperpolarization that follows action potential (AHP) in CAI neurons in the hippocampus, is significantly reduced and PKC is persistently translocated from the cystosol to the membrane, under high frequency stimulation (LTP) (LoTurco et al., 1988).

These effects might result from a persistent potentiation of PLC activity as a consequence of muscarinic receptor activation in depolarized cell membranes. More experiments should clarify the possible connection of enhanced IP-formation and potentiated neurotransmitter release within the context of long term potentiation in the cortex (Bindman et al., 1987; Artola and Singer, 1987).

CONCLUSIONS

Our results suggest three major conclusions: First, muscarinic agonists and depolarizing agents (of which KCl is the most effective) act in concert to induce potentiation of phosphatidyl inositol turnover and neurotransmitter release. Second, the simultaneous presence of both a depolarizing agent and a receptor agonist is obligatory for eliciting the potentiatory effect, which is blocked by receptor specific antagonist. Third, potentiation induced by K^+ + CCh persists in consecutive stimulation periods although the muscarinic receptor is already desensitized in the second stimulation interval.

Synergy of IP-accumulation in correlation with synergy of neurotransmitter release induced by muscarinic receptor activation and membrane depolarization suggests a possible role for PLC in the bifurcating control of neurotransmitter release and a possible involvement of PLC and membrane potential in mediation of LTP.

REFERENCES

Adams PR, Brown AD and Constanti A (1982) Pharmacological inhibition of the M-current. J. Physiol. 332:223-262.
Akers RF, Lovinger D, Colley P, Linden D and Routtenberg A (1986) Translocation of protein kinase C activity may

mediate hippocampal long term potnetiation. Science 231:
587-589.

Artola A and Singer W (1987) Long term potentiation and NMDA
receptors in rat visual cortex. Nature 330:649-651.

Baird JG and Nahorski SR (1986) Potassium depolarization
markedly enhances muscarinic receptor stimulated inositol
tetrakisphosphate accumulation in rat cerebral cortical
slices. Biochem. Biophys. Res. Commun. 141:1130-1137.

Berridge MJ, Downs CP and Hanley MR (1982) Lithium amplifies
agonist-dependent phosphatidylinositol responses in brain
and salivary glands. Biochem. J. 206:587-595.

Bindman LJ, Meyer T and Pockett SJ (1987) Long term
potneitation in rat-neocortical slices, produced by
repetitive pairing of an afferent volley with
intracellular depolarizing current. J. Physiol.
386:900-922.

Bliss TVP and Lomo TJ (1973) Long lasting potentiation of
synaptic transmission in the dentate area of the
anaesthesized rabbit following stimulation of the
perforant path. J. Physiol. 232:331-356.

Bondy CA and Gainer H (1988) Activators of protein kinase C
potentiate electrically stimulated hormone secretion from
the rat's isolated nerohypophysis. Neurosci. Lett. 89:
97-101.

Clapp LH, Vivaudou MB, Walsh JV Jr and Singer JJ (1987)
Acetylcholine increase voltage activated Ca^{2+} current in
freshly dissociated smooth muscle cells. Proc. Natl. Acad.
Sci. USA 84:2092-2096.

Cohen-Armon M, Garty H and Sokolovski M (1987) G-protein
mediates voltage regulation of agonist binding to
muscarinic receptors: Effects of receptor Na^+ channel
interaction. Biochem. 27:368-374.

Diamant S, Lev-Ari I, Uzielly I and Atlas D (1988) Muscarinic
agonists evoke neurotransmiiter release: Possible roles
for phsophatidylinositol breakdown products. J.
Neurochem., in press.

Eva C and Costa E (1986) Potassium ion facilitation of phosphoinositide turnover activation by muscarinic receptor agonists in rat brain. J. Neurochem. 46: 1429-1435.

Fisher SK and Agranoff BW (1987) Receptor activation and inositol lipid hydrolysis in neuronal tissue. J. Neurochem. 48:999-1017.

Fisher SK, Klinger PD and Agranoff BW (1981) Enhancement of the muscarinic synaptosomal phospholipid labeling effect by the ionophores A-23187. J. Neurochem. 37:968-977.

Hebb DO (1949) The organization of behavior (Wiley, New York).

Higashida H and Brown DA (1986) Two phosphatidylinositide metabolites control two K^+ currents in a neuronal cell. Nature 323:333-338.

Lisman JE (1985) A mechanism for memory storage insensitive molecular turnover: A bistable autophosphorylating kinase. Proc. Natl. Acad. Sci. USA 82:3055-3057.

LoTurco JL, Coulter DG and Alkon DL (1988) Enhancement of synaptic potential in rabbit CA1 pyramidal neurons following classical conditioning. Proc. Natl. Acad. Sci USA 85:1672-1676.

Lynch MA, Clements MP, Errington ML and Bliss TVP (1988) Increased hydrolysis of phosphatidyl-inositol-4,5-bis-phosphate in long term potentiation. Neuroscience Lett. 62:123-129.

Malenka RC, Madison DV and Nicoll RA (1986) Potentiation of synaptic transmission in the hippocampus by phorbol esters. Nature 321:175-177.

Miller SG and Kennedy MB (1987) Regulation of brain type II Ca^{2+}/cal modulin dependent protein kinase by autophosphorylation: A Ca^{2+} triggered molecular switch. Cell 44:861-870.

Minc-Golomb D, Levy Y, Kleinberger N and Schramm M (1987) D[^{3}H]Aspartate release from hippocampus studied in a multiwell system controlling factors and postnatal development of release. Brain Res. 402:255-263.

Acknowledgement

We thank Ms. R. Zonenshine and Mr. J. Yaffe for excellent assistance with the assays. This research is funded by the H. L. Leuterbach Fund (D.A.).

PURIFICATION AND LOCALIZATION OF KAINATE BINDING PROTEIN IN PIGEON CEREBELLUM

A.U. Klein and P. Streit
Brain Research Institute
University of Zürich
August-Forel-Strasse 1
CH-8029 Zürich
Switzerland

Excitatory amino acid (EAA) receptors are pharmacologically well characterized (Watkins and Evans, 1981; Foster and Fagg, 1984; Cotman et al., 1987; Mayer and Westbrook, 1987; Watkins and Olvermann, 1987), but virtually unknown at the molecular level. Kainic acid (KA), a heterocyclic L-glutamate analogue, defines as agonist a certain type of these receptors. The results of binding studies have indicated that brain tissues of lower vertebrates are relatively rich in kainate binding sites (Henke and Cuénod, 1980; London et al., 1980) and, thus, might serve as sources for molecules related to EAA receptors. In particular, Henke and collaborators (Henke et al., 1981) have described a high abundance (B_{max} = 118 pmol/mg protein) of kainic acid binding sites with relatively low affinity (K_d = 330 nM) in the pigeon cerebellum and have localized them autoradiographically with tritiated kainate in the molecular and Purkinje cell layers. From this tissue, kainate binding activity has been solubilized by means of the non-ionic detergent Triton X-100 (Dilber et al., 1983). - In the present contribution, we report on a monoclonal antibody (mAb) used to purify a kainate binding protein (KBP) from pigeon cerebellum and to localize it immunohisto-chemically. Furthermore, we describe the characteristics of polyclonal antibodies to purified KBP in immunohistochemical and immunoblotting experiments. Part of this work has been published previously (Klein et al., 1988 a,b).

Monoclonal antibodies precipitating KA binding activity from detergent extracts could be expected to serve as useful reagents for purification procedures. Therefore, we focused on the production and detection of such antibodies. The following procedures were used: Washed homogenates were prepared from cerebella or optic tecta of pigeons (Columba livia, 450 - 550 g, killed by decapitation) as described by Henke and collaborators (Henke et al., 1981; Klein et al., 1988 a,b) and used either fresh or after storage at -80 °C. In some experiments, the protease inhibitors phenylmethylsul-fonylfluoride (PMSF, 0.1 mM), Trasylol (200 units/ml; Bayer) and ethyleneglycol-bis-aminoethylether-N,N,N',N'-tetraacetic acid (EGTA, 1 mM) were included in the preparation. Before extraction with detergent, homogenates were washed once more

NATO ASI Series, Vol. H29
Receptors, Membrane Transport and Signal Transduction
Edited by A. E. Evangelopoulos et al.
© Springer-Verlag Berlin Heidelberg 1989

by centrifugation, resuspended to a protein concentration of 0.5 - 0.7 mg/ml and shaken in presence of Triton X-100 (TX-100) at 0.25% for 15 min. The extract was cleared by ultracentrifugation and used immediately or stored at -80 $^{\circ}$C. KA binding in homogenates was measured as described by Henke and collaborators (Henke et al., 1981). Binding in Triton extracts or in material resulting from purification procedures was determined by the method of Stephenson and collaborators (Stephenson et al., 1982). Test samples were incubated with 25 nM tritiated KA ([^{3}H]-KA: 5 Ci/mmol, Amersham, or 60 Ci/mmol, New England Nuclear) for 1 min. [^{3}H]-KA binding measured in the presence of 100 μM unlabeled KA was defined as non-specific binding.

It was interesting to note that a hybridoma secreting appropriate antibodies resulted from an immunization procedure which had been modified from that described by Matthew and Patterson (1983). Thus, since low affinity KA binding sites had not been detected in the pigeon optic tectum (Henke and Cuénod, 1980), a female BALB/c mouse was first intraperitoneally injected with detergent extract from tectal homogenate (1mg protein; emulsified with Freund's adjuvant, FA). Two days later, a suppression of the immune response to common and potentially dominant antigens in tectal extract was attempted by the application of the cytostatic drug cyclo-phosphamide (Endoxan, Asta; 1 g/kg body weight, intraperitoneally injected). One week later, this sequence of injections was repeated, and the mouse then rested for 2 weeks. Six intraperitoneal injections of cerebellar extract (1 mg protein emulsified with FA) were administered over a period of one year. Four days after the last injection, the spleen cells and P3U1 myeloma cells were fused (Galfré et al., 1977; Kennett, 1980). Precipitation of kainate binding activity was the criterion to detect and select a primary hybridoma and clones derived from it by the following assay system (all steps at 4 $^{\circ}$C): Spent culture medium (500 μl) was added to Protein-A-Sepharose CL-4B (9-10 mg in 150 μl; Pharmacia) which had been coated with rabbit anti-mouse IgG antibodies (Miles), and the mixture was shaken overnight. The beads were washed with Tris buffer (50 mM, pH 8) and incubated with cerebellar detergent extract (300 μl; containing 5% BSA, 0.15 M NaCl; pH 8) for 6 hours. Following another washing step, they were suspended in Tris buffer (400 μl, 50 mM) at pH 7.4, before KA binding was measured (25 nM [^{3}H]-KA, 1 min incubation time). Non-specific binding was deter-mined as mentioned above. A hybridoma secreting appropriate antibodies was cloned repeatedly, and large amounts of monoclonal antibody, mAb anti-KBP-1, were produced in ascitic fluid. This hybridoma secreted IgG1 and was the only stable cell line with the desired characteristics which could be developed from a total of 6 experiments and from 2 cases in which the immunization procedure involving partial immunosuppression (Matthew and Patterson, 1983) had been tried. However, it could not be determined whether this type of procedure was the decisive factor in the successful development of appropriate monoclonal antibodies because the various immunization schedules used in the present study had not been planned to allow a systematic and comparative study.

For immunoaffinity chromatography, cerebellar detergent extract was passed through a column packed with Affigel-10 (Bio-Rad) to which mAb anti-KBP-1 had been linked. After thorough washing, elution was accomplished at pH 11.0 with 50 mM diethyl-amine (in 50 mM Tris/0.25% TX-100). The eluate was neutralized immediately with 1 M Tris/0.25% TX-100 (pH 7.4) and, following equilibration of the column at pH 7.4, the eluted material was subjected to another cycle of the chromatographic procedure. - After these two chromatographic cycles, binding activity was increased 87-fold relative to homogenates (Table 1).

Material	Total protein	Total specific binding	Specific activity	Purification factor	Yield
	mg	pmol	pmol/mg		
Washed homogenate	157±5 (159)	3486±28 (3476)	22.1±0.7 (21.9)	1 (1)	100 (100)
Detergent extract	125±4 (114)	6545±250 (6314)	52.4±2.0 (55.5)	2.4 (2.5)	188 (182)
Pellet	29±2 (27)	209±14 (189)	7.2±0.5 (7.0)		
Immunoaffinity purified material	0.034±0.003 (0.030)	67±7 (56)	1912±65 (1880)	87 (86)	1.9 (1.6)
Material not retained	126±6 (112)	202±30 (135)	1.6±0.3 (1.2)		

Table 1. Purification of a kainic acid binding site from pigeon cerebella. Data are averages from three experiments in which 32 pigeon cerebella each were used. Values in parenthesis are from an experiment performed in the presence of protease inhibitors. Binding activity was determined with [^{3}H]KA at a concentration of 25 nM.

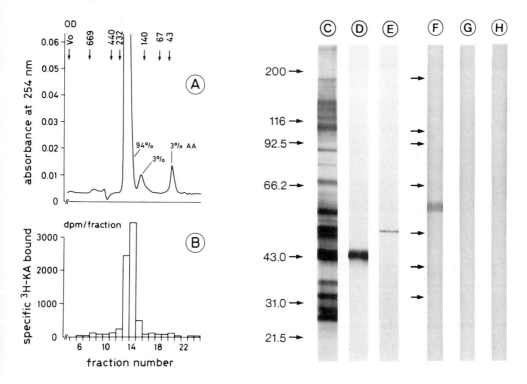

Figure 1. A,B: Gel filtration FPLC (on Superose 12) of material immunoaffinity purified by monoclonal antibody anti-KBP-1. A: A Peak at M_r = 220,000 predominates in the absorbance profile. Molecular weights (in kD) of standards at top of figure. B: Most KA binding activity is concentrated in fractions corresponding to major peak in A. 100 μl aliquots from 1 ml fractions were assayed for specific [^{3}H]KA binding. - C-E: SDS-PAGE of (C) detergent extract from pigeon cerebellum (3.5 μg) and (D,E) immunoaffinity purified material (approx. 0.3 μg) in 0.25% Triton X-100 (D) or following dialysis (E). Single major bands appear at M_r = 43,000 in D but at M_r = 50,000 in E. - F-H: Immunoblotting with polyclonal antibodies raised against immuno-affinity purified KBP on SDS extracts of pigeon cerebellum (F) or optic lobe (G) following SDS-PAGE and electrophoretic transfer to nitrocellulose membrane. A single major band at M_r =50,000 is labeled in cerebellar material (F). No signal is detected in extract of optic lobe (G) or if only the secondary peroxidase labeled antibody is applied (H). - Markers indicate position of standards (in kD).

Immunoaffinity purified samples prepared in the presence or absence of protease inhibitors showed the same characteristics as assayed by gel filtration chromatography and by polyacrylamide gelelectrophoresis (PAGE). Fast protein liquid chromatography (FPLC) was performed with purified and concentrated (Centricon 10, Amicon) material on a Superose 12 column (Pharmacia; HR 10/30 column) which had been equilibrated with Tris/TX-100 and calibrated with protein standards. A peak at M_r = 220,000 predominated (94%) in the absorbance profile (Fig. 1A). The fractions corresponding to this peak also contained virtually all the binding activity applied to and eluting from the gel filtration column (Fig. 1B). In addition, immunoaffinity purified material was resolved by sodium dodecylsulfate (SDS)-PAGE (Laemmli, 1970; Laemmli and Favre, 1973) under reducing conditions, and the 7% gels were silver-stained (Merril et al., 1981). A single major band migrating at M_r = 43,000 (broad band) was found in samples which contained 0.25% TX-100 (Fig. 1D) but at M_r = 50,000 (sharp band) following dialysis (Fig. 1E). The subunits of the purified protein exhibiting KA binding activity, thus, seemed to be homogeneous in apparent molecular weight. From frog brain, Hampson and Wenthold (1988) isolated recently a KBP which was resolved by SDS-PAGE at M_r = 48,000 but which was found to elute with M_r = 570,000 in gel filtration chromatography. Moreover, Gregor and collaborators (Gregor et al., 1988) reported even more recently on a KBP in chicken cerebellum with a M_r = 49,000 in PAGE under denaturing and reducing conditions and with an apparent molecular weight of 390,000 daltons in gel filtration chromatography. Only comparison of results from future structural investigations will determine the relation between these various KA binding proteins. Furthermore, since an effector function has not been described yet in association with these KA binding sites, their exact relationship with physiologically and pharmacologically defined KA receptors remains to be elucidated.

Immunohistochemistry allowed to determine the tissue distribution of material recognized by our antibodies. Pigeons or chickens were transcardially perfused with an aldehyde fixative (1% formaldehyde, or a mixture of 0.5% formaldehyde and 0.5% glutaraldehyde). A standard peroxidase anti-peroxidase (PAP) method (Sternberger, 1979) with a mouse monoclonal PAP complex was used to stain cryostat sections. In the cerebellar cortex of pigeons, staining obtained with mAb anti-KBP-1 was strong in the molecular layer and extended between and occasionally around the perikarya of Purkinje cells (Fig. 2A). The granule cell layer, the cerebellar white matter and the optic tectum, on the other hand, were free of any staining. The same staining pattern was observed in the cerebellum of chickens (Fig. 2C) and, moreover, with polyclonal antibodies raised in mice (Lacy and Voss, 1986) against immunoaffinity purified material (Fig. 2D,E). The fact that these immunohistochemical staining patterns matched the autoradiographic labeling pattern described by Henke and collaborators (Henke et al., 1981) strongly indicated that the KBP isolated from pigeon cerebellum represented the low affinity KA binding site characterized by these authors.

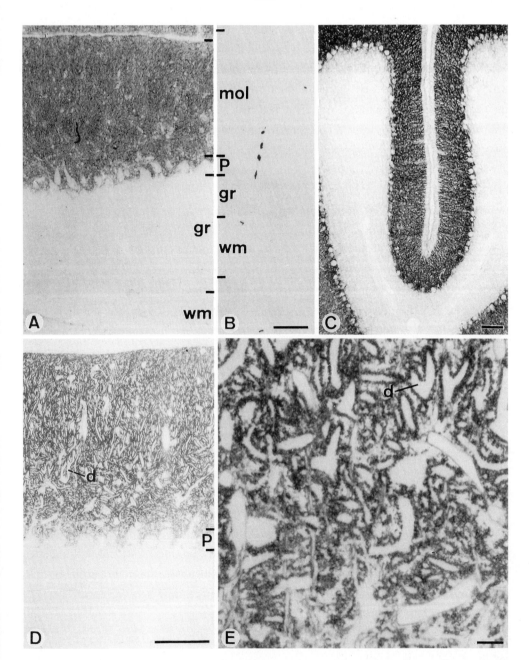

Figure 2. A-C: Immunohistochemical staining of cryostat sections from pigeon (A,B) or chicken (C) cerebellum (perfusion fixation with mixture of 0.5% glutaraldehyde and 0.5% formaldehyde; PAP staining). A: Monoclonal antibody anti-KBP-1 labels elements in molecular layer (mol) and as well as between and around Purkinje cell perikarya (P) in cerebellar cortex. No staining in granule cell layer (gr) and white matter (wm). B: Tissue is unstained if this antibody is omitted in staining procedure. C:

Our polyclonal antibodies were not only able to precipitate KA binding activity but, in contrast to mAb anti-KBP-1, they also labeled a single band at an apparent molecular weight of 50,000 daltons (Fig. 1F) in immunoblotting experiments (Towbin et al., 1979) which were performed on SDS extracts of pigeon cerebellum. No signal was detected on tectal material (Fig. 1G). Again in contrast to mAb anti-KBP-1, the polyclonal antibodies to KBP produced a labeling pattern in postembedding immunohisto-chemistry which was performed as described in detail by Liu and collaborators (Liu et al., 1988). For these experiments, pigeons were transcardially perfused with aldehyde mixtures (up to 1% formaldehyde and 1% glutaraldehyde). Semithin (1 µm) sections were cut from osmicated brain slices embedded in epoxy resin, etched, and following osmium removal stained with the polyclonal antibodies to KBP by means of a PAP method which involved silver intensification of the diaminobenzidine reaction product. A striking feature of the staining pattern in these semithin sections were small, almost circular elements which outlined what seemed to be Purkinje cell dendrites (Fig. 2D). Labeling also extended between the perikarya of Purkinje cells but could not be detected in the granule cell layer, the cerebellar white matter nor the optic tectum. Although the resolution in semithin sections was much better than in cryostat sections it was still limited and, thus, did not allow to define the labeled elements although dendritic spines of Purkinje cells or parallel fiber terminals appeared to be likely candidates.

In conclusion, mAb anti-KBP-1 allowed us to isolate and partially characterize a kainate binding protein from the pigeon cerebellum. This KBP is located in the molecular layer of the cerebellar cortex and represents the major kainate binding site in this structure as indicated by the striking similarity between our immuno-histochemical labeling patterns obtained with mono- and polyclonal antibodies and previously reported autoradiographic labeling patterns obtained with tritiated kainate. The first step, thus, has been taken towards a thorough characterization at the molecular level of a binding protein related to excitatory amino acid receptors.

This work was supported by grants 3.389.86 and 3.390.86 of the Swiss National Science Foundation, by the Dr. Eric Slack-Gyr Foundation and by the Ernst Göhner Foun-dation.

Staining pattern in chicken cerebellum is similar to that obtained in the pigeon (A). - D,E: Postembedding immunohistochemistry on semithin section of pigeon cerebellum (perfusion fixation as above). Staining pattern in D is generally the same as in A but immunoreactivity is mostly outlining what seem to be dendritic profiles (d). E: At higher magnification, labeling is resolved as small almost circular elements. - Bars: 100 µm in B-D, 20 µm in E. Same magnification in A and B.

Cotman CW, Monoghan DT, Ottersen OP, Storm-Mathisen J (1987) Anatomical organization of excitatory amino acid receptors and their pathways. Trends Neurosci 10:263-265

Dilber A, Henke H, Cuénod M, Winterhalter KH (1983) Characterization of the low affinity kainic acid binding site solubilized from pigeon cerebellum. Soc Neurosci Abstr 9:260

Foster AC, Fagg G (1984) Acidic amino acid binding sites in mammalian neuronal membranes: Their chararcteristics and relationship to synaptic receptors. Brain Res Rev 7:103-164

Galfré G, Howe SC, Milstein C, Butcher GW, Howard JC (1977) Antibodies to major histocompatibility antigens produced by hybrid cell lines. Nature 266:550-552

Gregor P, Eshhar N, Ortega A, Teichberg VI (1988) Isolation, immunochemical characterization and localization of the kainate sub-class of glutamate receptor from chick cerebellum. EMBO J 7:2673-2679

Hampson DR, Wenthold RJ (1988) A kainic acid receptor from frog brain purified using domoic acid affinity chromatography. J Biol Chem 263:2500-2505

Henke H, Cuénod M (1980) Specific [^{3}H]kainic acid binding in the vertebrate CNS. In Littauer UZ, Dudai Y, Silman I, Teichberg VI, Vogel Z (eds) Neurotransmitters and their Receptors. John Wiley New York pp373-390

Henke H, Beaudet A, Cuénod M (1981) Autoradiographic localization of specific kainic acid binding sites in pigeon and rat cerebellum. Brain Res 219:95-105

Kennett RH (1980) Fusion centrifugation of cells suspended in polyethylene glycol. In Kennett RH, McKearn TJ, Bechtol KB (eds) Monoclonal Antibodies. Hybridomas: A New Dimension in Biological Analyses. Plenum Press New York pp365-367

Klein AU, Niederöst B, Winterhalter KH, Cuénod M, Streit P (1988a) Monoclonal antibody to kainate binding protein in pigeon cerebellum. Eur J Neurosci Suppl 223

Klein AU, Niederöst B, Winterhalter KH, Cuénod M, Streit P (1988b) A kainate binding protein in pigeon cerebellum: purification and localization by monoclonal antibody. Neurosci Lett in press

Lacy MJ, Voss EW (1986) A modified method to induce immune polyclonal ascites fluid in BALB/c mice using Sp2/O-Ag14 cells. J Imm Meth 87:329-339

Laemmli UK (1970) Cleavage of structural proteins during the assembly of the head of bacteriophage T4. Nature 227:680-685

Laemmli UK, Favre M (1973) Maturation of the head of bacteriophage T4. J Mol Biol 80:575-599

Liu CJ, Grandes P, Matute C, Cuénod M, Streit P (1988) Glutamate-like immunoreactivity revealed in rat olfactory bulb, hippocampus and cerebellum by monoclonal antibody and sensitive staining method. Histochem in press

London ED, Klemm N, Coyle JT (1980) Phylogenetic distribution of [^{3}H]kainic acid receptor binding sites in neuronal tissue. Brain Res 192:463-476

Matthew WD, Patterson PH (1983) The production of a monoclonal antibody that blocks the action of a neurite outgrowth-promoting factor. Cold Spring Harbor Symp Quant Biol 48:625-631

Mayer ML, Westbrook GL (1987) The physiology of excitatory amino acids in the vertebrate central nervous system. Prog Neurobiol 28: 197-276

Merril CR, Goldman D, Sedman SA, Ebert MH (1981) Ultrasensitive stain for proteins in polyacrylamide gels shows regional variations in cerebrospinal fluid proteins. Science 211:1437-1438

Stephenson FA, Watkins AE, Olsen RW (1982) Physicochemical characterization of detergent solubilized γ-aminobutyric acid and benzodiazepine receptor proteins from bovine brain. Eur J Biochem 72:248-254

Sternberger LA (1979) Immunocytochemistry. John Wiley New York

Towbin H, Staehelin T, Gordon J (1979) Electrophoretic transfer of proteins from polyacrylamide gels to nitrocellulose sheets: Procedure and some applications. Proc Natl Acad Sci USA 76:4350-5354

Watkins JC, Evans RH (1981) Excitatory amino acid transmitters. Ann Rev Pharmacol Toxicol:165-204

Watkins JC, Olverman HJ (1987) Agonists and antagonists for excitatory amino acid receptors. Trends Neurosci 10:265-27

THE NOREPINEPHRINE ANALOG META-IODO-BENZYLGUANIDINE (MIBG) AS A SUBSTRATE FOR MONO(ADP-RIBOSYLATION).

C. Loesberg, H. v. Rooij and L.A. Smets
Department of Experimental Therapy
The Netherlands Cancer Institute
Plesmanlaan 121
1066 CX Amsterdam
The Netherlands

Introduction

Human tumors of neuro-ectodermal origin such as pheochromocy-toma, neuroblastoma and medullary thyroidoma, have often preserved the capacity of neuronal cells to recapture extracellular norepinephrine. Meta-iodo-benzylguanidine (MIBG; Wieland et al., 1980), is a functional analog of norepinephri-ne, derived from the neuron-blocking agents bretyllium and guanethidine which competes with biogenic amines for uptake and storage into chromaffin tissues. Radioiodinated [^{131}I]-MIBG has been successfully applied for the scintigraphic visualization of adrenergic tissues (Wieland et al., 1981) and neuroendocrine tumors such as pheochromocytoma (Sisson et al., 1981; Shapiro et al., 1985), neuroblastoma (Treuner et al., 1984; Hoefnagel et al., 1985) and many other tumors of the APUD series (Von Moll et al., 1987). During the clinical application of [^{131}I]-MIBG with radiotherapeutic intent, patients receive milligram amounts of a potentially active amine. We therefore investigated pharmacological properties of this novel radiopharmaceutical.

Cell Biological Effects

A number of reports have indicated that catecholamines can be deleterious to cells _in vitro_ with a clear preference for cells of neural origin. For instance, L-DOPA is selectively toxic for melanoma cells (Wick et al., 1977). Mouse melanoma cells are killed by 2,4-hydroxy-phenylalanine due to hydroxy-lation by the tissue-specific enzyme tyrosinase, eliciting a

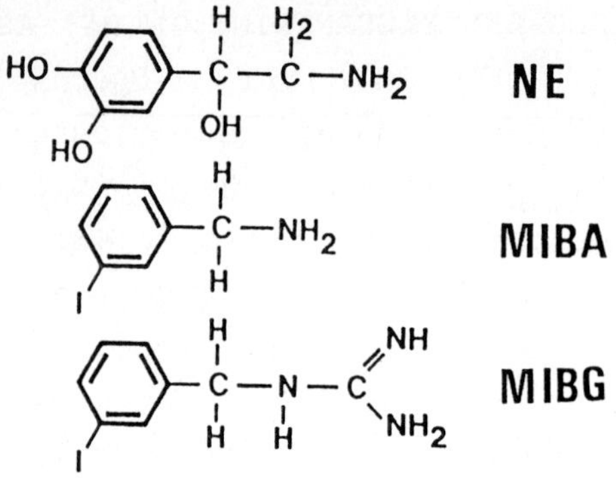

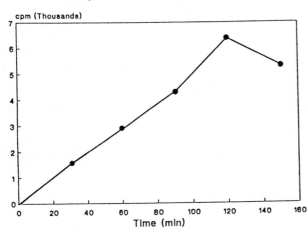

Fig. 1. Structure of norepinephrine (NE), meta-iodo-benzylamine (MIBA) and meta-iodo-benzylguanidine (MIBG)

Ribosylation of MIBG by cholera toxin

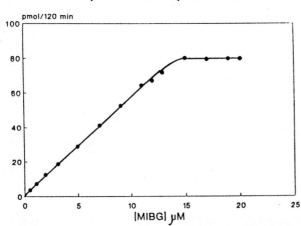

Fig. 2. Kinetics of formation of ADP-ribosyl-^{125}I-MIBG from ^{125}I-MIBG and NAD by cholera toxin.

Ribosylation of MIBG by cholera toxin

Fig. 3. Rate of formation of ADP-ribosyl-^{125}I-MIBG from ^{125}I-MIBG and NAD by cholera toxin at different ^{125}I-MIBG concentrations.

cascade of auto-oxidations and the generation of reactive oxygen radicals (Morrison et al., 1985). Similarly 6-hydroxy-dopamine is accumulated and hydroxylated by neuroblastoma cells (Bruchelt et al., 1985). Since MIBG could have similar effects on neuroblastoma or pheochromocytoma cells, clinical responses obtained with [^{131}I]-MIBG at high dose might in part be attributable to antitumor effects of the drug per se. For this reason its possible antiproliferative and cytotoxic effects were comparatively investigated in cell lines derived from neural crest tumors and control tissues, as well as in some tumors established from those cell lines.

MIBG appeared to be cytotoxic in a large panel of histogeneti-cally different cell lines without preference against tumor cells of neural origin. In view of its cytostatic action in vitro, MIBG was also tested for possible antitumor activity in vivo against N_1E115 neuroblastoma as a model for neural crest-derived tumors and L1210 lymphoma as an unrelated control in mice. It was demonstrated that MIBG had antitumor effects on both tumors in non-toxic schedules of drug administration (Smets et al., 1988).

Since MIBG is a structural and functional analog of natural norepinephrine, it was investigated if MIBG acted as a false hormone. It was shown, however, that cytotoxicity was unrelated to catecholamines. Pharmacological intervention data indicated that MIBG exerted its cytotoxic effects in the cytoplasma in its native form. To arrive at a conclusion regarding the active part of the molecule in cytotoxic activity, the mono-amine precursor of MIBG, meta-iodoben-zylamine (MIBA, see fig. 1) was tested as well. MIBA had only small biological effects both in vitro and in vivo.

Apparently, the guanidino-group was essential for both cytotoxicity and antitumor effects of MIBG. Since MIBG resembles the arginine acceptor for the enzyme mono (ADP-ribosyl) transferase (Ueda and Hayaishi, 1985) it was speculated that MIBG might exert these effects by interfering with this type of enzymes. In fact MIBG is very similar to some synthetic substrates of cholera toxin, also a mono(ADP-ribosyl) transferase that uses arginine as putative substrate.

(ADP-ribosyl) transferases.

(ADP-ribosyl) transferases are enzymes that utilize NAD as a donor for the ribosylation of proteins. The field of (ADP-ribosylation) reactions is emerging rapidly. At present two different classes can be discriminated, namely mono(ADP-ribosylation) and poly(ADP-ribosylation), although their coexistence in eukaryotes may involve a close functional link. Mono(ADP-ribosylation) predominates by far (i.e. ›10x) over poly(ADP-ribosylation), particularly in the extracellular compartments. Initially poly(ADP-ribosyl) transferases were characterized as being of nuclear origin and hence termed "nuclear(ADP-ribosyl) transferases", while mono(ADP-ribosyl) transferases were thought to be of extranuclear origin. At present it is clear that both types of enzymes can be found in nuclei as well as extranuclear (Althaus, 1987).

The two types of (ADP-ribosyl) transferases differ not solely in the number of the (ADP-ribose) molecules transferred to a protein, but also in the chemical nature of the (ADP-ribosyl)-protein bond. All purified mono (ADP-ribosyl) transferases catalyze the formation of N-glycosidic bonds between (ADP-ribose) and arginine (or guanidino-containing analogues) in vitro (Moss and Vaughan, 1980; Yost and Moss, 1983; Ritter et al., 1983; Tanigawa et al., 1984; Soman et al., 1984). In contrast, poly(ADP-ribose) transferases catalyze an O-glycosidic linkage (Ueda et al., 1975, 1985; Nishizuka et al., 1969; Reeves et al., 1981; Burzio, 1982; Adamietz, 1982).

The physiological function of the cellular mono(ADP-ribosyl) transferases is not known at present (for a review see Richter, 1988), partly due to the fact that no specific inhibitors for these transferases are known. The best characterized mono(ADP-ribosyl) transferases are evidently the microbial toxins. Three classes of bacterial toxins are described, based on the nature of the acceptor moiety in the protein: Diphtamide-, arginine- and cystein-specific (ADP-ribosylation). The last two reactions are known to interfere with signal transduction of G-protein linked receptors via ribosylation of G_s (cholera toxin catalyzed) and G_i (pertussis toxin catalyzed) respectively. It is conceivable that endogenous mono(ADP-ribosyl) transferases ribosylate similar substrates as those bacterial toxins albeit in a physiologically controlled manner.

Far more is known about the function of poly(ADP-ribosyl) transferases (for review see Althaus, 1987). Inhibitors of the poly(ADP-ribosyl) transferase activity have appeared extremely important tools in the elucidation of the biological function of these enzymes. A large number of inhibitors have become known in the past decade (Althaus, 1987), amongst which 3-aminobenzamide (Sims et al., 1982). These and other reports have demonstrated the importance of poly(ADP-ribosylation) as a major pathway of NAD catabolism in tissues and cells subjected to alkylations or ionizing radiation-damage. Lowering of NAD-pools results from activation of the enzyme poly(ADP-ribose) polymerase, an enzyme involved in the repair of breaks DNA breaks.

Effects on ADP-ribosylation

To test the hypothesis that MIBG may exert part of its cytostatic effects via interference with mono(ADP-ribosylating) enzymes, it was investigated first whether MIBG was a substrate for cholera toxin. Cholera toxin was tested, in triplicate, according to Mekalanos et al. (1979) with minor modifications. The reaction mixture (100 μl) consisted of 0.14 M sodium phosphate buffer (pH 7.0), 28 mM dithiothreitol, Triton X-100, 0.006% (w/v), 6 mM NAD and [^{125}I]-MIBG (usually 70,000 dpm, specific activity 40 mCi/mmol). The reaction was started by the addition of 40 μg activated cholera toxin and stopped by the addition of 7.5 ml of 0.01 M ice-cold sodium borate, pH 8.5. Subsequently the reaction mixture was passed through three stacked 2.4 cm DEAE-paper discs (Whatman DE81) which were then washed with 100 ml of the borate buffer and counted by liquid scintillation. Specific MIBG ribosylation was calculated from total binding to DEAE-filters minus binding in the absence of NAD. In this way it was demonstrated that MIBG served as an acceptor for cholera toxin-catalyzed

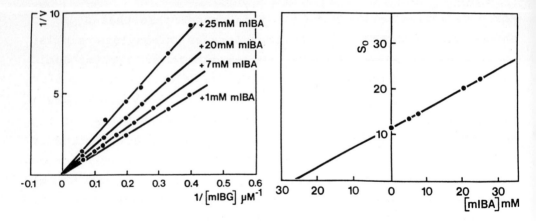

Fig. 4A. Effects of meta-iodo-benzylamine (MIBA) on the initial rate of mono (ADP-ribosylation) at different concentrations of ^{125}I-MIBG with NAD concentration of 6 mM. Fig. 4B. Secondary plot of the slopes (S_0) from Fig. 4A vs the MIBA concentrations.

Fig. 5. the activation of poly(ADP-ribose)polymerase activity by DNAse of permeabilised S49 lymphosarcoma cells in the absence and presence of 3-aminobenzamide (10 mM) and MIBG 3 or 5 μM).

(ADP-ribose) transfer reaction. This (ADP-ribosylation) of MIBG was linear for 120 minutes (Fig. 2).

Michaelis-Menten analysis of the enzymatic reaction, measuring initial rates of MIBG ribosylation at various concentrations of $[^{125}I]$-MIBG and NAD, demonstrate a K_m of 6.5 μM. A representative curve at 6 mM NAD is shown in Fig. 3.

To cheque the specificity of this reaction the inhibition by MIBA was also tested. The inhibition constant of MIBA (K_i) was determined by measuring enzyme substrate curves for $[^{125}I]$-MIBG in the presence of different amounts of MIBA (Fig. 4). The Ki calculated from the secondary plots of these data (Fig.4B) was 27 mM.

In conclusion it was demonstrated that MIBG was an excellent substrate for cholera toxin. In fact, MIBG is a synthetic substrate with the highest affinity for cholera toxin described as yet.

These results with cholera toxin are very suggestive for the possibility that MIBG is also a good substrate for cellular mono(ADP-ribosyl) transferases. Endogenous linkage of NAD to plasmamembrane proteins of human erythrocytes was found to be strongly inhibited by MIBG, probably by competition for binding the enzymes. Next it was tested if the action of MIBG was specific for the mono(ADP-ribosyl) transferase reaction. Poly(ADP-ribose) polymerase activity was assayed in S49-lymphosarcoma cells. Cells were permeabilised to › 90% with 15 μg/ml digitonin according to Trypan blue exclusion. These cells were cooled to 4°C, collected by centrifugation in the cold and washed once with an icecold saline solution (pH 7.4), 400 g KH_2PO_4, 350 mg Na HCO_3, 8 g NaCl and 1 g per liter glucose. The cells were resuspended in 1 ml cold hypotonic buffer (pH 7.8), 9 mM HEPES, 4.5 % (w/v) dextran, 1 mM EGTA, 4.5 mM $Mg-Cl_2$ and 5 mM dithiothreitol. After 30 minutes incubation at 4°C, cells were diluted in an isotonic buffer (pH 7.8), 40 mM HEPES, 130 mM KCl, 4 % (w/v) dextran, 225 mM sucrose, 2 mM EGTA, 2.3 mM $MgCl_2$ and 2.5 mM dithiothreitol. Finally, poly(ADP-ribose) polymerase activity was measured after deoxyribonuclease (DNAse I) addition (100 μg/ml, Sigma, type IV) to obtain an optimal number of DNA-breaks and thus a maximal polymerase stimulation (Farzaneh et al., 1982). Maximal enzyme inhibition was obtained using 3-aminobenzamide as an inhibitor (Sims et al., 1982). Results are given in fig.5. MIBG only marginally inhibited poly(ADP-ribose) polymerase.

In conclusion, MIBG is a high affinity substrate for the N-linked mono(ADP-ribosyl) transferase activity of cholera toxin, but does not inhibit cellular poly(ADP-ribose) polymerase. It exerts cell biological effects in micromolar concentrations, is metabolically very stable and is able to diffuse through the plasmamembranes. MIBG might turn out a valuable tool to interfere specifically with mono (ADP-ribosylation) of cellular proteins and to elucidate the largely unknown physiological role of this posttranslational modification.

REFERENCES:

Adamietz P (1982) Acceptor proteins of poly(ADP-ribose). In: Hayaishi O, Ueda K. (eds) ADP-ribosylation reactions. Biology and medicine. Academic Press New York, pp 77-101

Althaus FR (1987) Poly-ADP-ribosylation reactions. In: Althaus FR, Richter C (eds) ADP-ribosylation of proteins. Enzymology and biological significance. Springer Verlag, Berlin, pp 3-122

Bruchelt G, Buck J, Girgert R, Treuner J, Niethammer D (1985) The role of reactive oxygen compounds derived from 6-hydroxydopamine for bone marrow purging from neuroblastoma cells. Biochem Biophys Res Commun 130:168-174.

Burzio LO (1982) ADP-ribosyl protein linkages. In: Hayaishi O, Ueda K (eds) ADP-ribosylation reactions. Biology and Medicine. Academic Press, New York, pp 103-1226

Farzaneh F, Zalin R, Brill D, Shall S (1982) DNA strand breaks and ADP-ribosyl transferase activation during cell differentiation. Nature 33, 362-366

Hoefnagel CA, Voûte PA, De Kraker J, Marcuse HR (1985) Total body scintigraphy with 131-I-metaiodobenzylguanidine for detection of neuroblastoma. Diagn Imag Clin Med

54:21-27

Mekalanos JJ, Collier RJ, Romig WR (1979) Enzymic activity of cholera toxin. I New method of assay and the mechanism of ADP-ribosyl transfer. J Biol Chem 254:5849-5854

Moss J, Vaughan M (1978) Isolation of an avian erythrocyte protein possessing ADP-ribosyl-transferase activity and capable of activating adenylate cyclase. Proc Natl Acad Sci USA 75:3621-3624

Moss J, Stanley SJ, Watkins PA (1980) Isolations and properties of an NAD- and guanidine dependent ADP-ribosyl-transferase from turkey erythrocyte. J Biol Chem 255:5838-5840

Nishizuka Y, Ueda K, Yoshihara K, Takeda M, Hayaishi O (1969) Enzymic adenosine diphosphoribosylation of nuclear proteins. Cold Spring Harbor Symp Quant Biol 34:781-786

Reeves R, Chang D, Chung S-C (1981) Carbohydrate modification of the high mobility group proteins. Proc Natl Acad Sci USA 78:6704-6708

Richter C, Winterhalter KH, Baumhuter S, Lotscher HR, Moser B (1983) ADP-ribosylation in inner membrane of rat liver mitochondria. Proc Natl Acad Sci USA 80:3188-3192

Richter C (1987) Mono-ADP-ribosylation reactions. In: Althaus FR, Richter C (eds) ADP-ribosylation of proteins. Enzymology and biological significance. Springer Verlag, Berlin. pp 131-226

Sisson JC, Marc SF, Valk TW, Gross MD, Swanson DP, Wieland DM, Tobes MC, Beierwaltes WH Thompson NW (1981) Scintigraphic localization of pheochromocytoma. New Engl J Med 305:12-17

Shapiro B, Copp JE, Sisson JC, Eyre PL, Wallis J, Beierwaltes WH (1985) 131-I-metaiodobenzylguanidine for the locating of suspected pheochromocytoma: Experience with 400 cases (441 studies). J Nucl Med 26:576-585

Soman G, Mickelson JR, Louis CF, Graves DJ (1984) NAD:guanidino group specific mono-ADP-ribosyltransferase activity in skeletal muscle. Biochem Biophys Res Commun 120:973-980

Tanigawa Y, Tsuchiya M, Imai Y, Shimoyama M (1984) ADP-ribosyltransferase from hen liver nuclei. Purification and Characterization. J Biol Chem 259:2022-2029

Treuner J, Feine U, Niethammer D, Mueller-Schauenburg W, Meinke J, Eibach E, Dopfer R, Klingebiel TH Grumbach ST (1984) Scintigraphic imaging of neuroblastoma with ·m-[131I]iodobenzylguanidine. Lancet i:333-334

Ueda K, Omachi A, Kawaichi M, Hayaishi O (1975) Natural occurrence of poly(ADP-ribosyl) histones in rat liver. Proc Natl Acad Sci USA 72:205-209

Ueda K, Hayaishi O (1985) ADP-ribosylation. Annu Rev Biochem 54:73-100

Ueda K, Hayaishi O, Oka J, Komura H, Nakanishi K (1985) 5'-ADP-3"- deoxypentos-2"-ulose. A novel product of ADP-ribosyl protein lyase. In: Althaus FR, Hiltz H, Shall S (eds) ADP-ribosylation of proteins. Springer Verlag, Berlin, pp 159-166

Von Moll L, McEwan AJ, Shapiro B, Sisson JC, Gross MD, Lloyd R, Beals E, Beierwaltes WH, Thompson NW (1987) Iodine-131 MIBG scintigraphy of neuroendocrine tumors other than pheochromocytoma and neuroblastoma. J Nucl Med 28:979-988

Wick MM, Byers L. Frei E (1977) L-dopa selective toxicity for melanoma cells in vitro. Science 1967:468-469.

Morrison ME, Yagi MJ, Cohen G (1985) In vitro studies of 2,4-dihydroxyphenylalanine, a prodrug targeted against malignant melanoma cells. Proc Natl Acad Sci USA 82:2960-2964.

Wieland DM, Wu JI, Brown LW, Manger TJ, Swanson DP, Beierwaltes WH (1980) Radiolabeled adrenergic neuronblocking agents: Adrenomedullary imaging with (^{131}I)-iodobenzyl-guanidine. J Nucl Med 21:349-353

Wieland DM, Brown LE, Tobes MC, Rogers WL, Marsh DD, Manger TJ, Swanson DP, Beierwaltes WH (1981) Imaging the primate adrenal medulla with [123] and [131] metaiodo-benzylguanidine: Concise communication. J Nucl Med 22:358-364

Yost DA, Moss J (1983) Amino acid-specific ADP-ribosylation. J Biol Chem 258:4926-4929

THE SYNAPTIC VESICLE VESAMICOL (AH5183) RECEPTOR CONTAINS A LOW AFFINITY ACETYLCHOLINE BINDING SITE

B. A. Bahr and S. M. Parsons
Department of Chemistry and the Neuroscience Research Program
Institute of Environmental Stress
University of California
Santa Barbara, California   93106
United States of America

NATO ASI Series, Vol. H29
Receptors, Membrane Transport and Signal Transduction
Edited by A. E. Evangelopoulos et al.
© Springer-Verlag Berlin Heidelberg 1989

INTRODUCTION

Synaptic vesicles isolated from <u>Torpedo</u> <u>californica</u> electric organ exhibit active transport of acetylcholine (ACh). The transport system is composed of at least three components: a vacuolar-type ATPase thought to pump protons into the vesicle, an ACh transporter which utilizes the proton-motive-force to drive secondary active transport of ACh, and a receptor for l-<u>trans</u>-2-(4-phenylpiperidino)cyclohexanol (vesamicol, formerly AH5183) (Bahr and Parsons, 1986a). When the vesamicol receptor (vesamicol-R) of VP_1 cholinergic synaptic vesicles (SVs) is occupied, ACh transport is blocked noncompetitively with the absence of any concomitant effect on the total vesicular ATPase activity (Anderson et al., 1983; Bahr and Parsons, 1986b). This phenomenon of transport blockage is potentially of regulatory importance to neurotransmission since vesicular storage of neurotransmitter is required for evoked release and subsequent transynaptic signal transduction.

Vesamicol was originally discovered as a neuromuscular blocking agent with unique pharmacological characteristics (Marshall, 1970). The agent inhibits evoked quantal release of ACh from many cholinergic nerve terminal preparations, presumably as a secondary effect of ACh storage block and possibly by inhibition of SV mobilization to sites of release (Marshall and Parsons, 1987). The drug also inhibits nonquantal release of ACh from the motor nerve terminal, suggesting that nonquantal release is mediated by vesicular ACh transporters incorporated into the cytoplasmic membrane as a result of exocytosis (Edwards et al., 1985). The vesamicol-R is a major component of the cholinergic terminal having a local concentration of about 8 μM binding sites distributed throughout the cytoplasm. Quantitative autoradiography has shown the distribution of the vesamicol-R in brain to be highly correlated with other cholinergic terminal markers (Marien et al., 1987). The vesamicol-R has an unknown relationship with the ACh transporter, and its function <u>in</u> <u>vivo</u> is uncertain. We report here experiments designed to elucidate the relationship

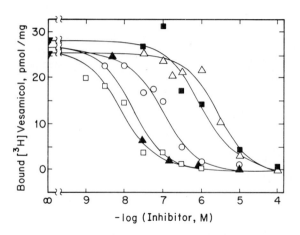

FIG. 1. Pharmacological characterization of the detergent solubilized vesamicol-R. Cholic acid-solubilized SVs (0.05 mg protein/ml) were equilibrated with 20 nM [^{3}H]vesamicol for 60 min at 23°C in the presence of the indicated concentrations of vesamicol analogues before the determination of bound ^{3}H by centrifugation-gel filtration (Bahr and Parsons, 1986a). A hyperbolic titration curve was fit to each data set by nonlinear regression analysis to obtain an IC$_{50}$ ± one standard deviation. The IC$_{50}$ values for [^{3}H]vesamicol binding inhibition by nonradioactive l-vesamicol (o), d-vesamicol (△), ±-trans-2-hydroxy-3-(4-phenylpiperidino)-trans-decalin (□), ±-trans-5-acetamido-2-hydroxy-3-(4-phenylpiperidino)tetralin (▲), or ±-N-(2-hydroxybutyl)-4-phenylpiperidine (■) were 90±20 nM, 2.8±0.8 µM, 8.8±2.4 nM, 17±7 nM, and 700±300 nM, respectively. The IC$_{50}$ values for ACh transport inhibition by these analogues were shown to be 20 nM, 0.5 µM, 10 nM, 50 nM, and 300 nM, respectively (Rogers et al., 1988).

between the vesamicol-R and a newly discovered low affinity ACh binding site that mediates the inhibition of [^{3}H]vesamicol binding (Noremberg and Parsons, 1988; Parsons et al., 1988).

PHARMACOLOGICAL COMPARISON BETWEEN THE VESICULAR AND DETERGENT-SOLUBILIZED VESAMICOL-Rs

The effects of detergent solubilization on the properties of the vesamicol-R were studied. The vesicular ACh transport pharmacology recently was well characterized with respect to vesamicol structure-function relationships (Rogers et al., 1988). Cholic acid solubilization of the vesamicol-R did not

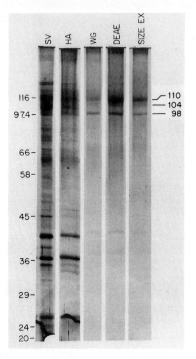

FIG. 2. Partial purification of the vesamicol-R. Samples from each chromatographic step were subjected to SDS-PAGE. The columns were hydroxylapatite (HA lane, 3 μg protein), agarose-wheat germ lectin (WG, 3 μg), DEAE ion-exchange (DEAE, 3 μg), and size exclusion (SIZE EX, 2 μg); SVs (3 μg) are shown in the SV lane. M_r standards ranged from 116 to 20 kDa. The partially purified receptor has prominent bands at 110 and 98 kDa, with suspected proteolytic fragments variably appearing at 104 and 42 kDa. Also, a diffuse glycoprotein-like material appears above 100 kDa.

change the degree of enantioselectivity expressed by the receptor (Figure 1). The saturable vesamicol-R is present on SVs at 250-400 pmoles binding sites per mg SV protein, or 5-8 sites/SV. The maximum [^{3}H]vesamicol binding was unchanged with solubilization in the presence of phosphatidylcholine (1 mg/ml) and glycerol (20%, wt./vol.) to stabilize the receptor. Various vesamicol analogues had similar apparent dissociation constants (IC_{50}), spanning almost four orders of magnitude, whether measuring their effects on intact or solubilized vesicles (Figure 1). Solubilization did not alter the characteristic first order dissociation of bound [^{3}H]vesamicol, although cholic acid (1%, wt./vol.) increases the dissociation rate by about 30% at 23°C (data not shown).

INHIBITION OF VESAMICOL BINDING BY ACh IS RETAINED THROUGH RECEPTOR PURIFICATION

The vesamicol-R was partially purified from Torpedo VP$_1$ SVs using a succession of chromatographic steps with a 14%

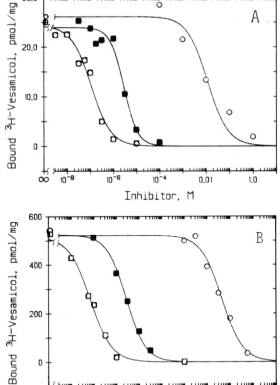

FIG. 3. Inhibition of [3H]vesamicol binding to cholate-solubilized SVs (A; 50 μg/ml) or partially purified vesamicol-R (B; 3.75 μg/ml). Samples were equilibrated with 20 nM [3H]vesamicol for 60 min at 23°C in the presence of the indicated concentrations of inhibitors. Bound [3H] was assessed by centrifugation-gel filtration. Nonradioactive l-vesamicol (□), d-vesamicol (■), and ACh (o) blocked binding to solubilized SVs with IC_{50} values of 90±20 nM, 2.8±0.8 μM, and 7.7±3.1 mM, respectively. The three inhibitors blocked binding to the purified receptor with IC_{50} values of 73±12 nM, 2.5±0.3 μM, and 37±9 mM, respectively. Endogenous ACh esterase was inhibited with 0.15 mM diethyl-p-nitrophenylphosphate prior to the experiment.

yield (Figure 2). The binding activity was increased about 40-fold for the native, detergent-solubilized vesamicol-R. It has a minimal apparent particle M_r of about 245 kilodaltons (kDa). The high affinity for [3H]vesamicol and the high degree of enantioselectivity were retained by the purified receptor (Figure 3). Exogenous ACh has only a small effect on [3H]vesamicol binding to intact vesicles but inhibits the binding to hyposmotically-shocked (Noremberg and Parsons, 1988) or solubilized SVs. This antagonistic effect by ACh was also retained by the vesamicol-R through purification (Figure 3). In contrast to the ATP dependence of [3H]ACh transport by SVs,

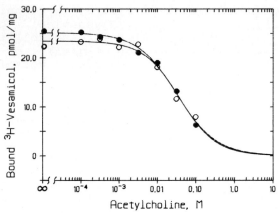

FIG. 4. ATP-independent ACh antagonism of [³H]vesamicol binding. Cholate-solubilized SVs (50 μg/ml) were equilibrated with 20 nM [³H]vesamicol and the indicated concentrations of ACh for 60 min at 23°C in the absence (o) or presence (•) of an ATP regeneration system (10 mM MgATP, 10 mM potassium phosphoenolpyruvate, and 30 μg/ml pyruvate kinase). Bound ³H was assessed and ACh esterase was inhibited as in Figure 3. ACh blocked binding with an IC_{50} of 36±6 mM in the absence and 32±4 mM in the presence of the ATP regeneration system.

ATP had no effect on the ability of ACh to inhibit [³H]vesamicol binding (Figure 4).

The competitive or noncompetitive nature of the interaction between ACh and vesamicol binding was probed as shown in Figure 5. [³H]Vesamicol binding could be completely inhibited by ACh at a low vesamicol concentration, but it was nearly completely insensitive at a 50-fold higher vesamicol concentration. Equations for competitive and noncompetitive equilibrium binding models were fit to the data in order to compare the apparent dissociation constants (K_{d-bi} and K_{d-ter}, respectively) for [³H]vesamicol with respect to the binary (receptor-vesamicol) and ternary (receptor-vesamicol-inhibitor) complexes. The noncompetitive model tends to fit the ACh data (Figure 5A) statistically better than the competitive model, but it was impossible to confirm the ACh inhibitory mechanism due to the experimental ACh concentration limit. The noncompetitive models fit to ACh antagonism data (IC_{50} = 5-35 mM) from three SV preparations had a K_{d-bi} range of 14-49 nM and a K_{d-ter} range of 100-500 nM. As a control for the

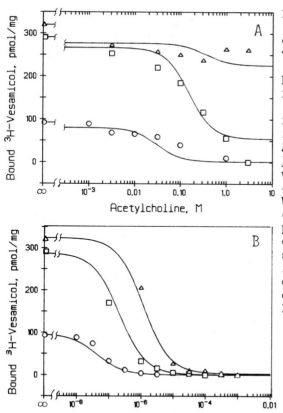

FIG. 5. The effect of [³H]vesamicol concentration on the ACh antagonism of the vesamicol-R. SV ghosts (0.13 mg/ml) were equilibrated with 20 nM (o), 200 nM (□), or 1 μM (Δ) [³H]vesamicol for 60 min at 23°C in the presence of the indicated concentrations of ACh (A) or nonradioactive l-vesamicol (B). Bound ³H was determined by applying 40 μl of each sample onto Whatman 1.2 cm GF/F filters coated 2 hr with 0.5% polyethyleneimine, then counted with liquid scintillation spectroscopy in 10 ml of aqueous counting cocktail. Endogenous ACh esterase was inhibited as in Figure 3.

behavior of a pure competitive inhibitor, a noncompetitive model fit to the nonradioactive l-vesamicol data (Figure 5B) had a K_{d-bi} and a K_{d-ter} of 35 nM and 0.5 mM, respectively.

DISCUSSION

The properties of the vesicular ACh transport- vesamicol-R system in vitro are complex and appear to involve regulatory phenomena. The vesamicol-R probably contains a binding site for ACh since the receptor can be partially purified with retention of its characteristic pharmacology including the ACh antagonism. ACh may be acting allosterically to block [³H]vesamicol binding. However, the ACh site in the vesamicol-R is not the same as the ACh transport site. The K_d

for ACh binding to the receptor is 5-35 mM, whereas the K_d for ACh binding to the transport site is 250±90 μM (n=6).

A minimal model which does not invoke a new ACh binding protein requires that both ACh binding sites be located in the ACh transporter. The low affinity site could correspond to the inwardly oriented transporter, which in the intact vesicle must release ACh in the presence of high concentrations of ACh. In vivo, such an internal ACh "signal" could control the conformation of the externally oriented vesamicol-R, thus reporting to the nerve terminal which vesicles are fully loaded with ACh and ready for release. This model suggests that the ACh transporter and the vesamicol-R are closely associated, but not necessarily within the same polypeptide. Presently, we are working toward a test of whether the vesamicol-R is the ACh transporter.

ACKNOWLEDGEMENTS

This work was supported by grant NS15047 from the National Institute of Neurological and Communicative Disorders and Stroke of the United States Public Health Service, and a grant from the American Muscular Dystrophy Association. We thank Catherine Fehlmann and Daniel Ray Oros for preparation of synaptic vesicles and Dr. Gary A. Rogers for synthesis of vesamicol and vesamicol analogues.

REFERENCES

Anderson DC, King SC, Parsons SM (1983) Pharmacological characterization of the acetylcholine transport system in purified Torpedo electric organ synaptic vesicles. Molec Pharmacol 24:48-54
Bahr BA, Parsons SM (1986a) Demonstration of a receptor in Torpedo synaptic vesicles for acetylcholine storage blocker l-trans-2-(4-phenyl[3,4-^{3}H]piperidino)cyclohexanol. Proc Natl Acad Sci USA 83:2267-2270
Bahr BA, Parsons SM (1986b) Acetylcholine transport and drug inhibition kinetics in Torpedo synaptic vesicles. J Neurochem 46:1214-1218

Edwards C, Dolezal V, Tucek S, Zemkova H, Vyskocil F (1985) Is an acetylcholine transport system responsible for nonquantal release of acetylcholine at the rodent myoneural junction? Proc Natl Acad Sci USA 82:3514-3518

Marien MR, Parsons SM, Altar CA (1987) Quantitative autoradiography of brain binding sites for the vesicular acetylcholine transport blocker 2-(4-phenylpiperidino) cyclohexanol (AH5183). Proc Natl Acad Sci USA 84:876-880

Marshall IG (1970) Studies on the blocking action of 2-(4-phenylpiperidino)cyclohexanol (AH5183). Br J Pharm 38:503-516

Marshall IG, Parsons SM (1987) The vesicular acetylcholine transport system and its pharmacology. Trends in Neurosci 10:174-177

Noremberg K, Parsons SM (1988). Regulation of the vesamicol receptor in cholinergic synaptic vesicles by acetylcholine and an endogenous factor. J Neurochem, in press.

Parsons SM, Noremberg K, Rogers GA, Gracz LM, Kornreich WD, Bahr BA, Kaufman R (1988) Complexity and regulation in the acetylcholine storage system of synaptic vesicles. Cellular and Molecular Basis of Neuronal Signalling (Synaptic Transmission), H Zimmermann (ed) Springer-Verlag, Berlin, 325-335

Rogers GA, Parsons SM, Anderson DC, Nilsson LM, Bahr BA, Kornreich WD, Kaufman R, Jacobs RS, Kirtman B (1988) Synthesis, acetylcholine storage blocking activities, and biological properties of derivatives and analogues of trans-2-(4-phenylpiperidino)cyclohexanol(vesamicol; AH5183). Submitted

PURIFICATION OF THE D-2 DOPAMINE RECEPTOR AND CHARACTERIZATION OF ITS SIGNAL TRANSDUCTION MECHANISM.

Zvulun Elazar, Gabriela Siegel, Hannah Kanety and Sara Fuchs.
Department of Chemical Immunology.
The Weizmann Institute of Science.
Rehovot 76100, Israel.

Dopaminergic pathways appear to be involved in many brain functions including motoric, mental and emotional states. Dopamine is also associated with controlling pituitary prolactin and α melanocyte-stimulating hormone release. Dopamine receptors have been divided into D-1 and D-2 subclasses according to their relationship to the adenylate cyclase system and their affinity for dopaminergic agonists and antagonists (reviewed in Seeman, 1980; Niznik, 1987). The D-1 receptor stimulates adenylate cyclase and is virtually insensitive to substituted benzamide neuroleptics, whereas the D-2 receptor inhibits adenylate cyclase activity and has picomolar or nanomolar affinities for all neuroleptics. The D-2 receptor has been solubilized from membranes of pituitary gland and striatum by use of various detergents, and its pharmacology is well defined. This receptor is a glycoprotein as indicated by its specific adsorption to and elution from wheat germ agglutinin-Sepharose. Photoaffinity labeling of the D-2 dopamine receptor has indicated that, in a variety of mammalian tissues and species, a 92-94 kDa polypeptide is probably the ligand binding unit of the receptor (Lew et al.,1985; Amlaiky and Caron, 1986; Kanety and Fuchs, 1988). In order to elucidate the molecular mechanisms by which dopaminergic signal transduction correlates with dopaminergic functions, it is essential to isolate and fully characterize this receptor.

In investigating the signal transduction mechanism of the D-2 receptor system, high and low affinity states of the receptor for dopaminergic agonists have been demonstrated (Sibley and Creese, 1983; Simmonds et al., 1986). Interconversion between the high and low affinity states was shown to be regulated by guanine nucleotides and to be sensitive to pertussis toxin , suggesting a receptor-Gi/Go coupling (Seeman et al., 1985 ; Niznik, 1987). Alternative pathways for signal transduction mediated by D-2 receptor , apart from inhibition of adenylate cyclase, were postulated . It was shown that in anterior pituitary dopaminergic inhibition of prolactin release is associated with Ca^{2+} mobilization (Delebeke and Dannies, 1985 ; Judd et al., 1985), an event possibly mediated by inhibition of phospholipase C activity. This regulation of prolactin release is blocked by pertussis toxin treatment (Cote et al. 1984), indicating the involvement of a G protein in this process. Several pertussis toxin sensitive G proteins may mediate these events (Katada et al., 1987). Recently Senogles et al. (1987) have identified a Go-related protein to be associated with the D-2 receptor in the anterior pituitary . In some neuronal systems D-2 receptor signal transduction has been shown to be cAMP independent (Stoof et al., 1986) and to involve inhibition of PI turnover and Ca^{2+} mobilization (Pizzi et al., 1987). In another neuronal system, the D-2 receptor has been shown to be associated with a decrease in voltage dependent Ca^{2+} currents (Harris-Warrick et al., 1988). The involvement of a pertussis toxin sensitive G protein has been suggested in these systems.

In this report we describe a purification procedure of the D-2 dopamine receptor from bovine striatum yielding a preparation which exhibits a single major band on SDS-polyacrylamide gel electrophoresis, with an apparent molecular weight of 92 kDa and a specific binding activity of 2490 pmoles spiperone per mg protein. Furthermore we have characterized two pertussis toxin sensitive G proteins which copurify with the D-2 dopamine receptor from striatal membranes following affinity chromatography on affi-gel-HGE. These G proteins are immunologically related to brain Gi and Go.

PURIFICATION OF THE D-2 RECEPTOR

The first step of the purification of the D-2 receptor was an affinity chromatography of CHAPS solubilized striatal membranes on an adsorbent of a dopaminergic antagonist. An analog of haloperidol, haloperidol glycine ester (HGE; Fig 1) (Kanety et al., 1988) was coupled to affi-gel 10. HGE inhibited the binding of spiperone to striatal membranes with a K_D value of 30 nM, in comparison with a K_D of 3 nM for haloperidol. Affi-gel-HGE was found to adsorb 70-80% of the spiperone binding activity and less then 5% of the total proteins from the CHAPS solubilized receptor preparation. Adsorption of the [^{3}H]-spiperone binding was inhibited by preincubation of the solubilized receptor preparation with spiperone or (+)-butaclamol, as well as by other dopaminergic antagonists and agonists (Fig 2). This profile demonstrates that the affinity matrix displays the predicted specificity for the D-2 dopamine receptor.

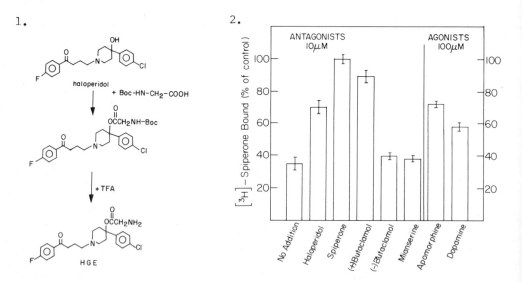

Fig. 1: Synthesis of haloperidol glycine ester.

Fig. 2: Specificity of adsorption of solubilized D-2 dopamine receptor to affi-gel-HGE (From Elazar et al., 1988a).

Approximately, 0.1% of the total protein applied to the affi-gel-HGE matrix was eluted with 10 μM spiperone (Table I; Elazar et al.,1988a). The eluted material (eluate I) had to be reconstituted by insertion into artificial phosphatidylcholine liposomes before assaying for [^{3}H]-spiperone binding activity. This step resulted in a 500 fold purification, and about 40% of the binding activity applied to the affi-gel-HGE was recovered.

TABLE I: Purification of D-2 dopamine receptor from bovine striatum

Step	Volume (ml)	Protein conc. (mg/ml)	Specific activity (pmole[^{3}H]-spiperone/ mg protein)	Purification (fold)	Recovery (%)
Membranes	25	10	0.116	1	100
CHAPS extract	25	2.45	0.20	1.72	42
Affi-gel-HGE (eluate I)	5	0.01	96.7[a]	833	17
Sephacryl	4.2	0.002	330[a]	2840	9.6
Sepharose-WGA (eluate II)	0.5	<0.0002	2490[a]	21460	0.9

[a] [^{3}H]-spiperone binding was measured following reconstitution into phosphatidylcholine liposomes.

Radioiodinated eluate I exhibited several protein bands on SDS-PAGE; however, mainly 92 kDa and 140 kDa protein bands were depleted following preincubation of the solubilized receptor with spiperone , prior to application on the affi-gel-HGE (Elazar et al., 1988a). The nature of the 140 kDa polypeptide is unknown. It might represent a complex of the 92 kDa protein with another polypeptide or a precursor of the D-2 receptor binding unit. Some of the other proteins which elute with spiperone at this step might represent functional components, e.g. G-proteins, which may be specifically coupled to the receptor complex (see below).

The affinity purified receptor preparation (eluate I) was further chromato-graphed on a Sephacryl S-300 column, resulting in an additional enrichment of the receptor binding activity (Table I). Fractions which contained the ligand binding activity and the 92 kDa protein band, were pooled, concentrated and loaded onto Sepharose-wheat germ agglutinin (Sepharose-WGA). Only a 92 kDa polypeptide was adsorbed to this adsorbent and was specifically eluted with N-acetylglucosamine . This eluted material was designated eluate II and when assayed for [³H]-spiperone binding activity, following reconstitution into artificial vesicles, it bound 2490 pmoles spiperone per mg protein (Table I and Elazar et al., 1988a). This binding was specifically inhibited by excess of either spiperone, haloperidol or (+)-butaclamol. In addition, samples of radiolabeled eluate II were shown to specifically readsorb to affi-gel-HGE (Fig. 3). Also, the pure receptor preparation (eluate II) comigrated on SDS-PAGE with a protein band which resulted from photoaffinity labeling of the D-2 dopamine receptor in striatal membranes with azido-haloperidol (Fig 3 ; Kanety and Fuchs, 1988).

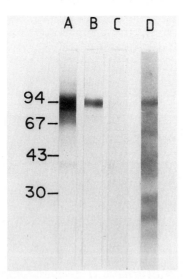

Fig. 3: Readsorption of pure receptor preparation on affi-gel-HGE. A, eluate II; B, eluate obtained following readsorption of eluate II on affi-gel-HGE; C, eluate obtained following readsorption of eluate II on affi-gel-ethanolamine; D, immunoblot of photoaffinity labeled dopamine receptor with anti-haloperidol antibodies (see Kanety and Fuchs, 1988).

ASSOCIATION OF TWO PERTUSSIS TOXIN SENSITIVE G-PROTEINS WITH THE D-2 RECEPTOR

In studying the signal transduction mechanism of the D-2 dopamine receptor it was demonstrated that the CHAPS solubilized receptor exhibits high and low affinity states for dopaminergic agonists. Guanine nucleotides and pertussis toxin can convert the solubilized receptor from a high affinity state to a low one (Elazar et al., 1988b). This suggests that the solubilized receptor is functionally coupled to one or more pertussis toxin sensitive G-proteins. Moreover, following affinity chromatography on affi-gel-HGE, the resulting receptor preparation (eluate I) exhibits, upon reconstitution, GTPase activity which is stimulated about 2 fold by the dopaminergic agonist N-propylapomorphine (NPA) (Fig 4A). This NPA-induced GTPase activity was inhibited by pertussis toxin with NAD (data not shown). NPA did not have any effect on the GTPase activity of eluate I that had not been reconstituted (Fig 4B). These data indicate that the affinity purified D-2 receptor preparation contains pertussis toxin sensitive G proteins, functionally coupled to the receptor .

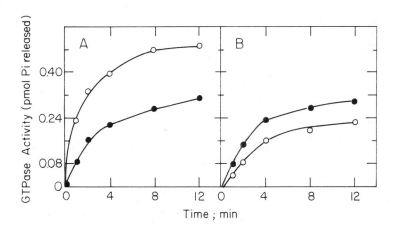

Fig. 4: GTPase activity of affinity purified D-2 receptor . Reconstituted affinity purified D-2 receptor (A) or non reconstituted preparation (B) were assayed for GTPase activity in the absence (●) or the presence (O) of 10 uM NPA.

In order to identify the G-proteins which copurify with the D-2 receptor upon affinity chromatography on affi-gel-HGE, eluate I was [^{32}P]-ADP ribosylated by pertussis toxin . Chromatography on SDS-polyacrylamide gel resulted in two labeled polypeptide bands of 39 kDa and 41 kDa (Fig 5). Preincubation of the solubilized membranes with spiperone prior to affinity chromatography , a treatment which has been shown to inhibit the binding of the receptor to the affinity column (see Fig. 2), resulted in a dramatic decrease in the amount of both labeled bands (data not shown). Similarly, pretreatment of the solubilized membranes with Gpp(NH)p also reduced the amount of the bound [^{32}P]-ADP. These data strongly suggests that two G proteins are specifically associated with the D-2 receptor and that this receptor-G complexes can be functionally dissociated by guanine nucleotides.

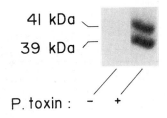

Fig. 5: Pertusis toxin mediated ADP ribosylation of affinity purified D-2 receptor preparation (eluate I).

Further characterization of the G-proteins which are associated with the D-2 receptor came from immunoblotting experiments employing specific anti G-protein antibodies, and from proteolytic fragmentation of the G-proteins. Immunoblotting experiments using specific anti-Gi and anti-Go antibodies (kindly provided by Dr. A. Spiegel) showed that the 41 kDa endogenous band is recognized exclusivly by anti-Gi antibodies while the endogenous 39 kDa band is recognized by anti-Go antibodies (data not shown). In addition, we have also compared peptide maps obtained from the endogenous G-proteins with those of brain specific Gi and Go proteins. For these experiments affinity purified

D-2 receptor and purified brain Gi/Go proteins were labeled with $[^{32}P]$-NAD and pertussis toxin. After electrophoresis of the labeled preparation on SDS-PAGE, the different bands were excised from the gel and digested in the stackinig gel of another SDS-PAGE, by either V-8 or bromelain for 1 hr and then were further electrophoresed and exposed for autoradiography. The peptide maps obtained from either V-8 or bromelain digestion indicate that the endogennous 41 kDa band is like brain Gi, whereas the 39 kDa band is like brain Go (Fig. 6).

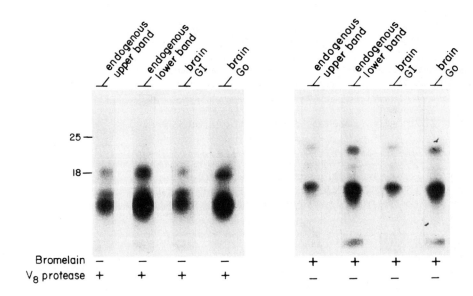

Fig. 6: Peptide map of $[^{32}P]$-ADP ribosylated affinity purified D-2 receptor and purified brain Gi/Go.

Acknowledgements: We wish to thank Mrs. S. Cannon for excellent technical assistance. This research was supported by grants from the Chief Scientist's Office, Ministry of Health, Israel, from Enosh and Harry Stern Foundations, from the Henry and Francoise Glasberg Foundation for Research on Cellular Diseases, and from the Rockfeller-Weizmann Collaboration Trust Fund.

REFERENCES

Amlaiky N, Caron MG (1986) J Neurochem 47: 196-204.

Cote TE, Frey EA, Sekura RD (1984) J Biol Chem 259: 8693-8698.

Delebeke D, Dannies PS (1985) Endocrinology 117: 439-446.

Elazar Z, Kanety H, David C, Fuchs S (1988a) Biochem Biophys Res Commun 156: 602-609.

Elazar Z, Siegel G, David C, Kanety H, Fuchs S (1988b) Soc Neurosci Abst 3778.2.

Harris-Warrick RM, Hammond C, Paupardin-Tritsch D, Homburger V, Rout B, Bockaert J, Gerschonteld HM (1988) Neuron 1: 27-32.

Judd AM, Koike K, Schettini G, Longin IS, Hewelett EL, Yosumoto T, MacLeod RM (1985) Endocrinology 117: 1215-1221.

Kanety H, Fuchs S (1988) Biochem Biophys Res Commun 155: 930-936.

Kanety H, Schreiber M, Elazar Z, Fuchs S (1988) J Neuroimmunol 18: 25-36.

Katada T, Oinuma M, Kusakabe K, Ui M (1987) FEBS Lett 213: 353-358.

Lew JY, Meller E, Goldstein M (1985) Eur J Pharmacol 113: 145-146.

Niznik HB (1987) Mol Cell Endocrinol 54: 1-22

Pizzi M, D'Agostini F, DaPrada M, Spano PF, Haefely WE (1987) Eur J Pharmacol 136: 236-269.

Seeman P (1980) Pharmacol Rev 32: 229-313.

Seeman P, Watanabe M, Grigoradis D, Tedesco JL, George SR, Svensson V, Nilsson JLG, Neumeyer JL (1985) Mol Pharmacol 28: 391-399.

Senogles SE, Benovic JL, Amlaiky N, Unson C, Milligan G, Vinitsky R, Spiegel AM, Caron MG (1987) J Biol Chem 262: 4860-4867.

Sibley DR, Creese I (1983) J Biol Chem 258: 4957-4965.

Simmonds SH, Strange PG, Hall AW, Taylor RJK (1986) Biochem Pharmacol 35: 731-735.

Stoof JC, Werkman TR, Lodder JC, De Vlieger ThA (1986) Trends Pharmacol Sci 7: 7-9.

DOWNREGULATION OF M1 AND M2 MUSCARINIC RECEPTOR SUBTYPES IN Y1 MOUSE ADRENOCARCINOMA CELLS

Nancy M. Scherer, Robert A. Shapiro, Beth A. Habecker, and Neil M. Nathanson.
Dept. of Pharmacology SJ-30,
Univ. of Washington,
Seattle, WA 98195.
USA

Muscarinic acetylcholine receptors are important in mediating neurotransmission in the central and peripheral nervous systems and regulate a broad spectrum of physiologic responses. Chronic exposure of cells to muscarinic agonists results in a decrease in receptor number and diminished coupling to effectors. We examined this phenomenon in 2 of the 5 described receptor subtypes: m1, one of the predominant subtypes expressed in brain, and m2, the major cardiac subtype. DNA clones for these receptors were transfected into the Y1 mouse adrenal cell line and a variant Y1 clone, Kin 8, in which cAMP-dependent protein kinase activity is greatly reduced. Transfected receptors could be expressed at high levels and were functional as determined by their ability to bind muscarinic ligands and stimulate phosphoinositide turnover (m1) or inhibit adenylyl cyclase (m2). We determined the susceptibility of the receptor subtypes to internalization after chronic exposure to the muscarinic agonist, carbachol, and as a consequence of activation of cAMP-dependent protein kinase and protein kinase C. These experiments suggest that the mechanisms for internalization of the m1 and m2 receptor subtypes differ. The m1, but not the m2, receptor was internalized in response to activation of protein kinase C and this internalization was dependent on the presence of a functional cAMP-dependent protein kinase. Activation of either protein kinase C or cAMP-dependent protein kinase did not mimic agonist-induced internalization of either m1 or m2. There was a large surplus of receptors for coupling to the effector enzymes in cells expressing several hundred fmol receptor per mg membrane protein. At low receptor numbers, agonist-induced desensitization was correlated with internalization of receptors.

Muscarinic acetylcholine receptors (mAChR) are anatomically widespread and functionally diverse. In the central nervous system, mAChR regulate long-term potentiation, a process considered integral to learning and memory (Williams et al., 1988) and are involved in arousal and control of movement. mAChR's mediate pre-ganglionic transmission in the autonomic nervous system and post-ganglionic transmission in the parasympathetic and some sympathetic neurons. Muscarinic stimulation decreases chronotropy and inotropy in the heart, increases smooth muscle contraction and increases secretion in glands innervated by the parasympathetic system (Nathanson, 1987).

As with many neurotransmitter receptors, the mAChR exists in multiple subtypes. Pharmacologically, brain, heart and glandular subtypes were identified based on their anatomical location and binding affinity for various muscarinic ligands. Until recently, it was not known whether this pharmacologic diversity arose from structural differences in the receptors or was

NATO ASI Series, Vol. H29
Receptors, Membrane Transport and Signal Transduction
Edited by A. E. Evangelopoulos et al.
© Springer-Verlag Berlin Heidelberg 1989

due to differences in the biochemical environment or cell type in which the receptor was expressed. Now, at least 5 distinct genes which code for mAChR subtypes have been identified (Kubo et al., 1986; Bonner et al., 1987, 1988; Peralta et al., 1987).

Current research aims to elucidate the function of the receptor subtypes. A cell or tissue often expresses more than one receptor subtype, complicating the correlation of function with structure. In addition, one mAChR subtype can couple to multiple effectors. The effectors to which mAChR's are known to couple include adenylyl cyclase, guanylyl cyclase, phospholipase C, and K^+, Ca^{2+}, and Cl^- channels (Nathanson, 1987). Transfection experiments show that the m1 and m3 subtypes stimulate phosphoinositide turnover, open a Ca^{2+}-dependent K^+ channel and inhibit a voltage-sensitive K^+ channel (the M current). The m2 and m4 subtypes inhibit adenylyl cyclase but do not affect these ion channels. The m5 receptor also stimulates phosphoinositide turnover (Ashkenazi et al., 1987; Bonner et al., 1988; Fukada et al., 1988; Peralta et al., 1988). This multiplicity of receptor subtypes has the potential advantage of permitting specificity of response to a common neurotransmitter. Specificity may occur at the tissue level, during ontogeny, or within a transduction pathway, as it has been suggested that m2 and m1 receptors may show preferential pre- and post-synaptic distributions (Barnard, 1988).

The mAChR's are part of a gene family of hormone and neurotransmitter receptors. These receptors consist of a single subunit containing 7 putative transmembrane regions and couple to effectors via G proteins. Members of this family identified to date include the opsin pigments (Nathans and Hogness, 1984), and the yeast α-factor (Nakayama et al., 1985), adrenergic (Dixon et al., 1986; Kobilka et al., 1987), acetylcholine (Kubo et al., 1986), substance K (Masu et al., 1987), serotonin (Fargin et al., 1988; Julius et al., 1988), angiotensin (Jackson et al., 1988) and cAMP receptors (Klein et al., 1988). It is important to understand the relationships between members of this gene family from both an evolutionary standpoint and a pragmatic one, as this knowledge will aid in the characterization and treatment of pathologic conditions.

Neurotransmitter receptors adapt to chronic stimulation by showing a diminished response to a constant stimulus. This process is called desensitization and often, though not necessarily, involves internalization of the receptors (Sibley et al., 1987b). In this paper we attempt to identify factors affecting internalization and desensitization of the m1 and m2 receptors. Receptors were transfected into the Y1 cell line which lacks endogenous mAChR's and a variant Y1 clone, Kin 8, in which the regulatory type 1 subunit of cAMP-dependent protein kinase (PK A) is defective resulting in kinase activity being reduced 600 fold relative to Y1 cells (Rae et al., 1979; Clegg et al., 1987). The Kin 8 clone was used to assess the contribution of PK A to internalization of the mAChR's. This system has several advantages. The receptor subtype is defined, and the receptors can be expressed at high levels which facilitates biochemical studies. Also, receptor level can be varied within a single subclone and the receptor subtypes are studied in a defined cell line thereby eliminating differences in receptor function that may arise from different biochemical environments.

Downregulation of several receptors in this gene family is correlated with phosphorylation of the receptor. Phosphorylation modulates the function of rhodopsin (Palczewski et al., 1988), the α_1 and β_2 adrenergic receptors (Bouvier et al., 1987, 1988; Sibley et al., 1987a), the cAMP receptor (Klein et al., 1988) and the yeast α-factor receptor (Reneke, 1988). Several studies suggest that phosphorylation also regulates the function of the muscarinic receptors. Phosphorylation of the m1 receptor by cAMP-dependent protein kinase (PK A) and protein kinase C (PK C) (Haga et al., 1988), and the m2 receptor by PKA (Rosenbaum et al., 1987) has been demonstrated *in vitro* . The mAChR isolated from porcine brain synaptic vesicles is phosphorylated by PK A and phosphorylation is associated with loss of agonist binding (Ho et al., 1987). In *in vivo* studies, agonist stimulation is correlated with activation of PKC and internalization of the mAChR in N1E-115 cells (Liles et al., 1986). In human salivary cells, PK C activation inhibits muscarinic-dependent Ca^{2+} fluxes (He et al., 1988). The chick cardiac receptor is phosphorylated *in vivo* in response to stimulation by carbachol and phosphorylation is correlated with a decreased affinity for agonist and attenuation of negative inotropy (Kwatra et al., 1987, 1988). In contrast to the *in vitro* study of the porcine m2 receptor, elevation of cAMP levels in chick atria did not result in phosphorylation of the cardiac receptor.

Although these studies suggest that phosphorylation may regulate the function of the mAChR, the role of phosphorylation *in vivo* of known receptor subtypes is unclear. *In vitro* studies may be faulted because of the high levels of kinases needed to achieve phosphorylation and non-physiologic conditions. The receptor type(s) present in tissues or cell lines is uncertain. Our approach has the advantage of determining the function of known receptor subtypes in a defined cell line.

In this paper we demonstrate that the m1 and m2 receptors: 1) differed in their susceptibility to internalization by PK C activation, 2) were not internalized by PK A activation, 3) differed in the kinetics of agonist-induced internalization and, 4) agonist-induced desensitization was correlated with internalization of receptors at low receptor levels. Agonist-induced internalization of m1 or m2 did not appear to involve PK A or C, suggesting that a specific kinase may be responsible, as has been demonstrated for rhodopsin and the β adrenergic receptors (Benovic et al., 1987; Palczewski et al., 1988).

Methodology

A genomic clone for m1 receptor from mouse brain and a cDNA clone for m2 receptor from pig atria were inserted into the ZEM 228 expression vector. Y1 and Kin 8 cells were grown in F-10 medium (Gibco) supplemented with 15% fetal calf serum, penicillin G (100 units/ml) and streptomycin sulfate (0.1 mg/ml) in 5% CO_2. The $CaPO_4$ method was used for transfection and individual colonies were subcloned. The vector contains a neomycin phosphotransferase resistance gene, allowing selection of transformants by growing the cells in the presence of 500 μM Geneticin. Expression of the receptor was driven by the metallothionein-1 promoter which was induced by exposure of the cells to 120 μM $ZnSO_4$ for 24 h.

Total receptor number was determined by binding of the muscarinic ligand, quinuclidinyl benzilate (^{3}H-QNB) to membrane homogenates, and surface receptors by N-methyl scopolamine (^{3}H-NMS) binding to intact cells. ^{3}H-NMS binding was determined on cells grown to 70 - 90% confluency in 24-well plates. Five wells were averaged per treatment and a sixth well contained 3 μM atropine to measure non-specific binding. Cells were exposed to ligands in F-10 media for 1h at 37°, rinsed with phosphate- buffered saline and incubated in 0.25 ml buffer containing 290 fmol ^{3}H-NMS for 4 h at 4°. Cells were scraped, filtered through GF/C paper, and counted. cAMP levels were determined by the method of Gilman (1970). Total inositol phosphates were isolated using ion exchange chromatography as described by Masters et al. (1984). Protein was determined by the method of Lowry et al. (1951).

Expression of Receptors

The m1 and m2 receptors were expressed in Y1 mouse adrenocarcinoma cells and a variant clone, Kin 8, in which the regulatory type 1 subunit of PK A is defective, resulting in a 600 fold decrease in kinase activity relative to Y1 cells. In the following report the receptor type is indicated first (M1 or M2) followed by the clone (#x) and a Y1 for wild type cells or "K" for the Kin 8 line. At least 2 clones were tested for each cell line to assess consistency of responses. Cell lines were generally tested at high and low receptor numbers by using cells exposed, or not exposed, to $ZnSO_4$. Figure 1 shows the receptor levels expressed by various clones as measured by ^{3}H-QNB and ^{3}H-NMS binding. When possible, clones were chosen for testing which expressed a similar number of receptors. The M1#7K clone expressed receptors at a very high level compared to the other clones and this will be shown to affect internalization. The m2 receptor was very poorly expressed in Y1 cells. Poor expression is unlikely to be due to the particular construct used because the same construct was well expressed in Kin 8 cells. It is tempting to speculate that expression of m2 is affected by PK A activity. Regulation might arise from a cAMP-responsive control element affecting transcription, as has been shown for other genes (Deutsch et al., 1988), instability of the mRNA, or PK A-dependent post-translational modification of the receptor.

A

B

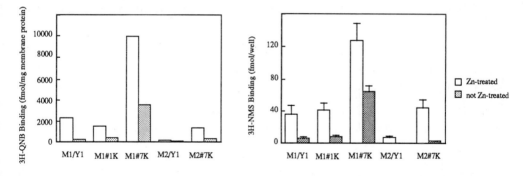

Figure 1 Typical receptor levels in Y1 and Kin 8 cells are shown as measured by the membrane-permeable muscarinic ligand, ^{3}H-QNB on membrane homogenates(A) or the impermeant ligand, ^{3}H-NMS on intact cells(B). Where indicated, cells were tested after exposure to 120 µM ZnSo$_4$ to increase receptor levels. QNB data represent duplicate determinations after subtraction of non-specific binding measured in the presence if 3 µM atropine. NMS data represent the mean of 10 to 30 experiments after subtraction of non-specific binding.

Transfected Receptors are Functional

Stimulation of the m1 receptor in Y1 cells resulted in a 30-40 fold increase in phosphoin-

ositde turnover, indicating m1 activates phospholipase C (figure 2). This increase was unaffected by incubating the cells for 18h with 75 ng/ml pertussis toxin which would ADP-ribosylate and inactivate G_i-type and G_o proteins. This suggests that the m1 receptor couples to phospholipase C by a pertussis toxin-insensitive G protein. Stimulation of the m1 receptor did not reduce forskolin-stimulated cAMP levels, instead, levels were increased. This may reflect Ca^{2+}/calmodulin-dependent activation of adenylyl cyclase subsequent to an inositol trisphosphate-induced increase in intracellular calcium (Shapiro et al., 1988).

Stimulation of the m1 receptor in Y1 cells decreased forskolin-stimulated cAMP accumulation by 30%. Inhibition was reversed by exposure of the cells to pertussis toxin, suggesting coupling was mediated by a G_i-type protein. We could not detect G_o in Y1 cells by immunoblotting (Shapiro et al., 1988), in agreement with the report of Heschler et al. (1988).

Transfected Receptors Can Be Internalized

We wished to determine whether chronic exposure to the muscarinic agonist, carbachol, or activation of PK A or PK C resulted in internalization of receptors as measured by ^{3}H-NMS binding. We also tested whether agonist-induced internalization involved activation of one of these kinases.

A   **B**

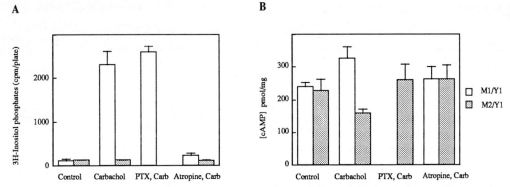

Figure 2 The ability of m1 and m2 receptors to mediate phosphoinositide turnover and cAMP accumulation in Y1 cells is shown. Assays were performed as described in the text. In the phosphoinositol experiment the receptor level of m1 was 215 fmol/mg membrane protein (cells not exposed to $ZnSO_4$) and m2 was 860 fmol/mg membrane protein (Zn-treated cells). In the cyclase experiment the receptor level of m1 was 1000 fmol/mg membrane protein (Zn-treated cells) and m2 was 214 fmol/mg membrane protein (Zn-treated cells). Where indicated, cells were exposed to 75 ng/ml pertussis toxin for 18 h prior to testing; or the muscarinic antagonist, atropine (1 μM), was added prior to stimulation by carbachol to determine whether the effect was specific. Data are presented as the mean ± std. dev. and representative of 2 to 6 experiments.

Agonist-Dependent Internalization

Exposure of cells to 1 mM carbachol for 1 h resulted in loss of about 50% of either m1 or m2 receptors from the surface of Y1 cells. The m2 receptor in Kin 8 cells was even more sensitive to agonist-induced internalization than in Y1 cells, with a loss of 85% of surface receptors after exposure to carbachol. In contrast, internalization of the m1 receptor was less in Kin 8 cells than Y1 cells, with a 34% decline in surface receptors. The m1 receptor in the M1#7K clone, expressed at about 5 fold greater level than the M1#1K clone, was not internalized in response to carbachol. This attenuation may reflect saturation of the internalization mechanism at high receptor levels. The M1#7K clone tested at lower receptor levels (cells not exposed to $ZnSO_4$) internalized receptors to a similar extent as in the M1#1K clone, supporting the interpretation that receptor number can limit the efficiency of internalization. A similar observation was made by Benveiste et al. (1988), who reported that over-expression of the EGF receptor in 3T3 cells decreased the rate of endocytosis of receptors. They proposed that endocytosis was limited by the number of clathrin vesicles, however Payne et al. (1988) reported that internalization of the yeast α-factor receptor in clathrin-deficient mutants still occurred at 30-50% of the level seen in the wild type. As the mechanism of receptor internalization is unknown, it is unclear why internalization is reduced in cells expressing large numbers of receptors.

Kinase-Dependent Internalization

The m1 and m2 receptors differed in their response to PK C activation by phorbol 12-myristate 13-acetate (PMA). There was no loss of surface receptors after 1h exposure to 10^{-7} M

PMA in M2/Y1 cells. In contrast, exposure of M1/Y1 cells to PMA resulted in a 30% loss of surface receptors. The biologically inactive 4α-phorbol-12,13 dideconoate was without effect. PK C activation also results in internalization of the β_2-adrenergic receptor (Kassis et al., 1985; Bouvier et al., 1987). Surprisingly, the PK C effect was greatly attenuated or blocked in Kin 8 cells, suggesting that PK A activity is required for PK C-dependent internalization. This result is particularly interesting in light of the report of Estensen et al. (1983) that PK C stimulated steroid secretion and plasminogen activator production in Y1, but not in Kin 8 cells.

Elevation of cAMP levels in Y1 cells by treatment with 8-(4-chlorophenylthiol)adenosine 3':5"-cyclic monophosphate or forskolin did not alter the number of either m1 or m2 receptors, suggesting that PKA activation can not directly induce internalization.

Agonist-induced Internalization was Independent of PK C Activation

The observation that agonist-induced internalization of m1 occurred in Kin 8 cells, although PK C-dependent internalization did not, suggests that the mechanisms of internalization are different. This interpretation was supported in further experiments. Downregulation of PK C by overnight exposure of cells to 10^{-7}M PMA did not attenuate carbachol-induced internalization. Moreover, the rate of internalization of m1 in Y1 cells after exposure to carbachol was faster than internalization due to PK C, with a $t_{1/2}$ of about 2 min compared to about 8 min. Internalization of the β_2-adrenergic receptor in response to agonist also appears occur by a different mechanism than PK C-induced internalization (Kassis et al., 1985).

The Kinetics of Agonist-induced Internalization of m1 and m2 Differ

Carbachol-dependent internalization was further characterized by examining the kinetics of internalization of the m1 receptor in Kin 8 cells and the m2 receptor in Y1 and Kin 8 cells. The m1 receptor was rapidly internalized in both the M1#1K and M1#7K clones, reaching a maximum within 5 min exposure to carbachol. In contrast, the m2 receptor was internalized more slowly, with the maximum reached after 30 min in Y1 cells and 15 min in Kin 8 cells. These data suggest that the m1 and m2 receptor subtypes are internalized by different mechanisms.

There appears to be a complex relationship between the rate and extent of receptor internalization and receptor number. We noted previously that the internalization mechanism was blocked in cells expressing high receptor numbers (5 000-10 000 fmol/mg membrane protein). However, the efficiency of internalization is not strictly dependent on receptor number. Receptors in the M1#1 clone are internalized to a greater extent in cells expressing low receptor numbers (not treated with $ZnSO_4$), than 3 times more receptor (Zn-treated). In contrast, receptors in Zn-treated M2#7 cells are internalized to a similar extent as receptors in cells not exposed to $ZnSO_4$ expressing one fourth as many receptors.

Differences in Receptor Number Affect Coupling to Effectors

Studies on cells expressing different receptor numbers due to treatment with $ZnSO_4$ or carbachol-induced internalization suggested that there is a large reserve of spare receptors for coupling to effectors in cells which express hundreds of fmol receptors per mg membrane protein. Coupling of m1 to phosphoinositide turnover after stimulation by the muscarinic agonist, pilocarpine, was less efficacious in cells not exposed to $ZnSO_4$ than in Zn-treated cells with about 3 times more receptors. The EC_{50} increased about 8 fold in cells expressing low compared to high receptor numbers. Pilocarpine was used in this study rather than carbachol because it is a less effective agonist and would be more likely to reveal differences in coupling. The maximal level of stimulation of phosphoinositide turnover by 0.1 mM pilocarpine was independent of receptor number. A change in the EC_{50} without a decrease in the maximal response was also observed for angiotensin II receptors in rat hepatocytes in which receptor number declined after exposure to agonist (Bouscarel et al., 1988).

Coupling of m2 receptors in Y1 cells to inhibition of adenylyl cyclase could be blocked by pre-treatment of the cells with 1 mM carbachol for 1 h, resulting in a decrease in surface receptor number from 62 to 26 fmol/ mg total cell protein. In M2#7K cells, carbachol pre-treatment resulted in a decrease in receptor number from 660 to 88 fmol/mg total cell protein and agonist-dependent inhibition of cAMP accumulation was greatly attenuated (89% of control versus 59% without pre-treatment). However, when the receptor number after carbachol-induced internalization was higher, 220 fmol/mg total cell protein, the decrease in maximal inhibition was less, 60% inhibition in control cells compared to 42% inhibition in cells without pre-treatment. There is little effect on inhibition of adenylyl cyclase by m2 stimulation until the receptor level is decreased to less than 100 fmol/mg total cell protein. These data suggest that desensitization of the responses results from receptor internalization, unlike the case for the β_2-adrenergic where desensitization may occur independently of internalization (Sibley et al., 1987a,b). Studies are currently underway to better clarify the relationship between receptor number and coupling.

Future Trends

The techniques of molecular biology have revolutionized the study of neurotransmission. Gene cloning is used to identify on a structural level receptors which had previously only been characterized pharmacologically. The deduced amino acid sequence of a polypeptide can even be determined before a function can be ascribed to it, such as the G21 clone (the 5-HT$_{1C}$ receptor, Fargin et al., 1988) and the *mas* oncogene product (the angiotensin receptor, Jackson et al., 1988).

The advantages of using a molecular biologic approach are numerous. Receptor function is often difficult to determine in tissues due to the multiplicity of receptor types, heterogeneity of cell types and, frequently, low level of expression of receptors. Transfection of cloned receptors into cultured cells lines allows one to study a known receptor at high levels of expression in a homogeneous cell population. Moreover, receptors can be selectively mutated to identify regions involved in ligand, G protein and allosteric regulator binding.

This approach should allow rapid progress in identifying the structure/function relationships of receptors and exploring the range of effectors that the receptor is capable of activating. Ultimately, it will be desirable to return to study receptors as they are natively expressed in cells. It is becoming increasingly apparent that post-translational modifications, such as glycosylation, phosphorylation and acylation, strongly affect receptor function. These modifications may be specific to cell type or developmental stage. Also, the very high receptor levels attained in some transfected cells may permit unnatural couplings of receptors to effectors. This may be the case for the m2 receptor which was found to stimulate phosphoinositde turnover when expressed at very high levels in CHO cells although coupling was not observed when receptors were present at lower levels (Ashkenazi et al., 1987) or in other cell lines (Shapiro et al., 1988).

One of the most important consequences of the application of molecular biology to the study of hormone and neurotransmitter receptors has been the identification of gene families. Four families are described to date: the multi-subunit receptors which include an ion channel in the receptor (including the nicotinic AChR, GABA$_A$, and glycine receptors; Schofield et al., 1987), single subunit receptors which are coupled through G proteins (including the acetylcholine, adrenergic receptors; Bonner et al., 1987), the growth factor receptor family exhibiting tyrosine kinase activity (including the insulin, epidermal growth factor, and platelet-derived growth factor receptors; Yarden, 1988), and the steroid and thyroid hormone family containing DNA-binding elements (including the androgen and thyroid hormone receptor; Evans, 1988). A comparison of the similarities and differences between members of these families will lead to an understanding of the basic principles and constraints of neurotransmission and growth regulation.

Supported by NIH Grant HL30639 and HL07312. N.M.S. is a fellow of the American Heart Association, Washington Affiliate.

REFERENCES
Ashkenazi, A., J.W. Winslow, E.G. Peralta, G.L. Peterson, M.I. Schimerlik, D.J. Capon, J. Ramachandran. (1987) An M2 muscarinic receptor subtype coupled to both adenylyl

cyclase and phosphoinositide turnover. Science 238: 672-675.

Barnard, E.A. (1988) Separating receptor subtypes from their shadows. Nature 335: 301-302.

Benovic, J.L., F. Mayor, C. Staniszewski, R.J. Lefkowitz and M.G. Caron. (1987) Purification and characterization of the β-adrenergic receptor kinase. J. Biol. Cem. 262: 9026-9032.

Benveniste, M., E. Livneh, S. Schlessinger, Z. Kam. (1988) Overexpression of epidermal growth factor receptor in NIH-3T3-transfected cells slows its lateral diffusion and rate of endocytosis. J. Cell Biol. 106: 1903-1909.

Bonner, T.I., N.J. Buckley, A.C. Young and M.R. Brann. (1987) Identification of a family of muscarinic acetylcholine receptor genes. Science 237: 358-360.

Bonner, T.I., A.C. Young, M.R. Brann and N.J. Buckley. (1988) Cloning and expression of the human and rat m5 muscarinic acetylcholine receptor genes. Neuron 1: 403-410.

Bouvier, M., L.M.F. Leeb-Lundberg, J.L. Benovic, M.G. Caron and R.J. Lefkowitz. (1987) Regulation of adrenergic receptor function by phosphorylation. J. Biol. Chem. 262: 3106-3113.

Bouvier, M., W.P. Hausdorff, A. DeBlasi, B.F. O'Dowd, B.K. Kobilka, M.G. Caron and R.J. Lefkowitz. (1988) Removal of phosphorylation sites from the β_2-adrenergic receptor delays onset of agonist-promoted desensitization. Nature 333: 370-373.

Clark, R.B., M.W. Kinkel, J. Friedman, T. J. Goka and J.A. Johnson. (1988) Activation of cAMP-dependent protein kinase is required for heterologous desensitization of adenylyl cyclase in S49 wild-type lymphoma cells. Proc. Natl. Acad. Sci. USA 85: 1442-1446.

Clegg, C.A., L.A. Correll, G.G. Cadd and G.S. McKnight. (1987) Inhibition of intracellular cAMP-dependent protein kinase using mutant genes of the regulatory type I subunit. J. Biol. Chem. 262: 13111-13119.

Deutsch, P.J., J.P. Hoeffler, J.L. Jameson and J.F. Habener. (1988) Cyclic AMP and phorbol ester-stimulated transcription mediated by similar DNA elements that bind distinct proteins. Proc. Natl. Acad. Sci. USA 85: 7922-7926.

Dixon, R. A., B.K. Kobilka, D.J. Strader, J.L. Benovic, H.G. Dohlman, T. Frielle, M. A. Bolanowski, C. D. Bennet, E. Rands, R. E. Diehl, R. A. Mumford, E.E. Slater, I. S. Sigal, M. G. Caron, R. J. Lefkowitz and C. D. Strader. (1986) Cloning of the gene and cDNA for mammalian β-adrenergic receptor and homology with rhodopsin. Nature 321: 75-79.

Estenesen, R. D., K. Zustiak, A. Chuang, P. Schultheiss and J. Ditmanson. (1983) Action of 12-O-tetradecanolylphorbol-13-acetate on Y1 adrenal cells apparently requires the regulatory subunit type I cyclic AMP-dependent protein kinase. J. Exptl. Path. 1: 49-60.

Evans, R. M. (1988) The steroid and thyroid hormone receptor superfamily. Science 240: 889-895.

Fargin, A., J.R. Raymond, M. J. Lohse, B. K. Kobilka, M. G. Caron and R.J. Lefkowitz. (1988) The genomic clone G-21 which resembles a β-adrenergic receptor sequence encodes the $5HT_{1A}$ receptor. Nature 335: 358-360.

Fukuda, K., H. Higashida, T. Kubo, A. Maeda, I. Akiba, M. Bujo, M. Mishina and S. Numa. (1988) Selective coupling with K^+ currents of muscarinic acetylcholine receptor subtypes in NG 108-15 cells Nature 334: 335-338.

Gilman, A.G. (1970) A protein binding assay for adenosine 3':5'-cyclic monophosphate. Proc. Natl. Acad. Sci. USA. 67: 305-312.

Haga, T., K. Haga, G. Berstein, T. Nishiyama, H. Uchiyama and A. Ichiyama. (1988)

Molecular properties of muscarinic receptors. TIPS 102 (suppl.): 12-18.

He, X., X. Wu and B. J. Baum. (1988) Protein Kinase C differentially inhibits muscarinic receptor operated Ca^{2+} release and entry in human salivary cells. Biochem. Biophys. Res. Comm. 152: 1062-1069.

Heschler, J., W. Rosenthan, K.D. Hirsch, M. Wulfern, W. Trautwein and G. Schulz. (1988) Angiotensin II-induced stimulation of voltage-dependent Ca^{2+} currents in an adrenal cortical cell line. EMBO J. 7: 619-624.

Ho, A, K. S., Q-L. Ling, R. Duffield, P.H. Lam and J. H. Wang. (1987) Phosphorylation of brain muscarinic receptor: evidence of receptor regulation. Biochem. Biophys. Res. Comm. 142: 911-918.

Jackson, R. R., L.A.C. Blair, M.M. Goedert and M.R. Hanley. (1988) The *mas* oncogene encodes an angiotensin receptor. Nature 335: 437-440.

Jones, S.V., J.L. Baker, T. I. Bonner, J.J. Buckley and M.R. Brann. (1988) Electrophysiological characterization of cloned m1 muscarinic receptors expressed in A9 L cells. Proc. Natl. Acad. Sci. USA. 85: 4056-4060.

Julius, D., A. B. MacDermott, R. Axel, and T.M. Jessell. (1988) Molecular characterization of a functional cDNA encoding the serotonin 1c receptor. Science 241: 558-564.

Kassis, S. T. Zaremba, J. Patel, and P.H. Fishman. (1985) Phorbol esters and β-adrenergic antagonists mediate desensitization of adenylate cyclase in rat glioma C6 cells by distinct mechanisms. J. Biol. Chem. 48: 913-933.

Klein, P.S., T.J. Sun, C.L. Saxe III, A.R. Kimmel, R.J. Johnson and P.N. Devreotes. (1988) A chemoattractant receptor controls development in *Dictyostelium discoideuim*. Science 241: 1467-1472.

Kobilka, B.K., H. Matsui, T. S. Kobilka, T.L. Yang-Feng, U. Francke, M. G. Caron, R.J. Lefkowitz and J.W. Regan. (1987) Cloning , sequencing, and expression of the gene coding for the human platelet α$_2$-adrenergic receptor. Science 238: 650-656.

Kubo, T., K. Fukuda, A. Mikoma, A. Maeda, H. Takahashi, M. Mishina, M. Haga, T. Haga, T. Hirose and S. Numa. (1986) Cloning, sequencing and expression of complementary DNA encoding the muscarinic acetylcholine receptor. Nature 323: 411-416.

Kwatra, M.M. and M.M. Hosey. (1986) Phosphorylation of the cardiac muscarinic receptor in intact chick heart and its regulation by muscarinic agonist. J. Biol. Chem. 261: 12429-12432.

Kwatra, M. M., E. Leung, A. C. Maan, K.K. McMahon, J. Ptasienski, R. D. Green and M.M. Hosey. (1987) Corrrelation of agonist-induced phosphorylation of chick heart muscarinic receptors with receptor desensitization. J. Biol. Chem. 262: 16314-16321.

Liles, W. C., D.D. Hunter, K. E. Meir and N.M. Nathanson. (1986) Activation of protein kinase C induces rapid internalization and subsequent degradation of muscarinic acetylcholine receptors in neuroblastoma cells. J. Biol. Chem. 261: 5307-5313.

Lowry, O.H., N.J. Rosebrough, A. J. Farr and R. J. Randall. (1951) J. Biol. Chem. 193: 265-275.

Masters, S.B., T.K. Harden, and J.H. Brown. (1984) Relationships between phosphoinositide and calcium responses to muscarinic agonists in astrocytoma cells. Mol. Pharmacol. 26: 149-155.

Masu, Y., K. Nakayama, H. Tamaki, Y. Harada, M. Kuno and S. Nakanishi. (1987) cDNA cloning of bovine substance-K receptor through oocyte expression system. Nature 329: 836-838.

Nakayama, N., A. Miyajima and K. Arai. (1985) Nucleotide sequences of STE2 and STE3, cell type-specific sterile genes from *Saccharomyces cerevisiae*. EMBO J. 4: 2643-2648.

Nathans, J. and D. S. Hogness. (1984) Isolation and nucleotide sequence of the gene encoding human rhodopsin. Proc. Natl. Acad. Sci. USA 81: 4851-4855.

Nathanson, N. M. (1987) Molecular properties of the muscarinic acetylcholine receptor. Ann. Rev. Neursci. 10: 195-236.

Palczewski, K., J. H. McDowell and P. A. Hargrave. (1988) Purification and characterization of rhodopsin kinase. J. Biol. Chem. 263: 14967-14073.

Payne, G. S., D. Baker, E. van Tuinen and R. Schekman. (1988) Protein transport to the vacuole and receptor mediated endocytosis by clathrin in heavy chain-deficient yeast. J. Cell Biol. 106: 1453-1461.

Peralta, E.G., A. Ashkenazi, J.W. Winslow, D.H. Smith, J. Ramachandran and D. J. Capon. (1987) Distinct primary structures, ligand-binding properties and tissue specific expression of four human muscarinic acetylcholine receptors. EMBO J. 6: 3923-3929.

Peralta, E.G., A. Ashkenazi, J.W. Winslow, D.H. Smith, J. Ramachandran and D. J. Capon. (1988) differential regulation of PI hydrolysis and adenylyl cyclase by muscarinic receptor subtypes. Nature 334: 434-437.

Rae, P.A., N. S. Gutmann, J. Tsao and B. P. Schimmer. (1979) Mutations in cyclic AMP-dependent protein kinase and corticotropin (ACTH)-sensitive adenylate cyclase affect adrenal steroidogenesis. Proc. Natl. Acad. Sci. USA 76: 1896-1900.

Reneke, T.I., K.J. Blumer, W.E. Courschesne and J. Thorner. (1988) The carboxy-terminal segment of yeast α-factor receptor is a regulatory domain. Cell 55: 221-234.

Schofield, P.R., M. G. Darlison, N. Fujita, D. R. Burt, F. A. Stephenson, H. Rodrigues, L. M. Rhee, J. Ramachandran. V. Reale, T. Glencorse, P.H. Seeburg, and E.A. Barnard. Sequence and functional expression of the $GABA_A$ receptor shows a ligand-gated receptor super-family. Nature 328: 221-227.

Shapiro, R.A., N.M. Scherer, B. A. Habecker, E. M. Subers and N. M. Nathanson. (1988) Isolation, sequence and functional expression of the mouse M1 muscarinic acetylcholine receptor gene. J. Biol. Chem. (in press).

Sibley, D. R., K. Daniel, C. D. Strader and R. J. Lefkowitz. (1987a) Phosphorylation of the β-adrenergic receptor in intact cells: relationship to heterologous and homologous mechanisms of adenylate cyclase desensitization. Arch. Biochem. Biophys. 258: 24-32.

Sibley, D. R., J. L . Benovic, M. G. Caron and R. J. Lefkowitz. (1987b) Regulation of transmembrane signalling by receptor phosphorylation. Cell 48: 913-922.

Williams, S. and D. Johnston. (1988) Muscarinic depression of long-term potentiation in CA3 hippocampal neurones. Science 242: 84-87.

Yarden Y. (1988) Growth factor receptor tyrosine kinases. Ann. Rev. Biochem. 57: 443-478.

UPTAKE OF GABA AND L-GLUTAMATE INTO SYNAPTIC VESICLES

Else M Fykse, Hege Christensen and Frode Fonnum
Norwegian Defence Research Establishment
Division for Environmental Toxicology
P. O. Box 25, N-2007 Kjeller
NORWAY

INTRODUCTION

Synaptic vesicles are supposed to play a central role in neurotransmission as the transmitter storing and releasing organelles of the nerve terminal. In the central nervous system γ-aminobutyric acid (GABA) and L-glutamate are the most important inhibitory and excitatory neurotransmitters, respectively (Krnjevic, 1970). There are, however, no evidence for any enrichment of GABA and L-glutamate in the synaptosomes or the synaptic vesicles compared to other subcellular fractions from brain tissue (De Belleroche and Bradford, 1973; Lahdesmaki et. al., 1977; Wood and Kurylo, 1984). The lack of evidence for an enrichment of GABA and L-glutamate has been attributed to the possible leakage of the amino acids during the isolation procedure. It has been shown that GABA and L-glutamate are taken up into isolated synaptic vesicles by an active transport mechanism (Philippu and Matthaei, 1975; Disbrow et al., 1982; Naito and Ueda, 1983; Fykse and Fonnum, 1988). The vesicular uptake is temperature dependent, and dependent on the presence of ATP and Mg^{2+} in the incubation medium. The vesicular uptake is not dependent on Na^+, and it is not inhibited by blockers of the glial and synaptosomal uptake. The vesicular uptake is therefore clearly different from that of glial cells and synaptosomes (Naito and Ueda, 1983, 1985; Fykse and Fonnum, 1988). The Km values determined, 5.6 mM and 1.6 mM for GABA and L-glutamate uptake respectively, indicate a low affinity uptake system for the vesicles (Naito and Ueda, 1985; Fykse and Fonnum, 1988). In contrast, the synaptosomal uptake is a high affinity uptake with the Km values in the micromolar range (Fonnum et.al., 1980). These results indicate the storage of GABA and L-glutamate in synaptic vesicles.

In the present article, the mechanisms of the uptake of GABA and and L-glutamate into synaptic vesicles isolated from rat brain are compared. Synaptic vesicles used in these experiments are isolated from rat brain by a subcellular fractionation procedure and fractionated on a discontinuous sucrose gradient (Whittaker et al. 1964; Stadler and Tsukita, 1984). The uptake experiments are performed as earlier described (Fykse and Fonnum, 1988; Fykse et al., 1989).

RESULTS AND DISCUSSION

Accumulation of GABA and L-glutamate by synaptic vesicles is highly dependent on temperature. The vesicular uptake requires ATP hydrolysis and Mg^{2+}. Without ATP in the incubation mixture the vesicular uptake of GABA and L-glutamate was 16 and 6 percent of control, and in the absence of Mg^{2+}, the uptake was 53 and 20 percent. Carbonylcyanid-m-chlorophenyl-hydrazone (CCCP), an inhibitor of proton pumps and an uncoupler of oxidative phosphorylation (Heytler and Prichard, 1962), inhibited the vesicular uptake of GABA and L-glutamate. In the presence of 10 µM CCCP the uptake was 37 and 20 percent of controls respectively. In contrast, oligomycine and ouabain, agents known to inhibit the mitochondrial H^+-ATPase and the plasma membrane Na^+/K^+-ATPase respectively, had no significant effect on the ATP dependent GABA and L-glutamate uptake (Naito and Ueda, 1988; Fykse and Fonnum, 1988). This indicates that neither mitochondrial nor plasma membranes could be responsible for the uptake. In agreement, ouabain inhibits the synaptosomal GABA uptake (Nichlas et. al., 1973). It is known that mammalian synaptic vesicles contain an ATP dependent proton pump (Stadler and Tsukita, 1984). The vesicular uptake of GABA and L-glutamate is supposed to be driven by a proton gradient generated by a Mg^{2+} ATPase.

In order to investigate the uptake mechanisms in more detail we have studied and compared the effect of different concentrations of nigericin, gramicidin, CCCP, N-etylmaleimide (NEM), N' N'-dicyclohexylcarbodiimide (DCCD) and valinomycin on a vesicle preparation that contains both GABAergic and glutamergic vesicles. The uptake experiments were done in the presence of 110 mM K-tartrate. The results are shown in figure 1 and the 50 percent inhibition (IC_{50}) values are given in table 1. These agents caused a potent inhibition of the vesicular GABA and L-glutamate uptake. The electroneutral K^+-H^+ or Na^+-H^+ exchanger nigericin caused a more potent inhibition of the L-glutamate uptake than the GABA uptake. In contrast, the channelformer gramicidin, which would allow free movement of H^+, K^+ and Na^+, caused a more potent inhibition of the GABA uptake. DCCD and the thiol reagent NEM, inhibitors of the Mg^{2+} ATPase, completely inhibited the vesicular uptake of GABA and L-glutamate. Nigericin, gramicidin and CCCP will destroy any pH gradient across the vesicle membrane. The K^+ carrier valinomycin did not inhibit the vesicular uptake. These results strongly support the hypothesis that the vesicular uptake is driven by a transmembrane pH gradient generated by a Mg^{2+}-ATPase. Uptake of [^{3}H] acetylcholine into synaptic vesicles isolated from Torpedo electric organ is also driven by a pH gradient generated by a Mg^{2+} ATPase (Koenigsberger and Parsons, 1980; Parsons and Koenigsberger, 1980; Anderson et. al., 1982). The same is true for uptake of [^{3}H] catecholamines into both synaptic vesicles and chromaffin granules (Seidler et. al., 1977; Toll and Howard, 1977; 1978). This mechanism seems to be general for uptake of neurotransmitters into synaptic vesicles. The vesicular Mg^{2+} ATPase is

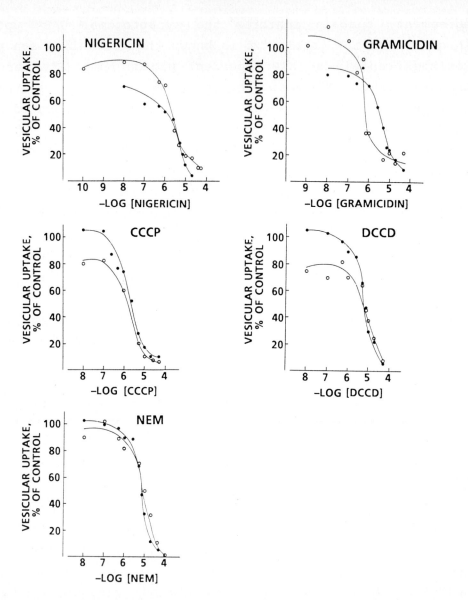

Figure 1. Effect of different inhibitors on the vesicular uptake of GABA and L-glutamate.
Crude synaptic vesicles were centrifuged on a discontinuous sucrose gradient for 2 hours at 4°C and 65000 x g in a contron TST 28.38 rotor. Synaptic vesicles were isolated from 0.4 M sucrose and incubated (about 0.1 mg protein) in 110 mM K-tartrate, 10 mM Tris-maleate (pH 7.4), 4 mM MgCl$_2$, 2 mM ATP and 1 mM [^{3}H] GABA or [^{3}H] L-glutamate (5 mCi/mmol). In the absence of inhibitors, the uptake was 342±38 (n=12) and 1399±134 (n=10) pmol/min/mg protein for GABA and L-glutamate. o GABA; ● L-glutamate. The figure is from Fykse et. al. J. Neurochem. 1989 (in press).

different from the mitochondrial H^+-and the plasma membrane
Na+/K+-ATPase. It belongs to a class of ATP driven ion pumps
very similar to that described in endosomes, lysosomes,
coated vesicles and plant vacuoles (Rudnick, 1986; Kanner
and Schuldiner, 1987).

Table 1. Inhibition of active GABA and L-glutamate uptake.

INHIBITORS	IC$_{50}$ values, (µM)	
	GABA	L-GLUTAMATE
NIGERICIN	2.0	0.3
GRAMICIDIN	0.8	3.2
CCCP	1.6	2.2
DCCD	5.4	6.4
NEM	7.2	7.2
VALINOMYCINE	>50	>50

The IC$_{50}$ values are calculated with a Multiple Drug Effect
Analysis Program (Chou and Chou, 1985). The data are from
Fykse et. al. J. Neurochem. 1989 (in press)

The Mg2+ ATPase of chromaffin granules is shown to be an
anion dependent proton pump (Moriyama and Nelson, 1987). We
therefore examined the effect of Cl^- on the vesicular GABA
and L-glutamate uptake. As shown in table 2, the uptake of
L-glutamate was highly stimulated, about 500 percent, by
addition of 5 mM Cl^-. In contrast, the uptake of GABA was
not affected. This is in agreement with the results of Naito
and Ueda (1985) on L-glutamate. The effect of low millimolar concentrations of Cl^- is still the only difference
described so far between GABA and L-glutamate uptake. In
these experiments, the synaptic vesicle preparation was run
through a Sephadex G-25 column (PD-10) to reduce the concentration of Cl^- in the incubation mixture. These results

are not excluding an effect of extremely low concentration of Cl⁻ on the uptake of GABA, since it is not possible to reduce the Cl⁻ concentration to zero. More experiments had to be done before this can be firmly stated.

Table 2. Effect of chloride on the vesicular uptake.

[KCl] (mM)	VESICULAR UPTAKE pmol/min/mg protein	
	GABA	L-GLUTAMATE
0	372±38 (7)	268±63 (7)**
5	368±45 (4)	1525±295 (6)
50	323±86 (4)	442±115 (6)
100	241±52 (4)*	204±61 (6)

Synaptic vesicles (0.4 M sucrose) were incubated as described in figure one. Data are mean±SEM (no of determinations). *p<0.05, **p<0.002 by student's t-test. Data are from Fykse et. al. J. Neurochem. 1989 (in press).

We have investigated the regional distribution of the uptake of GABA and L-glutamate (table 3). A crude vesicle fraction was isolated from cerebral cortex, cerebellum, medulla and the subcortical areas from rat brain. The ratio between L-glutamate and GABA uptake differed in different brain areas, which indicates that the GABA and L-glutamate were taken up into different vesicle populations. This suggestion is also supported by the fact that the uptake of GABA is not affected by high concentrations of L-glutamate and the L-glutamate uptake is not inhibited by high concentrations of GABA. (Naito and Ueda, 1985; Fykse and Fonnum, 1988). The distribution of the vesicular uptake was in agreement with the distribution of the sodium dependent synaptosomal uptake.

Table 3. Regional distribution of the vesicular and synaptosomal uptake of GABA and L-glutamate.

REGION	SYNAPTOSOMAL AND VESICULAR UPTAKE L-GLU/GABA	
	Vesicles	Synaptosomes
Cerebral cortex	6.2	5.4
Cerebellum	3.9	4.4
Medulla	1.3	2.0
Subcortical areas	2.1	2.7

Data are from 5 and 6 different experiments for the synaptosomal and vesicular uptake respectively.

CONCLUSION

Synaptic vesicles isolated from rat brain accumulate GABA and L-glutamate into different populations of vesicles. The uptake of both GABA and L-glutamate has been shown to be driven by a transmembrane pH gradient generated by a Mg^{2+} ATPase. The uptake of L-glutamate is highly dependent on low concentrations of Cl^-, while the uptake of GABA was not affected. The observed difference between these uptake systems may reflect the different charge of the GABA and L-glutamate molecules, or perhaps different properties of their Mg^{2+} ATPase proton pumps.

REFERENCES

Anderson DC, King SC, and Parsons SM (1982) Proton gradient linkage to active uptake of [3]-H acetylcholine by Torpedo electric organ synaptic vesicles. Biochemistry 21: 3037-3043
Chou J and Chou TC (1985) Dose effect analysis with computers. Elsevier-Biosoft, Cambridge

De Belleroche JS and Bradford HF (1973) Amino acids in synaptic vesicles from mammalian cerebral cortex: a reappraisal. J. Neurochem 21: 441-451

Disbrow JK, Gershten MJ, and Ruth JA (1982) Uptake of L[^{3}H]glutamic acid by crude and purified synaptic vesicles from rat brain. Biochem. Biophys. Res. Com 108: 1221-1227

Fonnum F, Lund-Karlsen R, Malthe-Sørensen D, Sterri S, and Walaas I. High affinity transport systems and their role in transmitter action. Cotman CW, Poste G, and Nicholson GL (eds) (1980) The cell surface and neuronal function, Elsevier/North-Holland Biomedical press, Amsterdam, pp. 171-183.

Fykse EM and Fonnum F (1988) Uptake of γ-aminobutyric acid by a synaptic vesicle fraction isolated from rat brain. J. Neurochem 50: 1237-1242

Fykse EM, Christensen H, and Fonnum F (1989) Comparison of the γ-aminobutyric acid and L-glutamate uptake into synaptic vesicles isolated from rat brain. J. Neurochem (in press)

Kanner BI and Schuldiner S (1987) Mechanism of transport and storage of neurotransmitters. CRC Critical reviews in biochemistry 22: 1-38

Koenigsberger R and Parsons SM (1980) Bicarbonate and magnesium ion-ATP dependent stimulation of acetylcholine uptake by Torpedo electric organ synaptic vesicles. Biochem Biophys. Res. Com 94: 305-312

Krnjevic K (1970) Glutamate and γ-aminobutyric acid in the brain. Nature 228: 119-124

Lahdesmaki P, Karpinnen A, Saarni H, and Winter R (1977) Amino acids in the synaptic vesicle fraction from calf brain: content and metabolism. Brain Res 138: 295-308

Moriyama Y and Nelson N (1987) The purified ATPase from cromaffine granule membranes is an anion dependent proton pump. J. Biol. Chem 19: 9175-9180

Naito S and Ueda T (1983) Adenosine triphosphate dependent uptake of glutamate into protein I-associated synaptic vesicles. J. Biol. Chem 258: 696-699

Naito S and Ueda T (1985) Characterization of glutamate uptake into synaptic vesicles. J. Neurochem 44: 99-109

Nicklas WJ, Puszkin S, and Berl S (1973) Effect of vinblastine and colchicine on uptake and release of putative transmitters by synaptosomes and on brain actomycin-like protein. J. Neurochem 20: 109-121

Parsons SM and Koeningsberger R (1980) Specific stimulated uptake of acetylcholine by Torpedo electric organ synaptic vesicles. Proc. Natl. Acad. Sci USA 77: 6234-6238

Philippu A and Matthaei H (1975) Uptake of serotonin, gamma-amino-butyric acid and histamine into synaptic vesicles of pig caudate nucleus. Naunyn-Schiedeberg's Arch. Pharmacol 287: 191-204

Rudnick G (1986) ATP-driven H+-pumping into intracellular organelles. Ann. Rev. Physiol 48: 403-413

Seidler F, Kirksey DF, Lau C, Whitmore WL, and Slotkin TS (1977) Uptake of (-)[^{3}H]norephineprine by storage vesicles prepared from whole rat brain: properties of the uptake system and its inhibition by drugs. Life Sci 21: 1075-1086
Stadler H and Tsukita S (1984) Synaptic vesicles contain an ATP dependent proton pump and show knob-like protrusions on their surface. EMBO J 3: 3333-3337
Toll L, Gundersen CB, and Howard BD (1977) Energy utilization in the uptake of catecholamines by synaptic vesicles and adrenal chromaffin granules. Brain Res 136: 59-66
Toll L and Howard BD (1978) Role of Mg^{2+}-ATPase and a pH gradient in the storage of catecholamines in synaptic vesicles. Biochemistry 17: 2517-2523
Whittaker WP, Michaelson JA, and Kirkland RJA (1964) The separation of synaptic vesicles from nerve endings particles (synaptosomes). Biochemical J 90: 293-303
Wood JD and Kurylo E (1984) Amino acid content of nerve endings (synaptomes) in different regions of brain: effect of gabaculline and isonicotinic acid hydrazide. J. Neurochem 42: 420-425

DEACTIVATION OF LAMININ-SPECIFIC CELL-SURFACE RECEPTORS ACCOMPANIES IMMOBILIZATION OF MYOBLASTS DURING DIFFERENTIATION.

Simon L. Goodman[§] Victor Nurcombe[+], and Klaus von der Mark[§].

MPI for Biochemie
Abt. Kühn
am Klopferspitz 18
Martinsried
8033
W. Germany

§ present address: MPI for Rheumatologie, Schwabachanlage 10, Erlangen, D8520, W. Germany: To whom correspondence should be addressed.

+MPI for Psychiatry, Abt. Thoenen, am Klopferspitz 18, Martinsried 8033, W. Germany

NATO ASI Series, Vol. H29
Receptors, Membrane Transport and Signal Transduction
Edited by A. E. Evangelopoulos et al.
© Springer-Verlag Berlin Heidelberg 1989

INTRODUCTION

During embryonal development, myoblasts move from the dermomyotome of the somites through the embryo before terminally differentiating and fusing in highly reproducible patterns to form static syncytial myotubes which integrate to produce the muscle analgen(Chevalier,1979;Christ et al.1983; Trinkaus,1984). Not all the migrated myoblasts go on to form myotubes; a reserve population of satellite cells is maintained in a quiescent state between the cell membrane of the myotube and a specialized ECM that surrounds it, the basement membrane (BM)(Vracko and Benditt, 1974; Timpl and Dziadek,1987). If the muscle is damaged, these cells are activated by unknown mechanisms to reiterate the developmental processes and rebuild syncytial muscle fibres(Gulati et al., 1983;Bischoff, 1986).

In vitro, myoblasts will not survive in culture unless they encounter ECM components, but their response to the components is highly selective. The protein laminin(LN; fig1), enriched in the BM, and the interstitial component fibronectin(FN), promote similar levels of myoblast attachment onto tissue culture substrates. However, LN stimulates myoblasts to proliferate, locomote, and differentiate into non-replicating myotubes(Kühl et al.,1985, Foster et al.,1987, Goodman et al., 1987,Risse et al., 1987,Öcalan et al, 1988), while fibronectin does not.

Laminin is a large multi domain glycoprotein ($M_r \approx 900,000$) and the many different effects that it induces in cells may not all be triggered by the same domains, nor need they operate over the same cell surface receptors. For example, LN has at least two distinct regions that can mediate cell attachment to otherwise non-adhesive substrates, E8, a 35nm length of the long arm and E1-4, the three short arms(Fig1)(Timpl et al.,1983.Edgar et al.,1984. Aumailley et al., 1987, Goodman et al., 1987).

Cellular interactions with ECMs can be mediated by specific receptors at the cell surface(Hynes,1987), however, although a number of putative LN receptors have been isolated from mammalian cells(reviewed in Timpl and Dziadek,1987; Risse et al.1987) they are still not well characterized – furthermore, the relationship between the presence of the 'receptor' (as detected with, typically, an antibody probe) and its biological activity, that is to say, whether it is able to bind its ligand at the cell surface, are seldom discussed. A reliable radio-ligand binding assay for LN and its subcomponents E8 and E1-4(Aumailley et al., 1987) allows the number of functional LN binding sites to be measured on the surface of living cells. This assay and cell attachment studies have strongly suggested that there are two distinct classes of cell receptors for E8 and E1-4(Aumailley et al., 1987, Goodman et al., 1987), each distinct from receptors for FN(Hynes, 1987.Gehlsen et al.1988).

The Structure Of Laminin

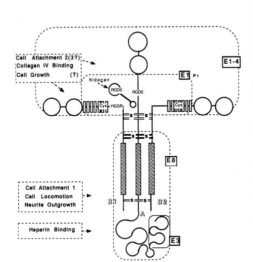

Figure 1. The structure of murine tumour laminin
Laminin has three disulphide linked polypeptide chains of apparent M_r ≈440,000 (A) and ≈220,000 (B1,B2), together with a closely associated molecule, nidogen, M_r ≈150,000. Its molecular domains include an extended stretch of α-helix in E8 (zig-zag pattern) and cysteine rich repeats in the B-chains('cys') of E1. Biological activities assigned to sub-fragments of the molecule are shown in the dotted boxes (left) and the position of active peptides RGDS and YIGSR are shown(see Timpl and Dziadek, 1987 for details).

What then happens to the motile abilities and the LN receptors of myoblasts when they differentiate? In principle differentiated myoblasts need no longer move; *in vivo* they will have reached their 'final resting place'. Similarly, the receptors for the ECM molecules that stimulate differentiation processes in muscle cells are not longer strictly 'necessary' when the cells have already differentiated. At least some growth factor receptors are removed functionally from cell surfaces when the cell no longer 'requires' the growth factor (eg. Lim and Hauschka, 1986). If interactions between ECM and ECM-receptors are directly triggering responses in myoblasts, then by analogy we might predict that the receptors would be somehow inactivated, either by clearing from the cell surface or by modification. We show here that there is a decrease in locomotory ability of muscle satellite cells as they differentiate and there is indeed an accompanying decrease in functional LN receptors.

MATERIALS AND METHODS

Cell culture and attachment assays(Kühl et al.,1985, Goodman et al., 1987,Öcalan et al.,1988) and radioligand binding assays (Aumailley et al.,1987) have been described elsewhere - the radio ligands all bound specifically and saturably. MM14dy is a differentiation-competent murine skeletal muscle satellite cell line derived from a post-crisis strain of a skeletal muscle culture. The creation of three differentiation states in MM14dy cells by nutrient depletion will be described elsewhere(Goodman et al. submitted). Video cell locomotion assays will be described in detail elsewhere(Goodman and Merkl, submitted, Goodman et al. submitted). Briefly, cells were plated onto 25 cm^{-2} flasks, pre-coated with LN or FN and recorded by time lapse video microscopy for at least 12h at 37°C. Cells were tracked using a computer controlled digitiser tablet and analysed by computer. Figure 2 shows the cell tracks as they appeared on the video-screen normallised to 700μm axes over a 5h period for 10-14 cells.

For immunoblots, sparse, intermediate or dense MM14dy cells were removed from the substrates by treatment with 0.05% EDTA in PBS, washed by centrifugation, and dissolved in reducing Lamelli sample buffer before loading 2.5x10^4 cells per track on 10% linear SDS gels. The gels were blotted onto nitrocellulose paper (Schleicher and Schuell; 0.22μM) in a dry blotting chamber (Biometra; using Towbin buffer without methanol; Towbin et al.1979; 20min at 300V , 12°C), blocked for 1h in PBS + 0.05% Tween-20 and incubated with a rabbit antiserum directed against the C-terminal 39 amino acids of the integrin β1 chain (Marcantonio and Hynes, 1988), followed by washing in PBS/Tween 20 and visualization of the bound antibody with peroxidase labelled goat anti-rabbit IgG (Biorad; 1:4000).

RESULTS

As MM14dy satellite cells differentiate they loose the ability to locomote over laminin substrates. *In vivo*, myoblasts become locomotory. We therefore studied the effect of the typical BM adhesion factor LN on myoblast locomotion by quantitative video-analysis (figure 2). Sparse MM14dy locomoted over LN with a mean speed of ≈60μM h^{-1}(Fig.2a). Intermediate density MM14dy locomoted more slowly than sparse populations, at ≈30μM h^{-1} (Fig2b), however, differentiated dense MM14dy moved at a mean speed of only ≈5μM h^{-1}, within the estimated background of the measurement technique, *ie*. they were essentially stationary (Fig2c). Primary myoblasts also locomoted, at a mean speed of ≈70μM h^{-1}(Fig2e). Myoblasts attach with similar kinetics and affinities to FN, but primary and sparse MM14dy populations were able to locomote only at ≈20μm h^{-1} over FN substrates(Fig2d,f)(Ocalan et al., 1988).

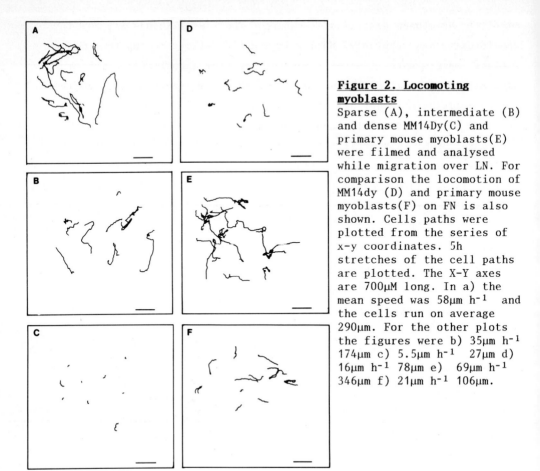

Figure 2. Locomoting myoblasts
Sparse (A), intermediate (B) and dense MM14Dy(C) and primary mouse myoblasts(E) were filmed and analysed while migration over LN. For comparison the locomotion of MM14dy (D) and primary mouse myoblasts(F) on FN is also shown. Cells paths were plotted from the series of x-y coordinates. 5h stretches of the cell paths are plotted. The X-Y axes are 700µM long. In a) the mean speed was 58µm h⁻¹ and the cells run on average 290µm. For the other plots the figures were b) 35µm h⁻¹ 174µm c) 5.5µm h⁻¹ 27µm d) 16µm h⁻¹ 78µm e) 69µm h⁻¹ 346µm f) 21µm h⁻¹ 106µm.

As MM14dy satellite cells differentiate laminin specific cell-surface binding sites are down-regulated. On preterminally differentiated sparse MM14dy the number of both LN and E8 binding sites per cell was ≈30,000 sites per cell, about double the number of sites found on primary murine myoblasts (table1). The number of sites dropped as the population differentiated, while their affinity remained constant at ≈2nM and scatchard plots of the data were linear and monotonic. (shown for laminin binding in Fig.3). The behaviour of the binding sites for E1-4 was more complex. Sparse MM14dy bound little E1-4 (≈1,500 per cell), but intermediate density MM14dy bound ≈5,500 molecules per cell, but both dense MM14dy and primary myoblasts bound undetectable numbers of E1-4 molecules (i.e. <500 per cell, the limit of resolution of the assay). The binding affinities agree with previously published values for LN, E8 and P1 (cf. E1-4) binding (Terranova et al, 1983, Albini et al.1986.Aumailley et al.,1987,Goodman et al. *manuscript in*

Table 1. The binding of Laminin and fragments to differentiating MM14 cells

Ligand	sMM14 no	Aff	iMM14 no	Aff	dMM14 No	Aff
LN	38,000	4.3	31,000	2.6	8,400	2.4
E8	34,000	2.9	25,000	2.1	n.d.	n.d.
E1-4	1,400	22.0	5,400	14.0	n.c.	n.c.

MM14dy were incubated in triplicate with iodinated LN, E8 or E1-4. The cell-bound and free radio-activity were separated by centrifugation and counted in a gamma-counter. **no**=number of molecules bound per cell. **Aff**=binding affinity (nanomolar)

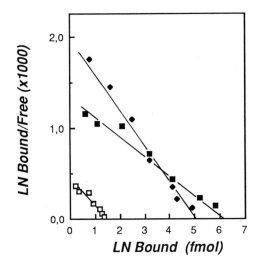

Figure 3. Binding of radio-labelled laminin to MM14dy.
Binding assays were performed on sparse(***closed squares***), intermediate(***diamonds***) and dense MM14dy (***open squares***) as described in table.1. The data were plotted according to Scatchard.

preparation). For each ligand concentration, non-specific binding was defined by incubating the cells with the same amounts of radioactivity plus 100 fold excess of unlabeled ligand. This non-specifically bound activity (usually 10-30% of the specific bound counts) was routinely subtracted from the total count to give the count used for further analysis.

The expression of integrin β1 chain does not alter as the satellite cells differentiate. Western blots were loaded for equal numbers of cells and blotted for the β1 chain of integrin. One broad band centered around 140kD was detected in each population; control pre-immune sera detected nothing(not shown). The intensity of the staining on the Western blot was relatively uniform in each state of cell differentiation(Fig.4).

a   b   c   **Mr (kD)**

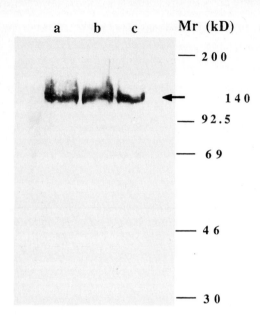

— 2 0 0

— 9 2 . 5

← 1 4 0

— 6 9

— 4 6

— 3 0

Figure 4. Western Blot of anti-integrin β1 MM14dy Myoblasts

2.5x10^4 cells in reducing sample buffer were resolved on a 10% gel, blotted onto nitrocellulose with Rainbow marker (Amersham) as blotting standards, and probed with a mono specific antibody against the integrin β1 chain. Sparse MM14dy(a), intermediate MM14dy(b) and dense MM14dy(c) are shown. The 140kD band is arrowed. Position of molecular weight markers in kilodaltons is also shown.

DISCUSSION

The mechanisms in the embryo which control how myoblasts migrate to their sites of terminal differentiation and halt and differentiate there are unknown, however, recent evidence points to the involvement of the specific receptors for the matrix components FN and LN[Horwitz et al. 1985,Hynes, 1987,Menko and Boettiger, 1987,Jaffredo et al.1988]. It has previously been noted that myoblasts respond differently to LN as opposed to FN and this is most striking in their locomotory behaviour[Kühl et al.1985,Öcalan et al., 1988].

Here we show that not only are myoblasts stimulated to locomote by LN, but there is also a correspondence between the stimulation and their expression of LN receptors. Differentiated myoblasts were unable to migrate over LN substrates, and this population had a drastically reduced number of LN binding sites; the affinity of the sites remained unchanged. The cells were not *immotile*, they produced ruffling lamellae and pseudopodia, but they couldn't translocate over the substrate.We hypothesize that there is a causal relationship between these observations: the lack of receptors inhibits cell locomotion over laminin.

Interestingly, the expression of the integrin β1 chain at the protein level was little changed as myoblasts differentiated. The integrins are a super family of cell surface recognition molecules with a common αβ heterodimeric structure whose specificity is modulated by permutating between some 10 α and 3 ß chains [Hynes, 1987]. Integrin ß1α5 is at least partially responsible for cell adhesion to fibronectin. An RGDS-independent laminin binding integrin has been isolated from RuGli glioblastoma cells[Gehlsen et al. 1988]; we have confirmed this observation (von der Mark et al. *in preparation*), but several other molecules especially of 68kD have also been implicated as LN receptors[reviewed in Risse et al. 1987]. Our Western blotting data suggest that integrin ß1 chain is *not* modulated at the protein expression level when the numbers of laminin receptors functionally decreases. We are investigating the expression of other integrins at the moment.

Although it is not yet known if myoblasts actually use either LN or FN during embryonal migration or during repair, pertur-bation studies from Jaffredo et al.[1988] indicate that they play a role. Such studies have used anti-bodies that react with a mixture of native integrins, or the RGDS-peptides, that can perturb more than one integrin-matrix interaction. The reagents are therefore not monospecific and interpretation of the data is not straightforward. However, satellite cells come into intimate contact with the LN-containing BM of the necrotic fibre during muscle repair. It is thus an interesting coincidence that the known cell biological events of the repair process, proliferation, migration and differentiation - are accelerated by LN and inhibited by FN. The comparative locomotory ability on LN and FN of myoblasts has to our knowledge not yet been compared *in vitro*(excepting our earlier work. [Öcalan et al.1988]). Avian myoblasts can move over FN[eg. Turner et al. 1983], but the rate of movement has not been reported.

Occupancy of the receptors for matrix molecules on the cell surface can control myoblast differentiation[Menko and Boettiger, 1987; Öcalan et al., 1988], yet there is little quantitative data available on the variations in the receptor number with the differentiation state of cells. FN receptors are instrumental during morphogenesis but less is known about LN receptors. However, migration of cranial neural crest cells can be altered by blocking LN-heparan sulphate proteoglycan complex[Bronner-Fraser and Lallier, 1988].

Our data suggests that the down-regulation of LN receptors may be important during myoblast differentiation and we suggest here that it influences their ability to locomote over LN substrates. These data may highlight aspects of the patterning mechanisms that myoblasts employ *in vivo* to control where and when they differentiate.

ACKNOWLEDGEMENTS

V.N is C.J. Martin fellow of the National Health and Medical Research
Council of Australia and the recipient of a fellowship from the Alexander
von Humbolt foundation. Laminin subfragments were the kind gift of Dr.
Deutzmann, Martinsried and the rabbit anti serum against the integrin β1
chain consensus cytoplasmic domain the generous gift of Drs. Marcantonio and
Hynes, MIT.

REFERENCES

Albini,A., Graf,J., Kitten,G.T., Kleinman,H.K., Martin,G.R.
Veillette,A. and Lippman,M.E.(1986). 17β-estradiol regulates and v-Ha-ras
transfection constitutively enhances MCF7 breast cancer cell interactions
with basement membrane. Proc.natn.Acad.sci. U.S.A.**83**,8182-8186.

Aumailley, M., Nurcombe, V., Edgar, D., Paulsson, M., and
Timpl,R.(1987). Cellular Interactions with laminin. Cell adhesion correlates
with two fragment specific high affinity binding sites. J. Biol. Chem. **262**,
11532-11538.

Bischoff,R.(1986). Proliferation of muscle satellite cells on intact
myofibres in culture. Dev. Biol.**115**, 129-139.

Bronner-Fraser,M. and Lallier,T.(1988). A monoclonal antibody against a
laminin-heparan sulphate proteoglycan complex perturbs cranial neural crest
migration *in vivo*. J. Cell Biol.**106**.1321-1329.

Chevalier,A.(1979). Role of the somitic mesoderm in the development of
the thorax in bird embryos. J.Embryol. exp. Morph. **49**,73-88.

Christ,B. Jacob,M., and Jacob,H.J.(1983). On the origin and development
of the ventrolateral abdominal muscles in the avian embryo.
Anat.Embryol.**166**,87-101.

Edgar, D., Timpl,R., and Thoenen H.(1984). The heparin-binding domain
of laminin is responsible for its effects on neurite outgrowth and neuronal
survival. EMBO. J. **3**, 1463-1468.

Foster, R.F., Thomson, J.M., and Kaufman, S.J.(1987). A laminin
substrate promotes myogenesis in rat skeletal muscle cultures: analysis of
replication and development using anti-desmin and anti-Brd-Urd monoclonal
antibodies. Dev. Biol. **122**, 11-20.

Gehlsen, K.R., Dillner,L., Engvall,E. and Ruoslathi,E. (1988). The
human laminin receptor is a member of the integrin family of cell adhesion
receptors. Science, **241**. 1228-1229.

Goodman, S.L., Deutzmann, R., and von der Mark,K.(1987). Two distinct
cell-binding domains in laminin can independently promote non-neuronal cell
adhesion and spreading. J. Cell Biol. **105**, 595-610.

Gulati, A.K., Reddi, A.H., and Zalewski,A.A.(1983). Changes in the
basement membrane zone components during skeletal muscle fiber degeneration
and regeneration. J.Cell Biol. **97**, 957-962.

Horwitz, A., Duggan, K., Greggs, R., Decker, C. and Buck,C.A. (1985).
The cell substrate attachment (CSAT) antigen has the properties of a
receptor for laminin and fibronectin. J.Cell Biol. **101**, 2134-2144.

Hynes, R.O.(1987). Integrins: a family of cell surface receptors.
Cell, **48**, 549-554.

Jaffredo, T., Horwitz, A.H., Buck, C.A., Rong, P.M. and Dieterlen-Lievre,F.(1988). Myoblast migration specifically inhibited in the chick embryo by grafted CSAT hybridoma cells secreting an anti-integrin antibody. Development, **103**, 431-466.

Kühl, U., Öcalan, M., Timpl, R., and von der Mark,K.(1985). The role of laminin and fibronectin in selecting myogenic versus fibrogenic cells from skeletal muscle in vitro. Dev. Biol. **117**, 628-635.

Kühl, U, Timpl, R., and von der Mark,K. (1982). Synthesis of type IV collagen and laminin in cultures of skeletal muscle cells and their assembly on the surface of myotubes. Dev. Biol. **93**, 344-354.

Lim, R.W., and Hauschka,S.D.(1984). EGF responsiveness and receptor regulation in normal and differentiation defective mouse myoblasts . J. Cell Biol. **98**, 739-747.

Marcantonio,E.E. and Hynes,R.O.(1988). Antibodies to the cytoplasmic domain of the integrin ß1 subunit react with proteins in vertebrates, invertebrates and fungi. J.Cell Biol **106**, 1765-1772.

Menko, A.S. and Boettiger,D.(1987). Occupation of the extracellular matrix receptor integrin is a control point for myogenic differentiation. Cell, **51**, 51-57.

Öcalan, M., Goodman, S.L., Kühl, U., Hauschka, S.D., and .von der Mark,K.(1988). Laminin alters cell shape and stimulates motility and proliferation of murine skeletal myoblasts. Dev. Biol. **125**, 158-169.

Risse, G., Dieckhoff, J., Mannherz,K.H., and von der Mark,K. (1987). The interaction of laminin with cell membranes. in Membrane Receptors, dynamics and energetics (*ed*. K.W.A. Wirtz, Plenum Publishing) *pp*173-180.

Terranova, V.P., Rao, C.N., Kalebic, T., Margulies, I.M., and Liotta,L.A.(1983). Laminin receptors on human breast carcinoma cells. Proc. natn. Acad. sci. USA. **80**, 444-448.

Timpl, R. and Dziadek,M.(1987). Structure, development and molecular pathology of basement membranes. Int. rev. Exp. Pathol. **29**, 1-112.

Timpl,R., Johansson,S., vanDelden,V., Oberbäumer,I. and Höök,M.(1983).Characterization of protease-resistant fragments of laminin mediating attachment and spreading of rat hepatocytes.J.Biol. Chem. **258**, 8922-8927.

Towbin, H., Staehelin,T. and Gordon,J.(1979). Electrophoretic transfer of proteins from polyacrylamide gels to nitrocellulose sheets: procedure and some applications. Proc. natn. Acad. Sci. US. **76**, 4350-4354.

Trinkaus,J.P.(1984). Cells into organs. The forces that shape the embryo. Prentice-Hall inc., Englewood Cliffs NJ.

Turner, D.C., Lawton, J., Dollenmeier,P., Ehrismann,R. and Chiquet.M.(1983). Guidance of myogenic cell migration by oriented deposits of fibronectin. Devl Biol. **95**. 497-504.

Vracko, R. and Benditt, E.P.(1974). Basal lamina scaffold anatomy and significance in orderly tissue structure. Amer. J. Pathol. **77**, 314-350.

SIGNAL TRANSDUCTION IN HALOBACTERIA

D. Oesterhelt and W. Marwan
Max-Planck-Institut für Biochemie
Am Klopferspitz 18A
8033 Martinsried
Federal Republic of Germany

Introduction

The search for photosynthetically efficient green light and
the avoidance of inefficient blue light or lethal ultra-
violet light allows halobacteria to survive by means of
photosynthesis in a natural habitat of brines and salt ponds
under strong sunlight. The entire photobiochemistry of these
archaebacteria is based upon the activation of retinal
proteins. Two light-driven ion pumps, bacteriorhodopsin as a
proton pump and halorhodopsin as a chloride pump, represent
light energy converters that power the energy-driven
processes of the cell (Lanyi, 1984). Two light sensors,
sensory rhodopsin and protein P_{480} (also called SR-II or
phoborhodopsin) mediate "colour" vision that aids the cell
in finding the optimal photosynthetic environment (Spudich &
Bogomolni, 1984; Takahashi et al., 1985; Marwan &
Oesterhelt, 1987).
The mechanical device mediating halobacterial motility is a
motor-driven flagellar bundle. Motion occurs by smooth
swimming (runs) of the cells interrupted by spontaneous
motor switching events (Alam & Oesterhelt, 1984). After a
switch the motor rotates in the opposite direction and the
cell reverses its direction but not necessarily taking the
original path. The resulting random walk keeps the cells
evenly distributed unless a light stimulus hits photoreceptor molecules, whereupon by transfer through a signal chain,

NATO ASI Series, Vol. H29
Receptors, Membrane Transport and Signal Transduction
Edited by A. E. Evangelopoulos et al.
© Springer-Verlag Berlin Heidelberg 1989

a prolongation (attraction) or shortening (repulsion) of a single run takes place. This results in stimulus-controlled behaviour.

Spontaneous switching of the flagellar motor

Rod-shaped species of halobacteria such as *Halobacterium halobium* carry at one, or both, of their polar ends flagellar bundles that are composed of right-handed helical filaments (Houwink, 1956; Alam & Oesterhelt, 1984). The bundles can be visualized by dark-field microscopy during certain stages of growth. Clockwise rotation (cw) of the flagellar bundle causes forward swimming (for a definition, see Macnab & Ornston, 1977), while counterclockwise (ccw) rotation reverses the direction of the cellular movement. Many Gram-negative eubacteria (such as *Escherichia coli*) show a so-called tumbling movement on a changing from ccw to cw rotation due to the dissociation of the formerly co-ordinated flagellar bundle (Macnab, 1976). In contrast, the halobacterial flagellar bundle remains tightly associated at all times and no tumbling occurs. The switch of the flagellar motor is seen as a short stoppage of the cell motion (Alam & Oesterhelt, 1984). No bias for cw or ccw rotation is observed. A stop is synonymous with a reversal or a switching event.

Monopolarly flagellated cells of *H. halobium* suspended in normal growth medium, were tracked in infrared light. Since no bias for runs by cw or ccw rotation of the flagella exists (see also Marwan et al., 1987) the sequence stop-run-stop served as a basis to measure the distribution of rotational intervals (time-span between two reversals, t_s). A broad and asymmetric frequency distribution of the intervals is found and shown in Figure 1(a). The broadness of the distribution curve is not due to individual differences of cells since qualitatively the same interval distribution is obtained when only one cell is observed for 300 events. Unlike in *E.coli*, the shortest intervals or runs are not the most frequent (probable), but the right part of

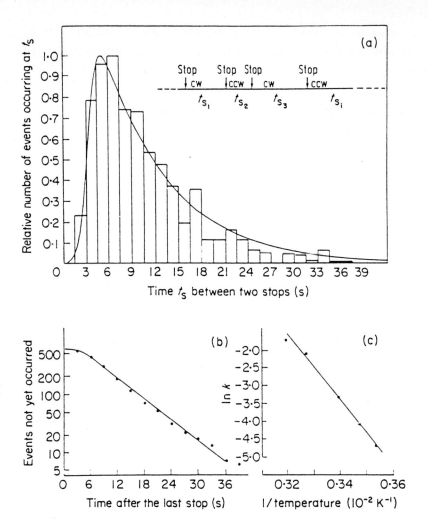

Figure 1. (a) Interval distribution at 34°C of run intervals t_s classified into bins of 1,5 s width. The total number of events (N_o) was 600 and 6 cells of strain Flx37 were used for the experiment. The drawn line is the result of a computer simulation of the kinetic model for the flagellar motor switch mentioned in the text and represents the first derivative (dB/dt) of the time-course of B.

(b) Number of events N_i that have not occurred at time t elapsed after the last stop plotted as a function of t. The Figure shows the integrated data from (a). The semilogarithmic plot has a constant slope after an initial phase of about 6 s.

(c) Temperature dependence of the first-order rate constant derived from experiments as in (a). k values were obtained from the plots as in (b) ($k = \ln 2/t_{1/2}$) and the activation energy was calculated. Taken from Marwan & Oesterhelt, 1987.

the distribution curve is exponential as in *E.coli* (Berg & Brown 1972). Figure 1(b) shows the semilogarithmic plot of the number of events still to be expected at time t after the last stop *versus* time. This exponential function suggests that the fraction of events dN occurring during an interval dt is proportional to the number (N) of events that have not yet occurred:

$$\frac{dN}{dt} = - k \, N. \tag{1}$$

This equation describes a first-order reaction and suggests that the switching of the flagellar motor occurs randomly during the exponential decay. In other words, a halobacterial cell behaves with respect to its flagellar motor like a radioactive atom with respect to the probability of its decay. The statistics of the run length distribution of a cell must therefore allow for an analytical description in terms of chemical kinetics. A first-order rate constant k can be calculated from the slope of the curve in Figure 1(b).

A simple switch between two states (A and B) of the flagellar motor representing cw and ccw rotation cannot explain the rise on the left side of the maximum in the distribution curve in Figure 1(a). It should also be noted that in *E.coli* it is possible that the shortest intervals are not the most frequent (Segall et al., 1982; Block et al., 1983). Therefore, *E.coli* and *H. halobium* may behave qualitatively in a similar fashion. A simple kinetic model that describes this phenomenon satisfactorily adds an intermediate state A_1 (B_1) to the transition from A (B) to B (A) (motor switching):

$$CW \longrightarrow CCW : \quad A \xrightarrow{k_1(1+A_1)} A_1 \xrightarrow{k_2} B$$

$$CCW \longrightarrow CW : \quad B \xrightarrow{k_1(1+B_1)} B_1 \xrightarrow{k_2} A$$

The conversion from A to A_1 is thought to be an autocatalytic process and the conversion of A_1 to B becomes the rate-limiting step in the course of the overall process. The occurrence of B is assigned to the visible motor switching from cw to ccw, whereas the transition from A to A_1 is not observable by monitoring the direction of movements. The transition from B to A has the same kinetics as A to B, since cw and ccw rotation have equal probability. The time-course of the probability for B (visible motor switching) to occur (equivalent to d[B]/dt) was calculated according to the model and plotted *versus* time in Figure 1(a). The values for $k_1 = 0,006$ s^{-1} and $k_2 = 0,115$ s^{-1} yielded the qualitative best fit to the experimental data of Figure 1(a) as judged by eye. The distribution of interval lengths was measured at temperatures between $10^{o}C$ and $40^{o}C$ and from the corresponding Arrhenius plot (Fig. 1c), an activation energy of 73 kJ/mol was calculated. This is equivalent to the average energy content of about 16 hydrogen bridges.

The photophobic transduction chain does not include a change of membrane potential.

In the dark, halobacteria carry out a random walk that is characterized by runs interrupted by spontaneous stops. After the onset of continuous illumination with blue light (step-up), the cells exhibit one additional stop, the so-called photophobic response. After such a stop the cell resumes its standard pattern of spontaneous reversals, even under continuous illumination, i.e. adaptation occurs within the average time between stops (Schimz & Hildebrand, 1985). If, instead of a step-up, an e.g. eight millisecond flash is applied to the cell, it takes more than one second until the flagellar motor responds with a stop (Fig. 2.). This is evidence that a light-independent step must be part of the photophobic tranduction chain.

Light-induced changes of the membrane potential as components of the signal tranduction chain have been discussed

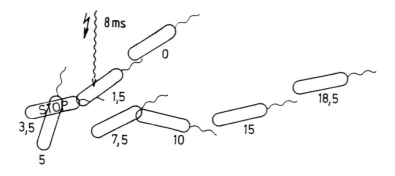

Figure 2. Photophobic reaction of *Halobacterium* after a blue light flash. The sequence of cell positions is redrawn from a video recording. After the flash (λ =480 nm, 8 ms duration) the cell continues to swim for more than a second until a reversal is observed. Taken from Marwan et al., 1987.

in the literature, but direct experimental evidence obtained by impaling a halobacterial cell with microelectrodes cannot be expected to be successful. A feasible experimental strategy, however, is the application of local light stimuli to a halobacterial cell during direct observation of the flagellar movement in the light microscope.

The smallest diameter of a light spot produced by a conventional light source through inversion of the optical path in the microscope used is about 2,3 μm. Standard halobacterial cells are about 3 μm in length and therefore experimentally unsuitable. By addition of aphidicolin, however, an inhibitor of the eukaryotic DNA polymerase to the nutrient broth, cells of up to 40 μm which contain no septa could be obtained (Forterre et al., 1984). These cells are either monopolarly or bipolarly flagellated and were used for microbeam irradiation (Oesterhelt and Marwan 1987).

Monopolar flagellated cells reacted with a photophobic response only when the flagella-carrying pole was irradiated. The response time, i.e. the time required after the stimulating flash for the flagellar motor to stop, was independent of the direction of rotation. No phobic response occurred when the middle or the end distal to the flagellated pole was irradiated.

Bipolarly flagellated cells behaved very similar to mono-polar cells. None of the flagellar motors reacted with a stop response when the cell was stimulated in the middle, but both ends were equally sensitive independent of the direction of flagellar rotation and of cellular transla-tional movement.

Figure 3 schematizes an experiment with a bipolarly fla-gellated cell which moved from left to right (1) before be-ing hit by a light flash in its middle part (2). The cell continued to move (3) until a flash was given to its back part (4), causing a stop (5) since both flagellar bundles excert an opposing force to the cell. After a second flash to the same pole (6), the cell started movement to the right again (7). The two flagellar bundles of a cell were not coordinated, since the second flagellar bundle did not respond to the stop of the irradiated first one.

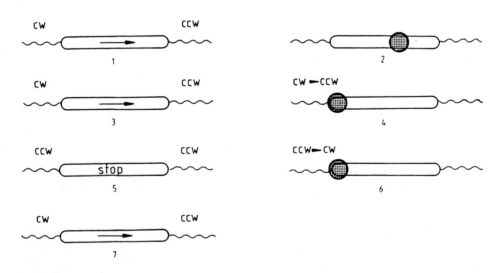

Figure. 3 Scheme of a microbeam irradiation experiment with a bipolar flagellated long halobacterial cell. The arrows indicate the translation direction of the cell, the dotted circles mark the position of the microbeam, and the numbers show the sequence of stimulation and response. Taken from Oesterhelt & Marwan, 1987.

By experiments of that kind on cells with lengths of 8 to 22 μm it could be shown that stimulation of one flagellated pole does not lead to a response of a second flagellar bundle beyond a distance of about 15 μm. This result was independent of the rotational mode of the flagellar bundle of the stimulated pole and independent of the movement of the cell. Furthermore, a flagellar motor does not stop if the stimulus is applied at a distance greater than 5 μm.

The short-range signalling in halobacteria excludes membrane potential changes as being directly involved, provided a potential electric signal does not decay over a path length of about 20 μm. According to the cable equation, a spatial decay of a membrane potential change to $1/e$ in halobacteria would occur only at a distance of 420 μm from the point of its origin.

In addition and independent of the result that membrane potential changes are not a component of the photophobic transduction chain, the photoreceptor could be localized at the poles of the cell.

Quantitative description of the behavior of the cells to blue light stimulation in the stochastic and kinetic range.

While the time span between spontaneous reversals of the flagellar motor (t_S) is about 20 seconds at $20^{\circ}C$, saturating flashes of blue light reduce this time to about 1-2 seconds (t_{min}), but not less (see also Fig. 2). This is evidence for rate-limiting dark reactions in signal formation (Marwan & Oesterhelt, 1987).

The time between the flash and the stop response of the cell is called the response time, t_R. It is a function of photon exposure (irradiance x time) and increases from t_{min} to about 2-5 seconds (t_{max}) upon weakening of the flashes. Further lowering the flash intensity led to the observation that a single cell does not always respond to a flash, thereby causing a split of the population of events into two groups with centers of frequency at t_{max} and t_S (Fig. 4).

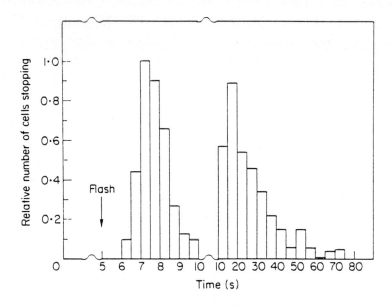

Figure 4. Distribution of response times upon a weak flash light stimulus (75 ms). Two groups can be distinguished. Light-stimulated events in the first group occur between 1 and 5 s after the flash, whereas the second group shows spontaneous reversals between 10 and 75 s after the last stop. The flash was applied 5 s after the last stop in order to assure that the cells were fully stimulatable. The total number of events were 570, irradiance was 1.1×10^{-4} mol $m^{-2}s^{-1}$, λ= 481 nm. Three cells of strain M407 were used. The same distribution was obtained with a single cell. Taken from Marwan & Oesterhelt 1987.

Clearly, at this range of photon exposure light reception and the photophobic response of the cell appear as stochastic processes. Thus, photon exposure regulates the ratio of probabilities for a response with t_{max} and for no photoresponse at all (t_s).

The dependence of the percentage of photophobic events on the photon exposure could then be measured and the results obtained analyzed with the help of Poisson statistics. This approach was used first by Hecht et al. (1942) to derive the minimal photon requirement in human vision and was recently applied to the *Chlamydomonas* photophobic response (Hegemann & Marwan, 1988).

In Figure 5 the experimental data are compared with

theoretical curves obtained for processes with different photon input requirement. The theoretical curves were calculated with the equation derived from the Poisson distribution: The probability $\hat{P}_n(\alpha)$ that n or more photons are effective is 1 minus the probability that less than n photons are effective.

$$\hat{P}_n(\alpha) = 1 - \sum_{i=0}^{n-1} \frac{\alpha^i e^{-\alpha}}{i!} \tag{2}$$

The average number of photons contributing to a photophobic response in many experiments is α. The value of α is proportional to the photon exposure F, the photoreceptor number R, the absorption cross-section σ and the quantum yield ϕ

$$\alpha = \text{const. F R } \sigma \phi \tag{3}$$

and

$$\log \alpha = \log F + \log (R \sigma \phi \text{ const.}). \tag{4}$$

Since a cell responds when the minimum number n or more photons are effective, the comparison of experimental and theoretical data requires plots which describe the probability that n = 1 photon or more, n = 2 photons or more, and so on induce a response as a function of the logarithm of photon exposure.

$\hat{P}_n(\alpha)$ is plotted for the n values 1 to 9 *versus log* α (Fig. 5, inset lower right) and experimental data for different or for individual cells fit only the curve for n=1, which is evidence that at minimum the photophobic response of the halobacterial flagellar motor is elicited by one absorbed photon.

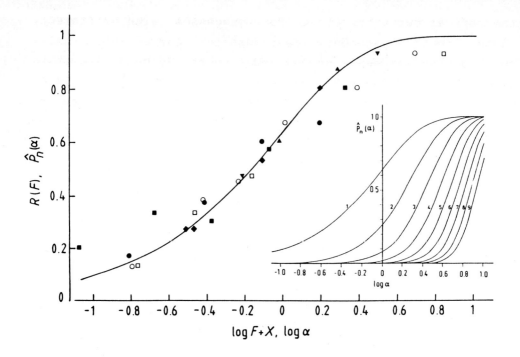

Figure 5. Minimal photon requirement for the photophobic response in *Halobacterium halobium* strain Flx37. Flashes of 50 ms were applied 2 s after a spontaneous reversal and the relative frequency of the response R(F) was plotted versus log photon exposure. Each cell was stimulated only once (○), each of the remaining symbols corresponds to one cell which had been repeatedly stimulated (about 15 times at each exposure). Each set of data was fitted using a least-squares procedure to one of the theoretical curves of the inset (lower right) by a linear shift on the abscissa. All data were fitting only the curve n=1, but with slightly different values for X, indicating slight variations of the photoreceptor concentration or the constant. At log α = 0 the photon exposure F was 8.12 x 10^{-6} mol m^{-2}, λ=481 nm. The results obtained were temperature independent. Taken from Marwan et al., 1988.

Further experiments were aimed at describing quantitatively the responses of halobacterial cells to blue light stimuli consisting of single and double flashes with varied time duration, intensity and dark interval. In these experiments the flash intensities were always such that response times between t_{min} and t_{max} were measured (see above). (A scheme of a double-flash experiment is shown in Fig. 6a.)

Qualitatively, the following observations were made.

(1) If a flash of a given duration, intensity and constant wavelength was split into two flashes of half the original duration, the response time, t_R, measured depended on the dark period (D) between the flashes.

(2) If two flashes of unequal time of duration but a constant dark period in between were applied, a longer t_R value was observed when the short flash was applied first, but a shorter t_R value was measured when the long flash was applied first (inversion experiment).

From the result of the inversion experiment, it seemed that the signal production depends on time, and that the time t_R, at which a critical signal concentration is reached depends on the length of time between the application of photons (irradiance I x flash duration τ) and the response. In other words, the sum of $I_1\tau_1 t_R$ and $I_2\tau_2(t_R-D)$ should describe the efficiency of a double flash in eliciting a stop response. Figure 6 represents a plot of data for the response time, t_R, measured as a response to double flashes of varying duration and dark intervals but constant intensity (dots and triangles). Clearly all data fit a straight line obeying the equation:

$$I\tau_1 t_R + I\tau_2(t_R-D) = t_{min} I(\tau_1 + \tau_2)+b, \qquad (5)$$

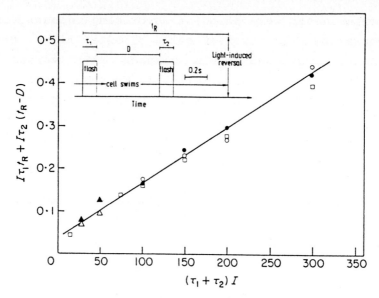

Figure 6 Experimental verification of an equation describing the response of a cell to single and double flashes. The response time t_R was measured as function of double flashes of varying time and dark intervals (D) but constant irradiance ($I=4.3\times10^{-4}$ mol m^{-2} s^{-1}, $\lambda=520$ nm) and $t_R I \tau_1 + I \tau_2 (t_R - D)$ plotted *versus* $\tau_1 + \tau_2$.
Symbols indicate: ($\bullet$) $\tau_1:\tau_2=3:1$, D=800 ms; ($\blacktriangle$) $\tau_1:\tau_2=1:1$, D=800 ms; (O) $\tau_1:\tau_2=1:3$, D=800 ms; ($\triangle$) $\tau_1:\tau_2=1:1$, D=50 ms. The response time to single flashes at constant irradiance ($I=4.3\times10^{-4}$ mol m^{-2} s^{-1}, $\lambda=520$ nm) was then measured and $t_R I \tau$ plotted *versus* τ ($\square$) ($I\tau_2(t_R-D)=0$). The numerical values are all given with the relative value for I of 1. Strain M407 was used. Taken from Marwan & Oesterhelt, 1987.

where t_{min} and b are constants. The equation describing the response to double flashes is then:

$$t_R = t_{min} + \frac{b}{I\tau} + \frac{\tau_2}{\tau} D \qquad (6)$$

The equation above with $\tau=\tau_1+\tau_2$ is converted into an equation describing the response time to single flashes by setting either D or τ_2 to zero:

$$t_R = t_{min} + \frac{b}{I\tau} \qquad (7)$$

As predicted by the equation, a plot of the response time t_R versus the inverse photon irradiance of a flash yields a straight line (Fig. 7).

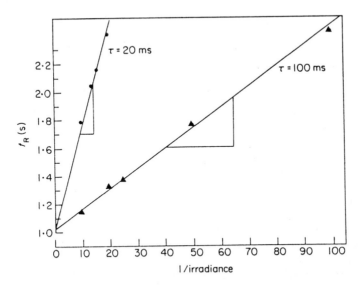

Figure 7. Variation of irradiance. The response time in seconds to flashes of constant duration and varying intensities was measured and plotted *versus* the reciprocal of intensity. Two curves are shown, for 20 ms and 100 ms flash duration. As predicted by eqn (7) the ratio of the slopes of the curves equals the reciprocal ratio of the flash times. A 1/I value of 10 corresponds to 8.63×10^{-4} mol m^{-2} s^{-1} at $\lambda=520$ nm, cells of the strain M407 were used and 90 events evaluated for each data point. Taken from Marwan & Oesterhelt, 1987.

Biosynthesis and isolation of the photoreceptor molecules

The blue light photophobic response in *H. halobium* was first reported by Hildebrand & Dencher (1975) and later a potential photoreceptor was analyzed in detail by Spudich & Bogomolni (1984). Sensory rhodopsin, as this photoreceptor was named, behaves like a photochromic pigment. It absorbs orange light at 587 nm (SR_{587}) and is converted into an intermediate state absorbing at 373 nm (SR_{373}). This intermediate was thought to induce the photophobic response of the halobacterial cell upon absorption of blue to near UV light (Spudich & Bogomolni, 1984). Two facts, however, could not be explained by this proposal. (1) The original action spectrum of Hildebrand & Dencher (1975) has a second maximum around 520 nm, which is red-shifted compared to the isosbestic point of SR_{587} and SR_{373} at about 410 nm (Bogomolni & Spudich, 1982). (2) Observation of cells with infrared light ($\lambda>780$ nm), revealed that blue light ($\lambda<530$ nm) alone causes a photophobic response, i.e. no orange background light is necessary to produce the blue light sensitivity (Marwan & Oesterhelt, 1987). The action spectrum of this background light-independent response is shown in Figure 8. It has a maximum at about 480 nm and was therefore called P_{480} (SR-II, phoborhodopsin). A similar observation was made independently by Takahashi et al., (1985). P_{480} must be a retinal protein since growth of the cells in the presence of 3 mM nicotine, which blocks retinal biosynthesis, prevents the blue light photophobic response altogether. Experimental results similar to those described above were also found for SR when photophobic behaviour of the cells was analyzed on a orange background light and single actinic light flashes of 366 nm were applied. The fact that in both cases equation (7) holds true indicates that both photoreceptors are linked to the motor by the same type of signal chain.

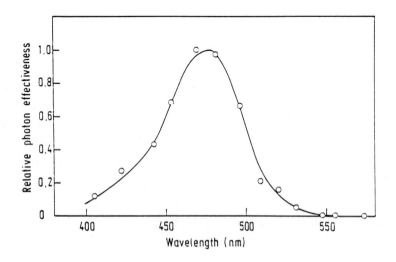

Figure 8. Action spectrum of the P_{480}-mediated photophobic response in *H. halobium* strain Flx37. Taken from Marwan & Oesterhelt, 1987.

The time course of cellular concentration changes of SR and P_{480}, however, is different. Figure 9 compares the sensitivity of the two photoreceptors during growth. SR is induced as are the ion pumps bacteriorhodopsin and halorhodopsin, but the P_{480} response is constant in the cells throughout the growth phase, except for very old cells in the stationary phase at more than 200 hours. Thus, very young cells exhibit only the P_{480} response, but in early stationary phase cells the SR response dominates before it declines to almost zero levels again. It could be shown, that the decrease of sensitivity in stationary state cells is not due to a decrease of the photoreceptor concentration but to a decrease of the efficiency of the signal transduction chain (Otomo et al., submitted).

The inducible sensory rhodopsin and the constitutive P_{480} together provide the cell with a perfect light control system in the search for orange light and the avoidance of blue light. While the blue light avoidance is important, independent of the bioenergetic basis on which the

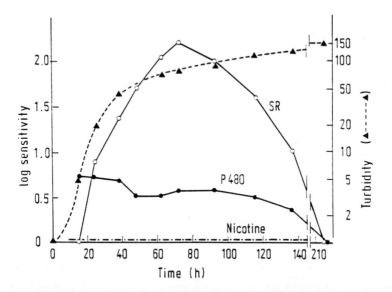

Figure 9. Time-course of blue light sensitivity mediated by P_{480} and sensory rhodopsin (SR) during growth of *H. halobium* Flx37. Taken from Marwan & Oesterhelt, 1987.

halobacterial cell lives, depletion of either arginine for fermentation or oxygen for respiration creates a situation where the photosynthetic apparatus must be induced and the active search for green to orange light by SR begins.

For their structures as members of the opsin family, and their function as prokaryotic photoreceptors in a photocatalytic signal chain, both, the P_{480} and the SR molecule deserve much biochemical interest. A comparison of the light sensors with the light-driven halobacerial ion pumps and with the eukaryotic rhodopsins should be of value for understanding the mechanism of photochemistry in retinal proteins.

Recently, Schegk & Oesterhelt (1988) succeeded in isolating one of the halobacterial photoreceptors, sensory rhodopsin, from a mutant, defective in BR and HR.

The assay for SR used for purification was based on its photochemical activity. Illumination with actinic light (550 nm) led to a photostationary mixture consisting of the initial state population, SR_{580}, which absorbed maximal at 580 nm, and an intermediate absorbing at about 380 nm (SR_{380}) which was characterized by a decrease at 380 nm. The kinetics of the reconversion of SR_{380} into SR_{580} after the actinic light was turned off were followed. Using this activity assay, the photoreceptor sensory rhodopsin was isolated from halobacterial cell membranes solubilized in lauryl-maltoside. In the presence of retinal, detergent and salt the native protein was obtained in pure form by sucrose density gradient centrifugation, hydroxyapatite chromato-graphy and gel filtration. The apparent mol. wt of the molecule was 24 kd if analyzed by SDS gel electrophoresis, and 49 kd by sedimentation and size-exclusion chromatogra-phic analysis, suggesting that it occurs as a dimer in vivo. The chromoprotein had an absorption maximum at 580 nm which was 8 nm blue-shifted compared to the membrane-bound state. The molecule was photochemically active and the action spectrum for formation of SR_{380}, the long-lived intermediate, coincided with the absorption spectrum. N-terminal sequencing of some peptides obtained by proteolytic cleavage of SR allowed for the isolation of the SR-gene (Blanck & Oesterhelt, in preparation), revealing the primary structure of the protein.

The primary structure of SR will be of great interest for the following reasons: assuming a seven α-helical motif as found for the retinal proteins thus far analyzed, a mol. wt of ~ 24 kd would not allow for large interhelical loops as found in bovine and other eukaryotic opsins. It is known that these loops interact with a component of the signal transduction chain, a G-protein, and also with protein kinases and phosphatases (Stryer, 1986). Thus, the *cis-trans* isomerization in rhodopsin apparently causes conformational changes in the helical connections mediating signal transduction and adaptation. In the ion pumps BR and HR, the helical segments form an intramolecular pore and have only very small interhelical loops because their function requires only the pore and retinal's *cis-trans* isomerization as an active switch to mediate ion translocation. SR as a member of the archaebacterial retinal protein family acting as a sensor should behave in the same way as rhodopsin but the mol. wt of its polypeptide chain falls into that of the pumps. Therefore, it will be of interest to see whether signal transduction is mediated by the loops or by the intramembrane part of the molecule.

References

Alam M, Oesterhelt D (1984) Morphology, function and isolation of halobacterial flagella. J Mol Biol 176:459-475

Berg HC, Brown DA (1972) Chemotaxis in *Escherichia coli* analysed by three-dimensional tracking. Nature 239:500-504

Block SM, Segall JE, Berg HC (1983) Adaption kinetics in bacterial chemotaxis. J Bacteriol 154:312-323

Bogomolni RA, Spudich JL (1982) Indification of a third rhodopsin-like pigment in phototactic *Halobacterium halobium*. Proc Natl Acad Sci USA 79:6250-6254

Forterre P, Elie C, Kohiyama M (1984) Aphidicolin inhibits growth and DNA synthesis in halophilic archaebacteria. J Bacteriol 159:800-802

Hecht S, Shlaer S, Pirenne MH (1942) Energy, quanta and vision. J Gen Physiol 25:819-840

Hegemann P, Marwan W (1988) Single photons are sufficient to trigger movement responses in *Chlamydomonas reinhardtii*. Photochem Photobiol 48:99-106

Hildebrand E, Dencher N (1975) Two photosystems controlling behavioural responses of *Halobacterium halobium*. Nature 257:46-48

Houwink AL (1956) Flagella, gas vacuoles and cell-wall structure in *Halobacterium halobium*; an electron microscope study. J Gen Microbiol 15:146-150

Lanyi J (1984) Bacteriorhodopsin and related light-energy converters. In: Ernster L (ed): Comparative biochemistry bioenergetics. Elsevier Amsterdam:315-335

Macnab RM (1976) Examination of bacterial flagellation by dark-field microscopy. J Clin Microbiol 4:258-265

Macnab RM, Ornston MK (1977) Normal-to-curly flagellar transitions and their role in bacterial tumbling. Stabilisation of an alternative quaternary structure by mechanical force. J Mol Biol 112:1-30

Marwan W, Alam M, Oesterhelt D (1987) Die Geisselbewegung halophiler Bakterien. Naturwiss 74:585-591

Marwan W, Hegemann P, Oesterhelt D (1988) Single photon detection in an archaebacterium. J Mol Biol 199:663-664

Marwan W, Oesterhelt D (1987) Signal formation in the halobacterial photophobic response mediated by a fourth retinal protein (P480). J Mol Biol 195:333-342

Oesterhelt D, Krippahl G (1983) Phototrophic growth of halobacteria and its use for isolation of photosynthetically deficient mutants. Ann Microbiol (Inst Pasteur) 134:137-150

Oesterhelt D, Marwan W (1987) Change of membrane potential is not a component of the photophobic transduction chain in *Halobacterium halobium*. J Bacteriol 169:3515-3520

Otomo I, Marwan W, Oesterhelt D, Desel H, Uhl R (1989) Biosynthesis of the two halobacterial light sensors P_{480} and sensory rhodopsin and variation in gain of their signal transduction chains. J Bacteriol (submitted)

Schegk ES, Oesterhelt D (1988) Isolation of a prokaryotic photoreceptor: sensory rhodopsin from halobacteria. EMBO J 7:2925-2933

Schimz A, Hildebrand E (1985) Response regulation and sensory control in *Halobacterium halobium* based on an oscillator. Nature 317:641-643

Segall JE, Manson MD, Berg H (1982) Signal processing times in bacterial chemotaxis. Nature 296:855-857

Spudich E, Spudich JL (1982) Control of transmembrane ion fluxes to select halorhodopsin-deficient and other energy-transduction mutants of *Halobacterium halobium*. Proc Natl Acad Sci USA 79:4308-4312

Spudich JL, Bogomolni RA (1984) Mechanism of colour discrimination by a bacterial sensory rhodopsin. Nature 312:509-513

Stryer L (1986) Cyclic GMP cascade in vision. Ann Rev Neurosci 9:87-119

Takahashi T, Tomioka H, Kamo N, Kobatake Y (1985) A photosystem other than PS 370 also mediates the negative phototaxis of *Halobacterium halobium*. FEMS Microbiol Lett 28:161-164

Wagner G, Oesterhelt D, Krippahl G, Lanyi JK (1981) Bioenergetic role of halorhodopsin in *Halobacterium halobium* cells. FEBS Lett 131:341-345

CONTROL OF BACTERIAL GROWTH BY MEMBRANE PROCESSES

K.van Dam, P.W.Postma, H.V.Westerhoff, M.M.Mulder and M.Rutgers
Laboratory for Biochemistry
University of Amsterdam
Plantage Muidergracht 12
1018 TV Amsterdam
The Netherlands

INTRODUCTION

Microorganisms have an enormous capacity to adapt to changing environmental conditions. They can often survive on many different carbon, nitrogen and energy sources by changing their enzymic makeup. Also they can slow down their metabolism to a very low level when one essential nutrient is exhausted and rapidly be reactivated when a fresh supply of that nutrient arrives.

During growth on a given combination of nutrients, the metabolism of a microorganism has to satisfy a number of criteria simultaneously. The most obvious are the restrictions imposed by the metabolic machinery: there must be redox balance and energy balance. Although these may seem trivial boundary conditions, the actual growth of a microorganism depends critically on the possibility to convert substrates into end products according to a metabolic scheme that leaves no redox equivalents or ATP.

In competitive situations, where other microorganisms share the use of some substrates, a further requirement for survival is the capacity to use the available nutrients with speed and efficiency. In some cases, the presence of more than one carbon source for growth may lead to a preferential order of use, resulting in the consumption of the 'easiest' substrate first.

In all cases, the microorganism senses the availability of nutrients primarily at the level of the membrane and, therefore, it is not surprising that adaptive responses occur to a large extent at the membrane. The entry and accumulation of substrates

NATO ASI Series, Vol. H29
Receptors, Membrane Transport and Signal Transduction
Edited by A. E. Evangelopoulos et al.
© Springer-Verlag Berlin Heidelberg 1989

303

across the membrane is highly regulated. Transport systems are usually involved in this entry and, by coupling to energy yielding reactions they can lead to the accumulation of the substrates.

Scavenging of the substrate that is most limiting in availability is of great importance, especially in competitive situations. In theory, one can envisage two ways in which the rate of uptake of any substrate can be enhanced: by increasing the affinity of the transport system or by increasing its maximal rate. In principle, the rate of a process is proportional to the free-energy change of that process. Therefore, the two ways of enhancing the uptake of a substrate can also be formulated as improvement of the kinetic capacity of the transporter or enlarging the free-energy change, for instance by coupling to an energy yielding reaction (such as ATP hydrolysis).

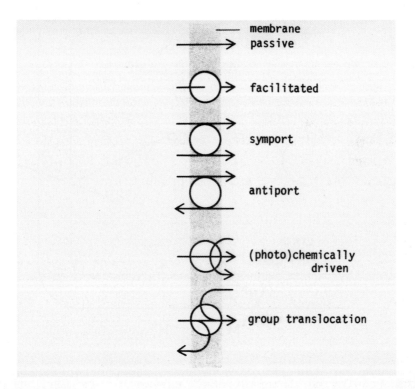

Fig.1 Classification of transport systems across biomembranes
We can schematically depict the different types of transport as in Fig.1. It should be emphasized that the transition between the different types of transport is sometimes rather arbitrary.

THE PEP:CARBOHYDRATE PHOSPHOTRANSFERASE SYSTEM (PTS)

As an example for the intricate coordination between different
uptake systems for carbohydrates, we will describe the regulatory
role of the PEP:carbohydrate phosphotransferase system (PTS; for
review see Postma and Lengeler, 1985). It will be shown that
understanding of the properties of this system leads to an
understanding how microorganisms can preferentially use one
particular carbon substrate (glucose) before it starts to convert
others (such as lactose). Lactose is one of the best studied
examples of a simple sugar-proton symport system; such a system
catalyzes the accumulation of a sugar until the gradient of sugar
is equivalent to that of the proton (in opposite direction).

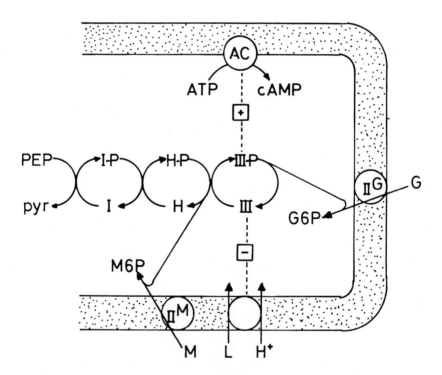

Fig.2 Scheme of the PEP:carbohydrate phosphotransferase system
and its regulating interactions. Enzyme I, I: HPr, H; Enzyme
III^{Glc}, III; adenylate cyclase, AC; glucose, G; mannose, M;
lactose, L.

The general scheme for the PTS is given in Fig.2. It is composed of two 'general' proteins, Enzyme I and HPr. These proteins can be phosphorylated and transfer the high-energy phosphate group of PEP to a next acceptor; this can be either an Enzyme II (for instance for mannitol) or an Enzyme III (for instance for glucose). In the latter case the phosphate group is further transferred to an Enzyme II in the membrane. The Enzymes II and III are specific for a particular carbohydrate. The Enzymes II ultimately transfer the phosphate group to the carbohydrate **in the process** of transmembrane movement. Thus, the carbohydrate ends up in the cell in its phosphorylated form. The PTS is a typical example of a group translocating system. It may be interesting to note that the phosphate group is bound to the different proteins at histidine residues, alternatingly at the N-1 or N-4 position.

Evidently, different PTS carbohydrates will compete with each other for the available phosphate groups and can inhibit the uptake of other PTS carbohydrates. However, it turns out that glucose inhibits also the uptake of non-PTS substrates. The mechanism of this inhibition is the central issue of this paragraph. To understand it, we first look at the phenotype of a number of mutants defective in the PTS (Fig.3).

Wild-type Salmonella typhimurium will grow on PTS carbohydrates and a number of other substrates. Growth on many of these other substrates is impossible in a strain in which Enzyme I has been deleted (ptsI). This is unexpected, since it would seem that growth of the non-PTS substrates does not require the presence of Enzyme I. Even less expected is the finding that a second deletion, namely that of Enzyme III^{Glc} (crr) leads to restoration of growth on some of the non-PTS substrates, namely maltose, melibiose and glycerol. Growth on citrate, succinate and xylose is not restored by the second mutation.

The common denominator in the last group of substrates is that the synthesis of the enzymes, required for their uptake and metabolism is stimulated by cAMP. Indeed, the addition of cAMP to the growth medium stimulates growth of a ptsI mutant on these substrates. It could, furthermore, be shown that the adenylate cyclase activity in the different mutants is correlated with the

	pts^+	$ptsI$	$ptsI\ crr$
PTS sugars	+	-	-
Maltose, melibiose, glycerol	+	-	+
Citrate, succinate, xylose	+	-	-
Galactose	+	+	+

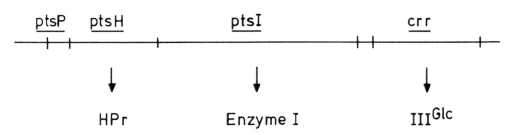

Fig.3 Growth behaviour of <u>Salmonella typhimurium</u> on different carbon substrates. Wild type, <u>pts</u>$^+$; mutant with Enzyme I defect, <u>ptsI</u>; mutant with Enzyme I and Enzyme III^{Glc} defect, <u>ptsI</u> <u>crr</u>. The scheme depicts the different gene products of the pts operon.

level of phosphorylation of Enzyme III^{Glc}: when phosphorylated Enzyme III^{Glc} is present, adenylate cyclase is stimulated.In the absence of Enzyme I, Enzyme III^{Glc} is completely dephosphorylated and at the same time adenylate cyclase activity is low. A low activity of adenylate cyclase is also found in the absence of Enzyme III^{Glc} (<u>crr</u> mutant).

These findings together lead to the postulate that phosphorylated Enzyme III^{Glc} is an activator of adenylate cyclase and, thereby, of the synthesis of all enzymes that have cAMP-dependent expression. The degree of phosphorylation of Enzyme III^{Glc} will depend on the rate of its phosphorylation and dephosphorylation. If phosphorylation is impeded because Enzyme I is inactive, there will be less P-III^{Glc}, leading to less activity of adenylate cyclase and, thus, less cAMP is present and synthesis of these enzymes is inhibited. One can predict that also enhanced dephosphorylation of P-III^{Glc}, for instance by adding α-methylglucoside (αMG, a non-metabolizable substrate for the glucose specific Enzyme II) will lead to inhibition of adenylate cyclase. Indeed, growth of <u>Salmonella</u> on citrate, succinate or xylose is inhibited by the presence of small amounts of αMG in the medium.

The effect of Enzyme III^{Glc} on substrates such as maltose, melibiose and glycerol (and lactose in the case of <u>E.coli</u>) is different. In these cases, further deletion of Enzyme III^{Glc} reverts the effect of deletion of Enzyme I alone. It would seem that Enzyme III^{Glc} itself has an effect on some step in the metabolism or uptake of these substrates. This effect can be directly demonstrated: uptake of for instance glycerol or maltose is inhibited by the presence of αMG. If Enzyme III^{Glc} is not present (<u>crr</u> mutant), this effect is also absent.

An attractive hypothesis to explain this effect is that the non-phosphorylated form of Enzyme III^{Glc} is an inhibitor of glycerol or maltose uptake, while the phosphorylated form has no effect. This hypothesis has been tested most directly for the case of the lactose carrier in <u>E.coli</u>. In <u>E.coli</u> lactose uptake is inhibited by αMG.

To study the interaction , Enzyme III^{Glc} has been purified and its interaction with the lactose transporter in <u>E.coli</u> has been studied directly. As shown in Fig.4, Enzyme III^{Glc} can bind to vesicles, that contain a high concentration of the lactose carrier. This binding is abolished if the Enzyme III^{Glc} is phosphorylated by the addition of the appropriate enzymes and PEP. The stoichiometry of binding indicates that a 1:1 complex between carrier and Enzyme III^{Glc} is formed. Furthermore, the binding is dependent on the presence of a substrate for the lactose carrier. In fact, the interaction between Enzyme III^{Glc} and lactose carrier has also been demonstrated with the purified carrier, incorporated in liposomes.

In the case of glycerol, the inhibitory interaction with Enzyme III^{Glc} turns out to be not at the level of the transporter, but with the first enzyme of intracellular metabolism, glycerol kinase. Also in this case, the binding of Enzyme III^{Glc} is

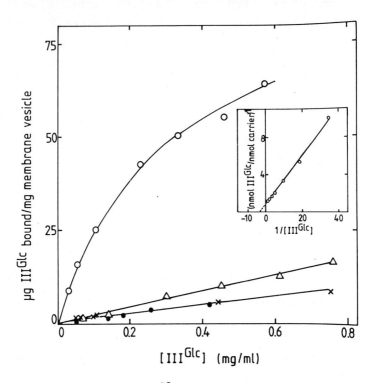

Fig.4 Binding of Enzyme III^{Glc} to membranes of <u>Escherichia coli,</u> <u>containing</u> the lactose carrier (Nelson et al, 1983). o, complete; $\triangle$, minus lactose; ●, plus PEP; x, membranes with less lactose carrier.

dependent on it being nonphosphorylated and on the presence of a substrate for the glycerol kinase. The requirement for a substrate of the target protein to be inhibited 'spares' Enzyme III^{Glc} for interaction with those systems that might be active. The many different functions of Enzyme III^{Glc} must of course influence each other. Since there is only a limited number of molecules in a cell, the most obvious effect will be that of saturation: if all Enzyme III^{Glc} molecules are involved in one process, they are not available for another. This can be shown nicely by the effect of different levels of induction of one uptake system on the extent to which it can be inhibited by nonphosphorylated Enzyme III^{Glc} (Fig.5). Uptake of maltose by

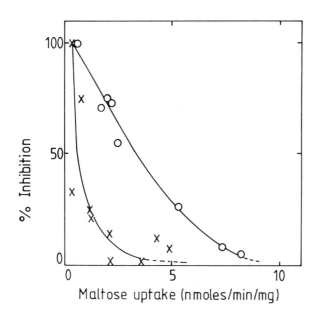

Fig.5 Inhibition of maltose uptake in <u>Salmonella typhimurium</u> by 2-deoxyglucose. The rate of maltose uptake was varied by varying the degree of induction. o, wild type; x, a <u>crr</u> mutant, containing much less Enzyme III^{Glc}.

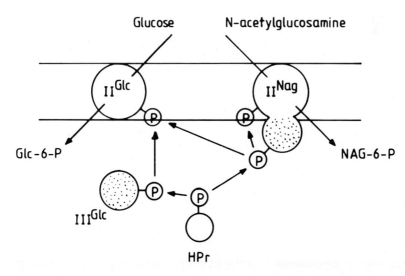

Fig.6 Proposed model for the transfer of phosphate groups in the PTS. In some cases (such as the Enzyme II for N-acetylglucosamine) the equivalent of Enzyme III is part of the membrane-bound Enzyme II.

S.typhimurium can be inhibited by the addition of 2-deoxyglucose, a non-metabolizable analog of mannose. This inhibition is almost complete in cells that have a low maltose uptake activity. However, the higher the maltose uptake activity, and therefore the number of transport molecules, the less it can be inhibited by 2-deoxyglucose. Apparently, under these conditions the number of Enzyme III^{Glc} molecules is not large enough to interact with all maltose transporters.

The effects of Enzyme III^{Glc}, described above, give a good explanation for the phenomenon known as 'diauxie'. An example of diauxie is the growth of Escherichia coli on glucose plus lactose. The cells will grow in two bursts: the first period of growth is completely on glucose, the second on lactose. It will be clear that during the growth on glucose, Enzyme III^{Glc} is continually dephosphorylated and, therefore, lactose can not enter the cells. At the same time, adenylate cyclase is not stimulated and the resulting lower level of cAMP causes a decreased synthesis of the metabolic enzymes. Only when all the glucose has been utilized will the cells readjust and start to use the lactose. This latter process takes some time, since the uptake and metabolic enzymes for lactose must first be induced. This is why the effect of glucose on uptake of lactose has also sometimes been called 'inducer exclusion'.

It must be noted that the phenomenon of inducer exclusion depends on the concentration of substrates. In a batch growth experiment the concentration of glucose will be high most of the time and, therefore, other substrates will be excluded from the cells. However, in a chemostat under carbon limited conditions the concentration of glucose will be so low that it can no longer effectively dephosphorylate Enzyme III^{Glc}, so that the steady-state degree of phosphorylation is higher and inhibition of uptake of other substrates diminishes or disappears altoghether. Thus, it is possible to grow Escherichia coli in the chemostat on a mixture of glucose and lactose, both of which are simultaneously consumed virtually completely.

The properties of Enzyme III^{Glc} represent an interesting example of regulation of metabolism by protein phosphorylation.It was found that there exist extensive homologies at the protein level between Enzyme III^{Glc} and some of the Enzymes II (Bramley and Kornberg, 1987; Rogers et al, 1988). These may be summarized as indicating that some Enzymes II contain a sequence of amino acids that can be recognized, and may function, as Enzyme III^{Glc}. In fact, it had been noted already that there was a membrane-bound Enzyme III^{Glc}-like activity in some strains of <u>Salmonella</u>. Thus, we arrive at the picture that there are two alternative routes for the transfer of phosphate from HPr to a carbohydrate via an Enzyme II: either via a specific Enzyme III or via <u>intra</u>molecular transphosphorylation within the Enzyme II itself (Fig.6) Furthermore, there may occur <u>inter</u>molecular transfer of phosphate groups, accounting for the membrane-bound residual Enzyme III^{Glc} activity.

EFFICIENCY OF BACTERIAL GROWTH: DISSIPATIVE PROCESSES

As mentioned before, growth of bacteria must occur in a balanced way; the substrates that enter the cell and the products that leave it (including new biomass) must be in redox balance. Furthermore, all ATP that is generated in catabolism must be used up somehow in anabolism or other dissipative processes.

We may consider a bacterium as a black box, that converts the free energy contained in metabolizable substrates into free energy contained in biomass and other products. Stucki (1980) has studied the properties of such an idealized energy converter. In particular, he has derived that such energy converters may be optimized towards certain output parameters, as a consequence of the constraints that thermodynamics sets to the system. The different output functions and their respective maximal efficiencies are listed in Fig.7. Each output function corresponds with a characteristic maximal efficiency.

Fig.7 MAXIMAL EFFICIENCY OF ENERGY CONVERTERS OPTIMIZED TOWARDS
    DIFFERENT OUTPUT PARAMETERS (STUCKI, 1980)

Output parameter	Maximal efficiency
Maximal rate (J_a)	0.24
Maximal power ($J_a \cdot \Delta G_a$)	0.41
Economic rate ($J_a \cdot \eta$)	0.54
Economic power ($J_a \cdot \Delta G_a \cdot \eta$)	0.62

We have studied the efficiency of growth of <u>Pseudomonas</u>
<u>oxalaticus</u> on different carbon substrates, by growing the
microorganism aerobically in the chemostat under carbon limited
conditions. The results are given in Fig.8 as a plot of the rate
of catabolism as a function of the rate of anabolism (=growth
rate). It is clear that the rate of catabolism is higher the more

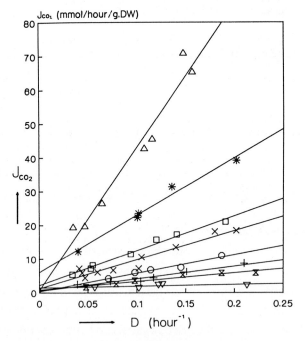

Fig.8 CO_2 production by <u>Pseudomonas oxalaticus</u> as a function of
the growth rate (J_a) in the chemostat. Oxalate, Δ ; formate, *;
glyoxylate, □ ; tartrate, x; citrate, o; acetate, +; fructose, X ;
ethanol, ∇ .

oxidized the growth substrate is. This is to be expected, because the more oxidized substrates will generate less ATP during their oxidation and also more ATP will be required to convert these substrates into biomass.

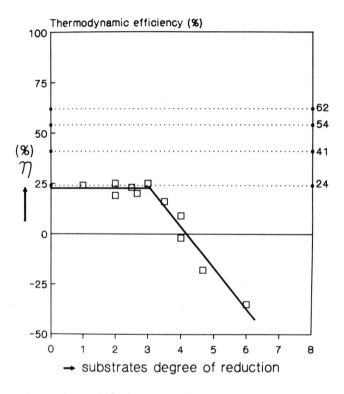

Fig.9 Thermodynamic efficiency of growth of P.oxalaticus as a function of the degree of reduction of the carbon substrate. The lines indicated in the figure correspond to the maximal efficiencies belonging to the output parameters defined by Stucki (see Fig.7)

Fig.9 represents the calculated thermodynamic efficiency of growth of P.oxalaticus as a function of the degree of reduction of the substrate. In the figure the efficiencies, corresponding with the different optimisation parameters of Stucki (1980), have been drawn. It is clear that with the most 'difficult' substrates the thermodynamic efficiency of growth approaches that of an energy converter optimized towards optimal efficiency at maximal rate. From the point of view of evolution this would seem quite reasonable: growth must be high to outcompete other microorganisms but, at the same time, with the highest achievable efficiency so as to produce as much offspring as possible to be able to survive periods of nutrient shortage.

Although the results, obtained by the black-box approach, are interesting, they do not make use of the available biochemical knowledge concerning the bacterial metabolism. We have developed a procedure that combines the use of the thermodynamic formalism with as much biochemical information as required to give a realistic description of the system. We called this formalism Mosaic Non-Equilibrium Thermodynamics (MNET; Westerhoff and van Dam, 1987). To show its use, we open the black box of the bacterium and consider the inside to be composed of three elements (Fig.10): catabolism, leading to generation of ATP, anabolism, leading to the utilization of ATP and an ATP 'leak', representing all other reactions that lead to the dissipation of ATP. Using this simple model, we can derive a number of equations that describe how anabolism and catabolism are related and what effects changes in the activities of the three elements would have on overall metabolism.

We asked ourselves the question: what kind of limitation describes best the situation of K^+-limited growth of Escherichia coli? Such a question could be answered, because the equations predict that catabolism depends much more strongly on the growth rate during catabolite-limited growth than under anabolite limitation. Furthermore, induction of extra ATP 'leak' should increase this dependency under catabolite limitation, but decrease it under anabolite limitation.

E.coli FRAG5 was grown in the chemostat with limiting amounts of K^+ in the medium (Mulder et al, 1986). A plot of ATP turnover (calculated from substrate utilization and product formation) versus growth rate gave a straight line, as expected (Fig.11). If the experiment was repeated in the presence of a certain amount of the uncoupler 2,4-dinitrophenol in the medium, ATP turnover was higher at all growth rates. This indicates that K^+-limited growth corresponds to a catabolite limitation.

E.coli FRAG5 differs from the parent strain E.coli FRAG1 in that

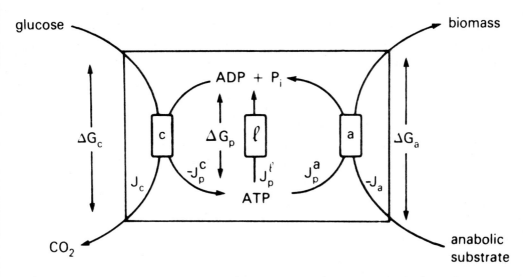

Fig.10 The model of bacterial metabolism, used in Mosaic Non-Equilibrium Thermodynamics (MNET; Westerhoff and van Dam, 1987)

it possesses only the low-affinity K^+-uptake system (Trk) but lacks the inducible high-affinity K^+ uptake system (Kdp). Fig.11 shows that FRAG1 behaves like FRAG5 plus 2,4-dinitrophenol as far as ATP turnover concerns. The explanation that we propose for this is that the presence of low- and high-affinity K^+ uptake systems simultaneously leads to a futile cycling of K^+ across the membrane. The high-affinity system would generate such a high gradient of K^+ that it would leak out again via the low-affinity system; the latter process is supposed to regenerate less ATP equivalents than the first has consumed. Thus, the presence of two different K^+ uptake systems simultaneously is equivalent to an ATP 'leak' in the cell.

One might wonder why the wild-type E.coli cells would maintain such a drain on their ATP supply. The answer to this question may be manyfold. However, two reasons can be given on the basis of experiments. The first is that the presence of two systems simultaneously gives the cells an advantage in reacting to a changing supply of K^+: FRAG1 has a much shorter time lag in

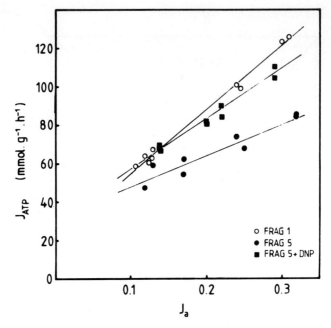

Fig.11 ATP turnover (J_{ATP}) in <u>Escherichia coli</u> cells as a function of growth rate (J_a) in the chemostat. FRAG1 contains both high- and low-affinity K^+ uptake systems, while FRAG5 lacks the former. DNP, 2,4-dinitrophenol (0.5 mM)

growth when fresh K^+ is supplied than FRAG5. Secondly, the possibility of modulating the efflux of K^+ may allow the modulation of the rate of dissipation of excess ATP through the futile cycle.

Interestingly, the rate of growth of <u>E.coli</u> under K^+-limited conditions appears to be determined by the <u>intracellular</u> concentration of K^+ and not by the extracellular concentration. This can be most dramatically illustrated by a batch experiment, in which the concentration of K^+ in the medium is followed (Fig.12). The cells grow exponentially until the extracellular concentration of K^+ drops below the detection level. However, growth does not stop immediately then, but continues for at least 15 min at the same rate. This is only possible because the presence of the second K^+ uptake system allows the rapid efflux of K^+. Thus, all cells can 'share' the available K^+. Growth

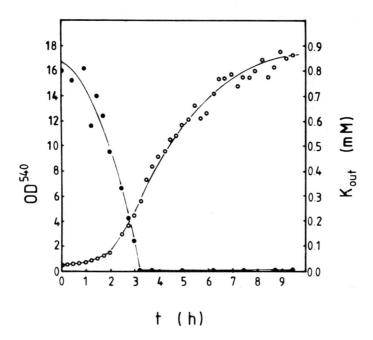

Fig.12 Growth of <u>E.coli</u> FRAG1 in batch culture with a limiting amount of K^+. Concentration of extracellular K^+, ● ; optical density at 540 nm, o.

continues at a decreasing rate until the intracellular concentration of K^+ has dropped to a critical level, because of dilution through the continued increase in intracellular volume. In Fig.13 the results are given of a calculation of the efficiency of growth of a microorganism as a function of the ratio of the free energies of anabolism and catabolism. An important result is that the point at which the optimum is reached depends on the type of dissipative process: a change in anabolic coupling has a different effect from a change in catabolic coupling (Westerhoff and van Dam, 1987). Thus, with the same overall loss of coupling, the optimal efficiency of growth may be reached at different external free energies. The opens the possibility that cells reach optimal efficiency under the conditions that are set by the growth medium. We have treated two examples of regulation of metabolism: via the PTS the relative contribution of different substrates to

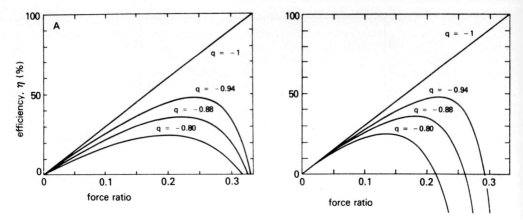

Fig.13 Variation of the thermodynamic efficiency of growth with the force ratio $(\Delta G_a/\Delta G_c)$, when the coupling coefficient (q) is varied by uncoupling catabolism from ATP production (left) or by uncoupling anabolism from ATP utilization (right)

metabolism could be varied, via the futile cycle of K^+ ions the ATP leak could be modulated. Such modulations are evidently very important for cells to adapt optimally to the prevailing conditions.

LITERATURE

Bramley HF, Kornberg HL (1987) Sequence homologies between proteins of the bacterial Pep;dependent sugar phosphotransferase systems: identification of possible phosphate-carrying histidine residues, Proc Natl Acad Sci US 84:4777-4780

Mulder MM, Teixeira de Mattos JM, Postma PW, van Dam K (1986) Energetic consequences of multiple potassium uptake systems in Escherichia coli, Biochim Biophys Acta 851:223-228

Nelson S, Wright JK, Postma PW (1983) The mechanism of inducer exclusion. Direct interaction between purified III[Glc] of the phosphoenolpyruvate:sugar phosphotransferase system and the lactose carrier of Escherichia coli, EMBO J 2:715-720

Postma PW, Lengeler JW (1985) Phosphoenolpyruvate:carbohydrate phosphotransferase system of bacteria, Microbiol Rev 49:232-269

Rogers MJ, Ohgi T, Plumbridge J, Soll D (1988) Nucleotide sequences of the Escherichia coli nagE and nagB genes: the structural genes for the N-acetylglucosamine transport protein of the bacterial phosphoenolpyruvate:sugar phosphotransferase system and for glucosamine-6-phosphate deaminase

Stucki JW (1980) The optimal efficiency and economic degrees of coupling of oxidative phosphorylation, Eur J Biochem 109:269-283

Westerhoff HV, van Dam K (1987) Thermodynamics and control of biological free-energy transduction, Elsevier, Amsterdam New York Oxford

CARBONYLCYANIDE-3-CHLOROPHENYLHYDRAZONE, A PROTOTYPE AGENT FOR THE SELECTIVE KILLING OF CELLS IN ACIDIC REGIONS OF SOLID TUMOURS

Ken Newell and Ian Tannock
Department of Medical Biophysics
Ontario Cancer Institute and University of Toronto
500 Sherbourne Street
Toronto, Ontario
M4X 1K9
Canada

Introduction

Poor vascularization resulting in the inadequate supply of nutrients (oxygen, glucose) and/or removal of catabolites (lactic acid) may produce regions of nutrient deprivation within solid tumours (Thomlinson and Gray, 1955; Tannock, 1968). Cells in a hypoxic microenvironment become dependent on glycolysis for energy metabolism. The resulting lactic acid production coupled with ATP hydrolysis may lead to the development of acidic conditions (Hochachka and Mommsen, 1983). It has been shown that the mean pHe[1] within both human and murine tumours (typically in the range 6.5-7.0) is on average 0.5 units lower than that of normal tissue (Wike-Hooley et al, 1984 . Hypoxic cells are resistant to radiation therapy and many anti-cancer drugs have limited penetration into tissue and/or have limited activity against slowly proliferating cells situated distant from blood

---

[1]Abbreviations used are: α-MEM, a minimum essential medium; BCECF-AM, 2',7'-bis-(2-carboxyethyl)-5(and-6)carboxyfluorescein acetoxymethyl ester; CCCP, carbonylcyanide-3-chlorophenylhyrazone; DIDS, 4,4'-diisothiocyanostilbene 2,2-disulfonic acid; FCS, fetal calf serum; Hepes, N-2-hydroxyethylpiperazine-N'-2-ethanesulfonic acid; MES, 2[N-morphilino]ethanesulfonic acid; pHe, extracellular pH; pHi, intracellular pH; Tris, tris(hydroxymethyl)aminomethane.

NATO ASI Series, Vol. H29
Receptors, Membrane Transport and Signal Transduction
Edited by A.E. Evangelopoulos et al.
© Springer-Verlag Berlin Heidelberg 1989

vessels. Thus, cells in nutrient deprived regions of solid tumours may be responsible for the failure of conventional therapy (Sutherland et al, 1982; Tannock, 1982). A major objective of our studies is to develop new therapeutic strategies which may utilize the acidic conditions within solid tumours to selectively kill nutrient-deprived cells.

Cells maintain pHi significantly higher than predicted by the electrochemical distribution of H^+ and HCO_3^- across cell membranes (Roos and Boron, 1981). The known mechanisms by which cells regulate pHi under acidic conditions are the amiloride sensitive Na^+/H^+ antiport and the DIDS senstive Na^+-dependent HCO_3^-/Cl^- exchanger (Grinstein et al, 1988; Cassel et al, 1988). Agents which can cause intracellular acidification by either (a) transporting H^+ equivalents into the cytoplasm or (b), by inhibiting either of the two pHi regulatory exchangers, may have therapeutic potential. Previous results from this laboratory revealed that conditions of hypoxia and low pHe (6.0-6.5) interacted to cause cell death in vitro whereas neither factor alone reduced cell viability (Rotin et al, 1986). Furthermore, it was shown that the electroneutral ionophore nigericin, which exchanges K^+ for H^+, was cytotoxic to cells only at pHe<6.5 in vitro (Rotin et al, 1987). To further examine the phenomenon of pH dependent cell killing, studies were undertaken with the protonophore CCCP.

Materials and Methods

Clonogenic cell survival experiments

The murine mammary sarcoma cell line EMT-6 and the human bladder cancer cell line MGH-U1 were grown in a-MEM supplemented with 0.1 mg/ml kanamycin and 5% FCS. Exponentially growing monolayers of cells were detached with 0.025% trypsin and 0.01% EDTA for experimentation. Cells at a concentration of 10^6 cells/ml were gassed with a humidified 5% CO_2/air mixture in glass vials at 37^oC as previously described (Whillans and Rauth, 1980). The pHe of a-MEM buffered with 25mM Hepes/bicarbonate showed a minimal acidification of 0.05-0.10 pH units over a 6 hour incubation period. After 30 minutes equilibration in various pHe, the appropriate volume of CCCP dissolved in 100% ethanol (0.02 M stock) was added to vials with control vials receiving equal volumes of 100% ethanol. Aliquots of cells were removed at appropriate time points, pelleted by mild centrifugation, resuspended in fresh a-MEM plus 5% FCS, diluted and plated in triplicate. Eleven days later colonies were stained with methylene blue and counted.

Determination of pHi

pHi was measured using the intracellularly trapped fluorescent indicator BCECF as described elsewhere (Grinstein et al, 1984). Cells were incubated with 2 μg/ml BCECF-AM for 30 minutes at 37°C in serum-free a-MEM after which cells were pelleted and resuspended in serum-free a-MEM at 6.0x10^6 cells/ml. For determination of the initial rate of acidification by CCCP, cells were placed in NMG media (10 mM glucose, 1 mM KCl, 1 mM MgCl$_2$ and 140 mM N-methyl-D-glucamine) buffered to various pHe's with Tris/MES. For studies of the effect of CCCP on long term pHi regulation cells were incubated in a-MEM plus 5% FCS buffered with Hepes/bicarbonate to the required pHe and were gassed with a humidified 5% CO$_2$/air mixture. Calibration between pHi and fluorescence was performed by 2 methods: (i) cells were ruptured with 0.05% Triton X-100 followed by direct titration with concentrated Tris and MES or (ii), cells were placed in either NMG or a-MEM in which Na$^+$ or NMG had been isoosmotically replaced with K$^+$ and exposed to the K$^+$/H$^+$ ionophore nigericin, which sets $H^+_i/H^+_e=K^+_i/K^+_e$, followed by titration with concentrated Tris or MES. When using Triton X-100 a correction factor was determined using the nigericin technique to account for red-shift upon release of the dye from cells (Thomas et al, 1979). Measurements of pHi were determined at 37°C in a Perkin Elmer LS3 fluorescence spectrophotometer.

Results

Effect of CCCP and low pHe on clonogenic cell survival

Incubation of EMT-6 cells at low pHe (6.1) in the presence of 5.0 μM CCCP led to a marked decrease in reproductive cell survival (Fig.1). Cells exposed to either low pHe (6.1) in the absence of the drug, or to physiological pHe (7.3) in the presence or absence of CCCP, had no decrease in colony-forming ability. The observed cytotoxicity was dependent on both the duration of exposure and on the concentration CCCP used. Furthermore, cytotoxicity of CCCP was strongly dependent on pHe, occurring only at pHe below 6.5 (Fig. 1). The above and all subsequent experiments were also performed with MGH-U1 cells and qualitatively similar results were obtained.

Effect of CCCP on regulation of pHi

Since CCCP has been shown to be capable of transporting H^+ equivalents through biological membranes (McLaughlin and Dilger, 1980), its effects on pHi regulatory mechanisms were examined. Fluorescent traces of pHi show that when EMT-6 cells were exposed to low pHe (6.15) the rate of acidification of cells was enhanced in the presence of CCCP, and that the rate of H^+ influx increased with increasing concentration of CCCP (Fig. 2). In the absence of CCCP, cells maintained their steady state pHi about 0.25 pH units above pHe in the acid range of pHe 6.0-6.7 (Fig. 3). However, the addition of CCCP caused equilibration of pHi and pHe in the acid range of pHe 6.0-6.7. CCCP did not appear to have any effect on steady state pHi in the range of pHe 6.7-7.4.

EMT-6 cells were also exposed to the combination of CCCP (7.5μM), amiloride and DIDS (0.1 mM), thereby inhibiting membrane exchangers that can lead to the net extrusion of protons. At physiological pHe 7.30 there was no apparent cytotoxicity for any combination of the three agents. At

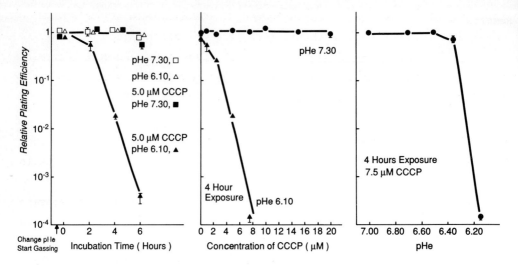

Figure 1 Time, dose and pHe dependent effects of CCCP on the colony forming ability of EMT-6 cells. Points represent mean and range of triplicate plates from two independent experiments.

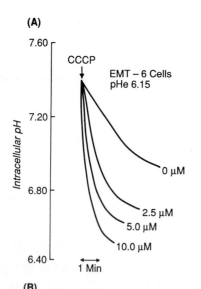

(A)

(B)

Concentration of CCCP (µM)	Fall in pH$_i$ in first minute (±S.D.)
0.0	0.17±0.03
2.5	0.46±0.02
5.0	0.53±0.09
10.0	0.67±0.11

Figure 2 Fluorescent traces showing change in pHi of cells exposed to CCCP at low pHe. Values represent mean and standard deviation from three experiments.

pHe 6.45, CCCP alone, or amiloride and DIDS combined were minimally cytotoxic whereas the combination of all three agents produced a large decrease in colony-forming ability (Fig 4A). Amiloride and DIDS increased the pHe dependent cytotoxicity of CCCP, and allowed killing of cells within the range of pHe (6.5-7.0) found in solid tumours (Fig. 4B).

Discussion

Results from this study show that it is possible to selectively kill malignant cells by using agents which prevent cells from maintaining pHi above pHe. In agreement with others who have studied the effects of ionophores on cell survival (Haveman, 1979; Rotin et al, 1987), we have found that the cytotoxicity of CCCP was limited to the range of pHe below 6.5. In the presence of low pHe 6.0-6.5 and CCCP, there was some variability in colony forming ability which was probably a result of the critical dependence of cell survival on pHe which varied slightly during the experiments, and on the dose of CCCP.

The primary cytotoxic effect of CCCP at low pHe probably relates to its ability to produce intracellular acidification. When CCCP was added to cells at low pHe, there was an immediate increase in the rate of H^+ influx into cells. This increased rate of acidification may have caused inactivation of biologically important molecules within cells. Furthermore, CCCP was shown to cause equlibration of pHi and pHe in the range of pHe 6.0-6.7. An increase in pHi has been implicated in mitogenesis for eukaryotic cells (Nucitelli and Heiple, 1982); therefore, CCCP which caused an equilibration of pHi and pHe thereby preventing an increase in pHi, might be expected to produce cytostatic if not cytotoxic effects. Further experiments need to be performed to determine if the pHe dependent cytotoxicity of CCCP is due entirely to equilibration of pHi and pHe, or if it is also due in part to some other secondary effects of CCCP. CCCP is a known uncoupler of oxidative phosphorylation (Heytler and Prichard, 1962) and low pH inhibits glycolysis (Trivedi and Danforth, 1966). Additional experiments have shown that cells become energy deprived

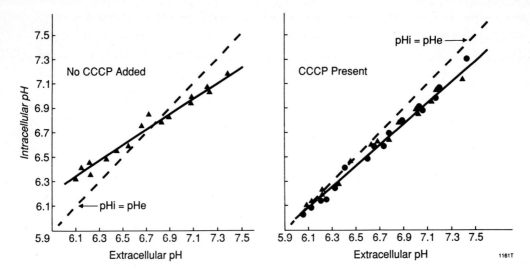

Figure 3 The relationship between pHe and pHi for EMT-6 cells in the presence (7.5 μM) or absence of CCCP. Measurements were taken after pHi had attained a constant level, generally after 30 minutes.

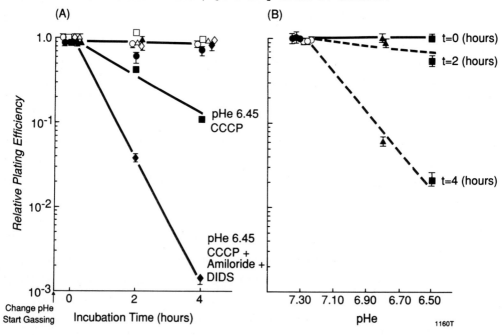

Figure 4 (A) Cell survival for EMT-6 cells exposed to CCCP alone, amiloride and DIDS, or CCCP plus amiloride and DIDS at pHe 6.45. Controls showing ~100% survival include exposure to the same combination of drugs at pHe 7.30 and exposure to pHe 6.45 alone. (B) Cell survival for EMT-6 cells exposed to various pHe in the presence of CCCP plus amiloride and DIDS. Points represent mean and range of triplicate plates.

when exposed to CCCP and low pHe, and energy deprivation may have resulted in the inhibition of the Na^+/K^+-ATPase, since we found a decrease in the normally large Na^+ gradient into cells. Given that Na^+/H^+ and Na^+-dependent HCO_3^-/Cl^- exchangers rely on a Na^+ gradient into cells for activity (Grinstein et al, 1988; Cassel et al, 1988), CCCP and low pHe may, in part, interact indirectly to produce intracellular acidification leading to cell death.

To selectively kill tumour cells in nutrient deprived regions, it will be necessary to design agents which have cytotoxic effects in the range of pHe 6.5-7.0. By combining CCCP with amiloride and DIDS it was possible to kill cells in the required range of pHe. Similar results using the ionophore nigericin in combination with amiloride and DIDS were obtained previously (Rotin, 1987). Amiloride and DIDS probably enhance the cytotoxicity of CCCP, and produce cell killing in the range of pHe 6.5-7.0, by preventing cells from extruding proton equivalents following the cytoplasmic acidification which is caused by CCCP. However, both amiloride and DIDS have a wide range of effects on cells besides the inhibition of ion exchange mechanisms, which may also contribute to cell killing when these agents are used in combination with CCCP.

CCCP, amiloride and DIDS may not be suitable agents for use in vivo; however, these compounds may be considered as prototype agents for the selective killing of cells in acidic regions of solid tumours. The results from this study suggest that the membrane bound exchangers responsible for the regulation of pHi may be useful targets for new forms of anticancer therapy.

References

Cassel, D., Scharf, O., Rotman, M., Cragoe, E.J., and Katz, M. (1988) Characterization of Na^+-linked and Na^+-independent Cl^-/HCO_3^- exchange systems in Chinese hamster lung fibroblasts. J. Biol. Chem., 263: 6122-6127

Grinstein, S., Rotin, D., and Mason, M.J. (1988) Na^+/H^+ exchange and growth factor-induced cytosolic pH changes. Role in cellular proliferation. Biochem. Biophys. Acta (in press)

Grinstein, S., Cohen, S., and Rothstein, A. (1984) Cytoplasmic pH regulation in thymic lymphocytes by an amiloride-sensitive Na^+/H^+ antiport. J. Gen. Physiol., 83: 341-369

Haveman, J. (1979) The pH of the cytoplasm as an important factor in the survival of in vitro cultured malignant cells after hyperthermia. The effects of carbonylcyanide-3-chlorophenylhydrazone. Europ. J. Cancer, 15: 1281-1288

Heytler, P.G., and Prichard, W.W. (1962) A new class of uncoupling agents - carbonyl cyanide phenylhydrazones. Biochem. Biophys. Res. Comm., 7: 272-275

Hochachka, P.W., and Mommsen, T.P. (1983) Protons and anaerobiosis. Science, 219: 1391-1397

McLaughlin, G.A., and Dilger, J.P. (1980) Transport of protons across membranes by weak acids. Physiol. Rev., 60: 825-863

Nuccitelli, R., and Heiple, J.M. (1982) Intracellular pH: Its measurement, regulation, and utilization in cellular functions. Alan R. Liss, Inc. New York 567-586

Roos, A., and Boron, W.F. (1981) Intracellular pH. Physiol. Rev., 61: 296-434

Rotin, D., Wan, P., Grinstein, S., and Tannock, I.F. (1987) Cytotoxicity of compounds that interfere with the regulation of intracellular pH: A potential new class of anticancer drugs. Cancer Res. 47: 1497-1504

Rotin, D., Robinson, B., and Tannock, I.F. (1986) Influence of hypoxia and an acidic environment on the metabolism and viability of cultured cells: Potential implications for cell death in tumours. Cancer Res., 46: 2821-2886

Sutherland, R.M., Eddy, H.A., Bareham, B., Reich, K., and Vanatwerp, D. Resistance to Adriamycin in multicellular spheroids. Int. J. Radiat. Oncol. Biol. Phys., 5: 1225-1230

Tannock, I.F. (1982) Response of aerobic and hypoxic cells in a solid tumour to Adriamycin and cyclophosphamide and interaction of the drugs with radiation. Cancer Res., 42: 4291-4926

Tannock, I.F. (1968) The relationship between cell proliferation and the vasculature system in a transplanted mouse mammary tumour. Br. J. Cancer, 22: 258-273

Thomas, J.A., Buchsbaum, A., Ziminiak, A., and Racker, E. (1979) Intracellular pH measurements in Ehrlich ascites tumour cells utilizing spectroscopic probes generated in situ. Biochemistry, 18: 2210-2218

Thomlinson, R.H., and Gray, L.H. (1955) The histological structure of some human lung cancers and the possible implications for radiotherapy. Br. J. Cancer, 9: 539-549

Trivedi, B., and Danforth, W.H. (1966) Effect of pH on the kinetics of frog muscle phosphofructokinase. J. Biol. Chem., 241: 4110-4114

Whillans, D.W., and Rauth, A.M. (1980) An experimental and analytical study of oxygen depeletion in stirred cell suspensions. Radiat. Res., 84: 97-114

Wike-Hooley, J.L., Haveman, J., and Reinhold, J.S. (1984) The relevance of tumour pH to the treatment of malignant disease. Radiother. Oncol., 2: 343-366

Ca^{2+} AND pH INTERACTIONS IN THROMBIN STIMULATED HUMAN PLATELETS

M.T. Alonso, J.M. Collazos and A. Sanchez
Departamento de Bioquímica y Biología Molecular y
Fisiología. Facultad de Medicina
Universidad de Valladolid
47005-VALLADOLID. SPAIN

INTRODUCTION

It has been reported previously that thrombin can elicit a cytoplasmic alkalinization through protein kinase C (PKC)-mediated stimulation of the Na^+/H^+ exchanger of the plasma membrane (Sanchez et al, 1988). This alkalinization is preceeded by a transient acidification, which is more prominent when the Na^+/H^+ exchanger activity is abolished. Intracellular pH drops have also been reported in growth factor-stimulated fibroblasts (Ives and Daniel, 1978) and TPA treated neutrophils (Grinstein and Furuya, 1986). The mechanistic interpretation of the observed acidification is not univoque: Whereas for neutrophils a metabolically driven stimulation of the hexose monophosphate shunt has been proposed, in fibroblasts the acidification has been attributed to a Ca^{2+}-dependent mechanism. We have followed the changes of (Ca^{2+}) and pHi in platelets loaded with fura-2 or BCECF and stimulated with thrombin in Ca^{2+}-free medium, in order to study possible interactions between the Ca^{2+} rise and the pH drop observed in this particular system.

METHODS

Platelets were prepared from freshly drawn human blood anticoagulated with 1/6 volume of ACD (2.5 g of sodium citrate, 1.5 g of citric acid, and 2.0 g of glucose in 100 ml of water). Platelet-rich plasma (PRP) was obtained by centrifugation of the whole blood for 5 min at 800 x g, and then centrifugated for 20 min at 350 x g. The platelet pellet was resuspended in a

NATO ASI Series, Vol. H29
Receptors, Membrane Transport and Signal Transduction
Edited by A.E. Evangelopoulos et al.
© Springer-Verlag Berlin Heidelberg 1989

solution containing (mM): NaCl, 150; KCl, 5; $MgCl_2$, 1; glucose, 10; Na-Hepes, 10; pH 7.40.

The fluorescent indicators were loaded into the cells by incubation at $3x10^8$ cells/ml with 4 μM of either fura-2 or BCECF in their acetoxymethylester forms for 45 min at 37ºC. The cell suspension was then diluted with two volumes of standard medium containing albumin (1 mg/ml) and ACD (2%) and centrifugated at 350 xg for 20 min to remove the extracellular dye. Cells were then resuspended in fresh standard medium to a final concentration of 10^8 cells/ml. Platelets were kept at room temperature in a capped plastic tube for up to 3 hours. Thromboxane synthesis was blocked by incubation of this suspension for 10 min at 37ºC with 250 μM aspirin. In some experiments the cells were kept at a higher density (10^9 cels/ml) in the usual Na^+ containing medium and diluted 10 times in the cuvette with either control, high K (NaCl replaced isoosmotically by KCl) or deoxiglucose (DOG) medium (10 mM DOG substituting glucose).

The intracellular Ca^{2+} concentration $(Ca^{2+})_i$ and the intracellular pH (pH_i) were monitored in an Hitachi 650-10s spectrofluorimeter using excitation/emission wavelengths of 340/500 nm (fura-2) and 505/530 nm (BCECF). Calibration of Ca^{2+} signal was performed in every cellular batch as described by Pollok et al (1986), using a value for Kd of 224 nM (Grynkiewiez et al, 1985). Extracellular dye leakage was estimated from the drop in the fluorescence observed after 1 mM EGTA addition, and corrected from the total fluorescence. The BCECF signal was calibrated before and after the experiment in high-K^+ medium containing 2 μM nigericin as described previously (Thomas et al, 1979). Dye leakage was estimated from the quick shift in the fluorescent signal observed after changing the pH_e of the suspension by aproximately 0.4 pH units. For every resting pHi value a fluorescence ratio of 505/430 nm was obtained at the begining of each experiment in order to correct for diferences among cell batches.

Aggregation and secretion were followed in some experiments to asses the normal platelet behavior. To perform this experiments, biolumisnescence of ATP released to the medium, and changes in the optical density of the suspension were continuously monitored in a chronolog-lumiaggregometer.

EIPA (ethylisopropylamiloride, a generous gift of Drs T. Friedrich and G. Burchardt (Max-Planck-Institüt für Biophysik, Frankfurt, FRG), CCCP (carbonyl cyanide chlorophenyl-hydrazone), ionomycin and nigericin were dissolved in ethanol at concentrations 1000 times higher than the respective final ones. Stock solutions (1 mM) of the acetoxymethyl esters of fura-2, and BCECF were made in dimethyl sulfoxide. Human thrombin was added from a 10 u/ml stock solution in Na^+-free medium. Alpha-cyano-4-hydroxycinnamic acid (CHC) was dissolved at 0.1 M and neutralized with KOH.

RESULTS AND DISCUSSION

Three different possibilities have been analized here to explain the observed thrombin-induced acidification: 1) Ca^{2+} displacement of H^+ from common binding sites; 2) increased intracellular H^+ generation; 3) H^+ entry into the cytoplasm either from the extracellular medium or from intracellular compartments. To test the first posibility we have studied the effects of increasing Ca^{2+} levels on the intracellular resting pH. Fig. 1 shows that the Ca^{2+} ionophore ionomycin can mimic the thrombin-induced transient acidification. The acidification is also followed in this case by an EIPA-sensitive recovery of pH_i that tends to restore the resting pH. This ionophore-mediated acidification is more prominent in EIPA-treated cells and it was enhanced by increasing ionomycin dose. This suggests that when H^+ cannot be extruded from the cytoplasm, this cell compartment becomes more acidic when the Ca^{2+} rise is more prominent.

Since the thrombin-induced acidification goes parallel with a $(Ca^{2+})_i$ rise (Sanchez et al, 1988), the above findings would suggest a Ca^{2+}-dependency of the acidification observed in both instances. However, Fig. 2 shows that thrombin can still promote an intracellular acidification when the Ca^{2+} stores had been previously emptied by a ionomycin pretreatment. In this particular instance, in which the intracellular Ca^{2+} is in fact dropping by the activity of membrane attached pumps, one migth speculate about H^+ being incorporated into the cytoplasm in exchange for Ca^{2+} through the normal operation of the Ca^{2+} pump (Carafoli and Longoni, 1986)

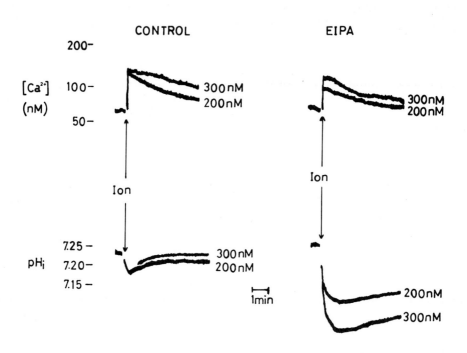

Fig. 1. Effects of ionomycin on $(Ca^{2+})_i$ (upper panel) and pH_i (lower panel) of human plateles. Ionomycin was added at the time marked by the arrow to give the final Ca^{2+} concentrations given in the figure (in nM). The incubation medium contained 1 mM EGTA. Representative experiments in control (left) or EIPA-treated cells (rigth) are shown.

To test that possibility we looked to the effects of manipulations of extracellular pH on both the Ca^{2+} buffering and the initial acidification observed after thrombin stimulation. Fig. 3 shows that changing pH_e from the normal physiological value to a more alkaline value by tritiation of extracellular medium, did not modify the initial acidification in EIPA-treated cells, although it became smaller in control conditions, suggesting a transtimulation of the exchanger in the predicted way (Grinstein and Rothstein, 1986).

333

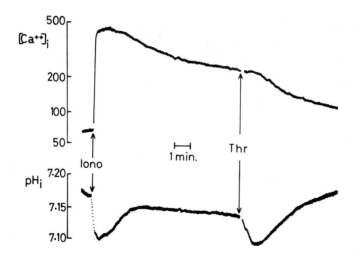

Fig. 2. Time course of the changes of $(Ca^{2+})_i$ (upper traces) and pHi (lower traces) induced by consecutive additions of ionomycin (100 nM), and thrombin (0.1 u/ml). 1 mM EGTA was added prior to any addition.

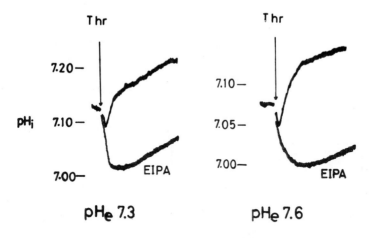

Fig. 3. Effects of extracellular pH on the changes of pHi induced by thrombin (0.1 u/ml). The experiments records of pHi changes following stimulation were performed at pHe values of

7.38 (left) and 7.68 (right) in the presence (lower trace) and absence (upper trace) of 40 μM EIPA. Alkalinization of the medium was performed inmediatly before thrombin addition by adding 2.5 mM of Tris Base. 1 mM EGTA was added before the experiments. EIPA treatment was performed independently for every cell batch during 1 min.

The thrombin-induced acidification can also be seen when the transmembrane proton gradients were collapsed with CCCP. Results shown in Fig. 4 indicated that: i) there is an inward H^+ gradient in resting platelets that promotes a marked intracellular acidification after CCCP treatment; ii) thrombin can still induce an EIPA-inhibitable alkalinization in CCCP-treated cells. This may reflect a H^+ redistribution following membrane potential changes or a clear predomination of the Na^+/H^+ exchanger mediated fluxes over the leak (CCCP); iii) When the Na^+/H^+ exchange is inhibited, thrombin can still promote an initial acidification. This suggests increased H^+ generation after stimulation since all the proton gradients are collapsed under these circumstances.

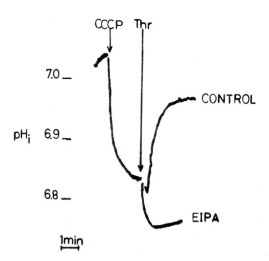

Fig. 4. Effects of CCCP (1 μM) and thrombin (0.1 u/ml) on intracellular pH. Experiments obtained in control cells and EIPA-treated cells are shown. The pH drop induced by CCCP was not affected by EIPA.

The most likely source for a thrombin-induced intracellular H^+ generation would be metabolic. To test that possibility glucose was substituted for the non-metabolizable analog deoxiglucose (DOG). Experiments were performed in EIPA-treated cells to look exclusively to thrombin-induced acidification (Fig. 5). DOG decreased by about 50% the initial acidification produced by all the thrombin doses studied, suggesting that at least part of the observed phenomenon has a metabolic origin.

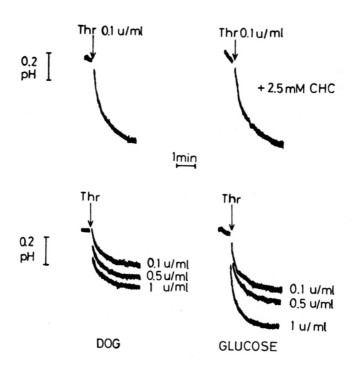

Fig. 5. Upper pannel: pHi records obtained after adding 0.1 u/ml of thrombin to an EIPA treated platelet batch in the absence (left) and the presence of 2.5 mM of (alpha-cyano-4-hydroxycinnamic acid (CHC) (right). Lower pannel: intracellular pHi drops obtained after adding thrombin at different doses (0.1, 0.5 and 1 u/ml) to EIPA treated platelet batches in medium either containing either 10 mM deoxiglucose (DOG) (left) or 10 mM glucose (right). To perform this experiment the platelet suspension was kept at a higher cell density and

diluted with either DOG- or glucose- containing medium
inmediatly before the experiment to avoid platelet starving.
Cell viability was assesed by showing that thrombin-induced
secretion was similar in DOG-suspended cells and their
corresponding controls (83% versus 100% of maximal releasable
ATP pool, for 6 diferent experiments .

Current hypothesis for metabolically driven acidification
includes H^+ generation during: a) glucose phosphorilation, b)
lactic acid production, and c) $NADH^+ + H^+$ generation in the
hexose monophosphate shunt. Since DOG can be phosphorilated the
first possibility can be excluded. Alpha-cyano-4-
hydroxycinnamic acid (CHC), a known inhibitor of lactic acid
transport (Halestrap and Denton, 1975) did not potentiate the
acidification induced by thrombin in control conditions (Fig.
5, upper panel). This suggests that the acidification is not
due to lactic acid accumulation after cell stimulation.
Increased operation of the pentose phosphate pathway has been
proposed to be responsible for the cytoplasmic acidification
observed in TPA-activated human neutrophils (Grinstein and
Furuya, 1986). Pentose phosphate pathway has been also
demostrated in platelets, where it is activated by thrombin
(Akkerman, 1978). Our results suggest that activation of the
pentose phosphate pathway is responsible for at least the DOG-
sensitive component of the acification induced by thrombin,
although other additional mechanism(s) cannot be excluded at
present.

REFERENCES

Akkerman JKN (1978) Regulation of carbohydrate metabolism in
    platelets. Thrombos Haemostas (Stuttg) 39:712-724
Carafoli E, Longoni S (1986) in Cell Calcium and the control of
    membrane transport. Mandel L and Eaton D (eds) Chap 2:21-29
    Society of General Physiologists. 40th Annual Symposium
Grinstein S, Furuya W (1986) Cytoplasmic pH regulation in
    phorbol ester-activated human neutrophils. Am J Physiol
    251:C55-C65
Grinstein S, Rothstein A (1986) Mechanisms of regulation of the
    Na^+/H^+ exchanger. J Membrane Biol 90:1-12
Grynkiewiez G, Poenie M, Tsien RY (1985) A new generation of
    Ca^{2+} indicators with greatly improved fluorescence
    properties. J Biol Chem 260:3440-3450
Halestrap AP, Denton RM (1975) The specificity and metabolic
    implications of the inhibition of pyruvate transport in
    isolated mitochondria and intact tissue preparations by

alpha-cyano-4-hydroxycinnamate and related compounds.
  Biochem J 148:97-106
Ives HE, Daniel O (1987) Interrelationship between growth
  factor-induced pH changes and intracellular Ca^{2+}. Proc Natl
  Acad SciUSA, 84: 1950-1954
Pollock WK, Rink TJ, Irvine RF (1986) Liberation of (^{3}H)
  arachidonic acid and changes in cytosolic free calcium in
  fura-2 loaded human platelets stimulated by ionomycin and
  collagen. Biochem J 235:869-877
Sanchez A, Alonso MT, Collazos JM (1988) Thrombin-induced
  changes of intracellular (Ca^{2+}) and pH in human platelet.
  Cytoplasmic alkalinization is not a prerequisite for calcium
  mobilization. Biochim Biophys Acta 938:497-500
Thomas JA, Buschsbaum RN, Zimniak A, Racker E (1979)
  Intracellular pH measurements in Ehrlich ascite tumor cells
  utilizing spectroscopic probes generated in situ.
  Biochemistry 18:2210-2218

J. Luirink
Department of Biology
De Boelelaan 1087
1081 HV Amsterdam
The Netherlands

Introduction

Gram-negative bacteria have developed special mechanisms to translocate proteins from the cytoplasm, across both the cytoplasmic and outer membrane and into the extracellular medium (Pugsley, 1988). One of these "export systems", the export of bacteriocins and more specifically the release of cloacin DF13 by *E. coli* cells, is under study in our laboratory. Bacteriocins are plasmid-encoded bacteriocid proteins which are synthesized without a cleavable signal peptide. The gene encoding cloacin DF13 is located in a mitomycin-C inducible operon with the genes encoding the immunity protein and the bacteriocin release protein (BRP) (Fig. 1).

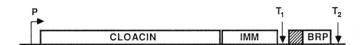

Fig. 1. Organization of the cloacin DF13 operon on pCloDF13. The (inducible) promoter P and terminators of transcription T_1 and T_2 are indicated. Imm, immunity protein; BRP, bacteriocin release protein.

The immunity protein forms an equimolar complex with the cloacin polypeptide and protects the producing cells against the lethal action of the bacteriocin. The BRP is essential for the release of the cloacin-immunity protein complex across both membranes of *E. coli* (De Graaf and Oudega, 1986). Moderate induction of BRP gene expression not only results in the release of the cloacin DF13 complex but also in an inhibition of colony formation on broth agar plates ("cell lethality") (Luirink et al., 1988). Full induction leads to a marked decline in culture turbidity, referred to as "lysis". Divalent cations such as Mg^{2+} and Ca^{2+}, in the culture medium, suppress these effects without affecting the release of cloacin DF13 (Luirink et al., 1986).

The molecular mode of action of the BRP is largely unclear, especially at the level of the cytoplasmic membrane. Functioning of the BRP in "lysis" and release of cloacin DF13 is

NATO ASI Series, Vol. H29
Receptors, Membrane Transport and Signal Transduction
Edited by A. E. Evangelopoulos et al.
© Springer-Verlag Berlin Heidelberg 1989

dependent on phospholipase A in the outer membrane. The phospholipase A is directly or indirectly activated by the BRP (Luirink et al., 1986). The accumulation of lysophosphatidylethanolamine, the product of the action of phospholipase A on phosphatidylethanolamine is thought to cause an increase in cell envelope permeability thereby facilitating the release of cloacin DF13 and certain periplasmic proteins from the cells.

BRP's are small, membrane bound proteins (molecular mass about 3 kDa), which are synthesized as precursor proteins with a typical lipobox around the signal peptidase processing site (Wu and Tokunaga, 1986). Site-directed mutagenesis and gene fusion experiments have shown that BRP's are indeed lipoproteins and that their acylation is at least in part indispensable for functioning and correct localization (Cavard et al., 1987; Pugsley and Cole, 1987; Luirink et al., 1987). In this report the structure-function relationships of the pCloDF13 encoded BRP are studied in more detail by using site directed mutagenesis to create carboxy-terminally truncated BRP's. Special attention was paid to the role of the signal peptide of the BRP which appeared to be extremely stable after cleavage by the prolipoprotein-specific signal peptidase II.

Results and Discussion

1. Expression of mutant BRP's

In a previous paper the subcloning of the BRP gene under control of the tandem *lpp/lac* promoter-operator complex was described (Luirink et al., 1988). Induction of BRP expression with IPTG resulted in "lysis", "lethality" and the release of cloacin DF13 depending on the concentration of the inducer. Using oligonucleotide-directed in vitro mutagenesis, four stopcodons were introduced at different positions in the part of the BRP gene encoding the mature portion of the BRP (Fig. 2). The resultant mutant BRP genes, encoding 4 (pJL17-5), 9 (pJL17-10), 16 (pJL17-17) and 20 (pJL17-21) amino-terminal amino acids of the mature BRP, were subcloned as the wild-type BRP gene (pJL17), and the structure and functioning of the mutant BRP's were studied upon IPTG induction. Furthermore, a mutant BRP gene was created in which the last codon, CAG, was changed into an AAC codon giving a substitution of glutamine at position +28 into an asparagine residue (pJL22, Fig. 2).

The expression of the described mutant BRP's was studied using a three-layer discontinuous SDS-PAGE system with tricine instead of glycine as the trailing ion (Schägger and von Jagow, 1987), which gives superior resolution of relatively small proteins. Cells of *E. coli* strain FTP4170 harboring the (mutant) BRP-plasmids were induced with IPTG and labeled with a mixture of [³H]-amino acids or with [³H]-palmitate in the presence or absence of globomycin (Fig. 3). Globomycin is a peptide antibiotic capable of blocking the action of

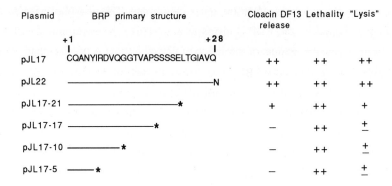

Plasmid	BRP primary structure	Cloacin DF13 release	Lethality	"Lysis"
pJL17	CQANYIRDVQGGTVAPSSSSELTGIAVQ	++	++	++
pJL22	————————————————N	++	++	++
pJL17-21	——————————————*	+	++	+
pJL17-17	———————————*	–	++	±
pJL17-10	——————*	–	++	±
pJL17-5	———*	–	++	±

Fig. 2. Primary structure and functioning of mutant BRP's. The position of a stopcodon is given by a *. The release of cloacin DF13 was determined using cells harboring both pJL25 and one of the indicated mutant BRP plasmids. The cells were grown in broth with 20 mM $MgCl_2$ to the early logarithmic growth phase. Mitomycin C (500 ng/ml) and various amounts of IPTG were added to induce synthesis of cloacin DF13 and the BRP, respectively. Cultures induced with suboptimal (non-lysing) concentrations of IPTG were selected for detection of cloacin DF13 release. Five hours after induction started, the cells from these cultures were collected by centrifugation and the amount of cloacin DF13 in pelleted cells and spent medium was determined using an ELISA. ++, > 50% release of cloacin DF13; +, about 20% release; -, < 5% release. "Lethality" was determined by spotting about 30 cells from a stationary culture on broth plates containing various amounts of IPTG. The plates were incubated at 37°C for 20 h. ++, complete inhibition of colony formation on plates containing > 60 μM IPTG. "Lysis" was determined by measuring the culture turbidity of cells harboring one of the mutant BRP plasmids. The cells were grown in broth to the early logarithmic growth-phase and induced with varying amounts of IPTG. ++, decline in culture turbidity of moderately induced cells (40 μM IPTG); +, decline only after strong induction (100-1000 μM IPTG); ±, modest decline only after strong induction (100-1000 μM IPTG).

signal peptidase II, which specifically cleaves acylated prolipoproteins. Labeling of cells harboring the wild-type BRP gene with a [3H]-amino acid mixture gave rise to three BRP-derived protein bands. The upper and middle band could be identified as the lipid-modified modified precursor- and mature form of the BRP respectively, on the basis of their relative mobility in the gel, incorporation of [3H]-palmitate and reaction with an antiserum raised against a synthetic peptide encompassing the complete mature BRP (not shown). Furthermore, the middle band disappeared upon globomycin treatment at the expense of intensity of the upper band indicating a precursor-product relationship between the two bands and processing by signal peptidase II. The lower band was identified as the surprisingly stable signal peptide of the BRP, since it could not be labeled with [3H]-palmitate, disappeared upon globomycin treatment and did not react with the antiserum mentioned above. The signal peptides of other BRP's have also shown to be extremely stable (Cavard et al., 1987; Pugsley and Cole, 1987).

All mutant BRP's were acylated and slowly processed by signal peptidase II. The extent of acylation and processing of the mutant BRP's was deduced from the relative fluorogram intensities of the various bands after labeling with ^{3}H-palmitate. The precursor forms of the pJL17-21 and pJL17-17 encoded BRP's as well as the mature form of the pJL17-10 encoded BRP were relatively weakly labeled. Acylated pJL17-5 encoded BRP could not be detected, probably because of its small size. The pJL22 encoded BRP could not be distinguished from wild-type BRP in all experiments described above (not shown).

Taken together, these data indicate that all mutant BRP's are expressed, acylated and processed albeit with different efficiencies.

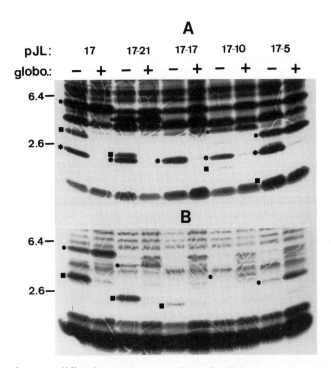

Fig. 3. Expression, modification and processing of wild-type and mutant BRP's. Cells harboring a (mutant) BRP plasmid were labelled with a [^{3}H] amino acid mixture (part A) or [^{3}H] palmitate (part B). The cells were grown in minimal medium ([^{3}H] amino acid labeling) or rich medium ([^{3}H] palmitate labeling) to the early logarithmic growth-phase. Ten minutes after induction of BRP expression with 1 mM IPTG, a [^{3}H] amino acid mixture (30 μCi/ml) of [^{3}H] palmitate (100 μCi/ml) was added. Growth was continued for 30 minutes in the presence or absence of 100 μg/ml globomycin. The labeled cells were spun down, washed with PBS ([^{3}H] amino acid labeling) or methanol to remove lipids ([^{3}H] palmitate labeling), suspended in sample buffer and subjected to Tricine-SDS-PAGE. The lower parts of the fluorograms are shown. The relative molecular mass of marker proteins are indicated at the left (kDa). o, lipid-modified BRP precursor; Δ, unmodified BRP precursor; □, BRP mature form; *, stable BRP signal peptide.

2. Functioning of mutant BRP's

To study whether the mutant BRP's are still functional in the export of cloacin DF13, cells harboring pJL17 or one of the mutant plasmids were complemented with a second plasmid, pJL25. This plasmid carries the genes encoding cloacin and its immunity protein under control of their own (inducible) promoter. Cells harboring this binary vector system were tested for the export of cloacin DF13 (Fig. 2). In contrast to the 'shorter' BRP's encoded by pJL17-17, pJL17-10 and pJL17-5, the mutant BRP's encoded by pJL22 and pJL17-21 were able to provoke the release of the cloacin DF13 complex.

Strong expression of the wild-type BRP results in "lethality" and "lysis" (see introduction). All mutant BRP's were tested for these "side" effects (Fig. 2). "Lethality" levels were equal for the wild-type BRP and all mutant BRP's tested. Significant "lysis" occurred when the pJL17-21 and pJL22 derived BRP genes were strongly expressed. Cells expressing other mutant BRP's were relatively resistant to IPTG induced lysis although some decline in culture turbidity was observed at higher induction levels. In all mutants tested, the observed "lysis" proved to be dependent on the presence of phospholipase A in the outer membrane (not shown). Furthermore, addition of 20 mM $MgCl_2$ to the culture medium inhibited "lysis" (not shown). Apparently, the molecular mechanism which underlies the decline in culture turbidity is unaltered in all mutant BRP's.

Taken together, these results indicate that the eight carboxy-terminal amino acids are not essential for proper functioning of the BRP in "lysis", "lethality", and release of cloacin DF13. This is in agreement with the finding of Toba et al. (1986) who demonstrated that the nine carboxy-terminal amino acids of the pColE2 or pColE3 encoded BRP's can be replaced by one glycine residue without affecting "lysis". The removal of eleven or more carboxy-terminal amino acids of the BRP totally blocks the release of cloacin DF13, whereas the "lysis" and "lethality" functions are still present upon induction, indicating that the amino-terminal part of the BRP is involved in these side effects.

The observation, that pJL17-5, encoding only four amino acids of the mature BRP is, at least in part, functional in "lysis" and "lethality" suggests that the stable signal peptide is somehow responsible for these processes. To study this possibility, the signal peptide and the mature BRP were genetically uncoupled in two ways. Firstly, the BRP signal sequence plus the amino-terminal cysteine residue of the mature BRP were fused to the gene encoding the mature portion of the periplasmic enzyme β-lactamase. Secondly, the (unstable) signal sequence of Braun's lipoprotein was fused to the gene encoding the mature portion of the BRP. Preliminary results using the constructed hybrid genes, indicate that both hybrid proteins are lipid modified and processed by signal peptidase II. Furthermore, the first construct caused "lysis" and "lethality" of host cells, but was not able to bring about the release of cloacin DF13. Further research on the structure, functioning and localization of these two constructs will probably provide more information on the role of the signal peptide in the functioning of the pCloDF13 encoded BRP.

Literature references

Cavard D, Baty D, Howard SP, Verhey HM, Lazdunski C (1987) Lipoprotein nature of the colicin A lysis protein: effect of amino acid substitutions at the site of modification and processing. J. Bact. 169: 2187-2194

Graaf FK de, Oudega B (1986) Production and release of cloacin DF13 and related colicins. Curr. Top. Microbiol Immunol 125: 183-205

Luirink J, Hayashi S, Wu HC, Stegehuis F, Graaf FK de, Oudega B (1988) Effect of a mutation preventing lipid modification of the pCloDF13 encoded bacteriocin release protein on the release of cloacin DF13 and on its subcellular localization. J Bact. 170: 4153-4160

Luirink J, van der Sande C, Tommassen J, Veltkamp E, Graaf FK de, Oudega B (1986) Mode of action of protein H encoded by plasmid CloDF13: effects of culture conditions and of mutations affecting phospholipase A activity on excretion of cloacin DF13 and on growth and lysis of host cells. J Gen. Microbiol. 132: 825-834

Luirink J, Watanabe T, Wu HC, Stegehuis F, Graaf FK de and Oudega B (1987) Modification, processing and subcellular localization in *Escherichia coli* of the pCloDF13-encoded bacteriocin release protein fused to the mature portion of β-lactamase. J Bact. 169: 2245-2250

Pugsley AP (1988) Protein secretion across the outer membrane of gram-negative bacteria. In "Protein Transfer and Organelle Biogenesis" (Das RC, Robbins PW eds) pp. 607-642. Academic Press, San Diego, California.

Pugsley AP, Cole ST (1987) An unmodified form of the ColE2 lysis protein, an envelope lipoprotein, retains reduced ability to promote colicin E2 release and lysis of producing cells. J. Gen. Microbiol. 133: 2411-2420

Schägger H, von Jagow F (1987) A tricine-sodium dodecylsulphate polyacrylamide gelelectrophoresis for the separation of proteins in the range from 1 to 100 kDa. Anal. Biochem. 166: 368-379

Toba M, Masaki H, Ohta T (1986) Primary structures of the ColE2-P9 and ColE3-CA38 lysis genes. J. Biochem. 99: 591-596

Wu HC, Tokunaga M (1986) Biogenesis of lipoproteins in bacteria. Curr. Top. Microbiol. Immunol. 125: 127-157.

BINDING OF A BACILLUS THURINGIENSIS DELTA ENDOTOXIN TO THE MIDGUT OF THE TOBACCO HORNWORM (MANDUCA SEXTA).

K. HENDRICKX, H. VAN MELLAERT[*], J. VAN RIE[*] and A. De LOOF
Zoological Institute
Catholic University Leuven
Naamsestraat 59
B 3000 Leuven
Belgium

Introduction.

Upon sporulation the gram-positive bacterium, Bacillus thuringiensis (B.t.), produces parasporal crystals. These crystals contain insecticidal proteins, called delta-endotoxins. Since more than thirty years B.t. preparations have been successfully used as biological insecticides. However their limited field stability and the rather narrow insecticidal spectrum of B.t. strains used in commercial formulations were important limitations to a more extensive use. With the successful expression of a B.t. delta endotoxin-gene in a tobacco plant (Vaeck et al., 1987), the possibility to apply B.t. delta-endotoxins in a totally new strategy for plant protection was demonstrated. This resulted in a new interest for the toxin from molecular biologists and biochemists.
Different strains of Bacillus thuringiensis produce delta-endotoxins with a different insecticidal spectrum. Up to now B.t. delta-endotoxins with specific toxicity towards lepidopteran, coleopteran and dipteran larvae have been described. Even within the group of strains toxic to Lepidoptera marked differences exist in toxicity towards different species of this order (Hofmann et al., 1988).
While it is now clear that the different strains contain delta-endotoxins with different primary structure, the question why insects exhibit a differential sensitivity has not been fully elucidated. Several insect related factors (midgut pH, midgut proteases, ...) were forwarded. Recent evidence now suggests that differences in receptor site population in the midgut are an important determinant (Hofmann et al., 1988). This is a first reason why identification of a B.t. delta-endotoxin

NATO ASI Series, Vol. H29
Receptors, Membrane Transport and Signal Transduction
Edited by A. E. Evangelopoulos et al.
© Springer-Verlag Berlin Heidelberg 1989

receptor would be extremely interesting. Furthermore the
molecular mode of action, i.e. the sequence of events between
the tentative receptor binding and the actual toxic effect is
poorly understood. Various authors agree that the permeability
of the midgut epithelial cells is altered (Himeno et al., 1985;
Gupta et al., 1985; English and Cantley, 1985), which then
leads to swelling and bursting of the cells through colloid
osmotic lysis (Knowles and Ellar, 1987). Still, essential
questions about the nature of the receptor/effector mechanism
and the precise changes in permeability remain. This paper
presents an initial study to characterize the receptor site for
Bt2, a recombinant delta-endotoxin from B.t. berliner 1715
(Hofte et al., 1986), in the midgut epithelium of the tobacco
hornworm (Manduca sexta).

Methods.

Isolation and activation of the recombinant Bt2-protoxin
expressed in E. coli is performed according to the method
described by Hofte et al. (1986). Purification of the active
Bt2-toxin and iodination with chloramin T was done following
the procedure of Hofmann et al. (1988). Brush border membrane
vesicles were prepared by a magnesium precipitation technique
(Wolfersberger et al., 1987) and binding was measured by a
rapid filtration method using Whatmann GF/F filters. Different
detergents were added to brush border membranes and incubated
for 30 minutes. The supernatant of a 105.000g centrifugation
(60 min) was assumed to contain the solubilized molecules.
Binding was tested by spotting the supernatant on
nitrocellulose strips. The strips were subsequently blocked
with bovine serum albumin. Total binding was measured by
incubating these strips with 1 nM iodinated toxin. For the non-
specific binding the labeled toxin was combined with 170 nM
unlabeled toxin.

Results and discussion.

Brush border membrane vesicles from midgut epithelial cells were prepared from the tobacco hornworm (Manduca sexta), a lepidopteran insect highly sensitive to Bt2 and from the grasshopper (Locusta migratoria), an orthopteran insect, non-sensitive to Bt2. For both we measured binding of ^{125}I-labeled Bt2-toxin (1nM) alone (total binding) and binding of ^{125}I-labeled toxin in combination with excess (170 nM) unlabeled Bt2-toxin (non-specific binding). No significant specific binding (total minus non-specific binding) was found in membrane preparations of Locusta migratoria, the non-sensitive insect (0.59 %). Also membranes prepared from pig kidney and mouse midgut showed no specific binding. In contrast 29 % of the labeled Bt2 was specifically bound to the membranes pepared from Manduca sexta.

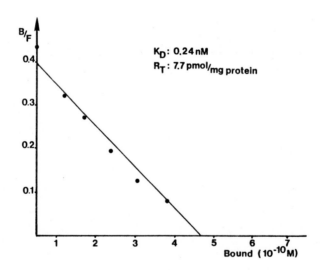

Fig. 1: Scatchard plot of Bt2-binding to brush border membranes from Manduca sexta. Displacement was done with different concentrations of cold Bt2. Binding was measured by filtration through Whatmann GF/F-filters.

An equilibrium binding experiment was performed with Manduca sexta membrane vesicles in order to determine essential characteristics of the toxin's interaction with its binding site. Increasing concentrations of unlabeled Bt2-toxin were incubated in combination with a constant amount of labeled toxin. Scatchard plot analysis of our binding data (fig. 1) indicated a single binding site for the Bt2-toxin. This binding exhibits high affinity (K_D = 0.24nM). The binding site concentration is 7.7 pmole/mg protein.

Thus when we compare the binding of Bt2 to membranes prepared from the sensitive insect, Manduca sexta, with binding to membranes from a non-sensitive insects, saturable binding was only found in the sensitive species. In addition the affinity of the interaction between the Bt2-toxin and its binding site is high and suggests a receptor-like interaction.

In further experiments we wanted to investigate the nature of the binding site. A first characterization was done by treating the membrane preparations with enzymes. Both pronase and proteinase K reduced the specific binding in a time dependent way (fig. 2 A,B). Appropriate controls were performed to confirm that the observed reduction in binding was not due to degradation of the labeled toxin itself. The contribution of glycoconjugates in binding was investigated by treating the membrane preparations with mixed glycosidases. Specific binding was also reduced in a time dependent way (fig. 2 C). The enzyme preparation was checked for the presence of proteinases and no proteinase activity was found.

These data suggest that the receptor is probably a glycoprotein, and that the sugar moiety is directly or indirectly involved in binding. Knowles et al., (1986) and Hofmann et al.(1988) previously also suggested that the receptor could be a glycoprotein. Dennis et al. (1986) demonstrated binding of a labeled B.t. delta-endotoxin to an insect glycosphingolipid, but they do not exclude that, in vivo, a glycoprotein is the real target.

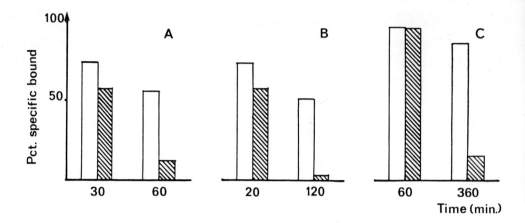

Fig. 2: Brush border membranes were incubated with enzyme at 37
        C and pH 7.4 (A: pronase, B: proteinase K, C: mixed
        glycosidase, 1mg/ml) for the indicated time. After
        treatment with pronase and glycosidases the membranes
        were separated from the enzymes by centrifugation to
        stop the reaction. In the case of proteinase K, PMSF
        was added. Binding was measured by the filtration
        method (see methods). The specific binding of a sample
        kept at 4 C during the incubation was taken as 100
        %.(open bars = without enzyme; striped bars = with
        enzyme).

Furthermore Knowles et al. (1984) suggested a lectin-like
binding of B.t.-delta-endotoxins since the toxicity of
Bacillus thuringiensis var. kurstaki toxin on CF1 cells was
inhibited by N-acetylgalactosamine (125mM) and N-
acetylneuraminic acid (125mM) as well as by wheat germ and
soybean agglutinin. In our experiments we tested the inhibitory
effect of the addition of different sugars (D-fucose, D-
glucose, D-mannose and N-acetyl-galactosamine), both to the
toxin or to the membrane preparation, in a concentration range
from 0.01 mM to 300 mM. None of them gave a reduction of the
binding over fifteen per cent. We also tested whether wheat
germ and soybean agglutinin (1mg/ml) had an inhibitory effect
in our bindingassay. The binding of Bt2 to brush border
membranes of the midgut epithelium of Manduca sexta was not
affected.

Our results from sugar and lectin addition indicate that the binding is not a pure lectin-type interaction. The difference with the results of Knowles et al. (1984) may be due to the choice of a different model system (insect midgut as opposed to an insect cell-line from non-midgut origin). The fact that the simple sugars tested do not interfere with the binding of Bt2 is not in conflict with a possible glycoprotein receptor for the Bt2-toxin.

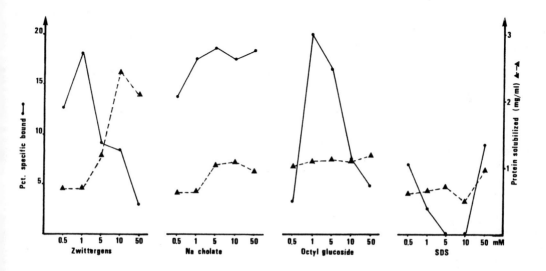

fig. 3: Solubilization of the binding site with detergents. The solubilized faction was spotted on 0.5cm^2 nitrocellulose strips after centrifugation at 105,000g for 1 hour. Binding was measured as indicated in the methods.

Purification of the receptor would be the most direct way to its characterization, identification and to a deeper understanding of the toxin's mode of action (see for example the identification of the GM1 ganglioside as the choleratoxin-receptor and the contribution of these findings to a better insight in the mode of action). A necessary step to purification is solubilization of the receptor. Solubilization, maintaining the integrity and activity, has often been found to be a difficult step. We have tried to solubilize the receptor by means of various detergents. Several detergents gave positive results as shown in fig. 3.

With zwittergens (N-dodecyl-N,N-dimetyl-ammonio-3-propane sulfonate) and octyl glucoside solubilized binding is found in a narrow concentration range around 1mM. Higher concentrations of the detergents strongly reduce activity. With sodium cholate on the other hand, binding in the solubilized fraction is found over a large concentration range from 1mM to 50mM. At the concentration of 1mM the highest ratio between binding activity and soluble protein is obtained. With sodium dodecylsulphate (SDS), as a negative control, very little saturable binding remained. In further experiments sodium cholate will be used for solubilization and we will attempt various purification methods. Apart from the further characterization of the receptor, the final aim is the isolation and partial aminoacid-sequencing of the protein part of the receptor. This should permit us in the future, to study the B.t. delta-endotoxin receptor using methods of molecular biology.

References.

Dennis RD, Wiegandt H, Haustein D, Knowles BH and Ellar DJ (1986) Thin layer chromatography overlay technique in the analysis of the binding of the solubilized protoxin of Bacillus thuringiensis var. kurstaki to an insect glycosphingolipid of known structure. Biomedical chromatography 1 31-37.

English LH, Cantley LC (1985) Delta-endotoxin inhibits Rb+ uptake, lowers cytoplasmic pH and inhibits a K+-ATPase in Manduca sexta CHE cells. J. membr. Biol. 85 199-204.

Gupta BL, Dow JA, Hall TA and Harvey WR (1985) Electron microprobe X-ray microanalysis of the effects of Bacillus thuringiensis var. kurstaki crystal protein insecticide on ions an electrogenic K+-transporting epithelium of the larval midgut in the lepidopteran, Manduca sexta, in vivo. J. Cell Sci. 74:137-152.

Himeno M, Koyama N, Funato T and Komona T (1985) Mode of action of Bacillus thuringiensis insecticidal delta-endotoxins on insect cells in vitro. Agr. biol. Chem. 49:1461-1468.

Hofmann C, Vanderbruggen H, Hofte H, Van Rie J, Jansens S and Van Mellaert H (1988) Specificity of Bacillus thuringiensis delta-endotoxins is correlated with the presence of high-affinity binding sites in the brush border membrane of target insect midguts. Proc. Natl. Acad. Sci. USA (in press).

351

Hofmann C, Luthy P., Hutter R and Pliska V (1988) Binding of
the delta-endotoxin from Bacillus thuringiensis to brush-
border membrane vesicles of the cabbage butterfly (Pieris
brassicae). Eur. J. Biochem. 173:85-91.
Hofte H, De Greve H, Seurinck J, Jansens S, Mahillon J, Ampe C,
Vandekerckhove J, Vanderbruggen H, Van Montagu M, Zabeau M
and Vaeck M (1986) Structural and functional analysis of a
cloned delta endotoxin of Bacillus thuringiensis berliner
1715. FEBS Eur. J. Biochem.
Knowles BH and Ellar DJ (1986) Characterization and partial
purification of a plasma mambrane receptor for Bacillus
thuringiensis var. kurstaki lepidopteran-specific delta-
endotoxin. J. Cell Sci. 83:89-101.
Knowles BH and Ellar DJ (1987) Colloid-osmotic is a general
feature of the mechanism of action of Bacillus thuringiensis
delta-endotoxin with different insect specificity. Biochem.
Biophys. Acta 924:509-518.
Knowles BH, Thomas WE and Ellar DJ (1984) Lectin-like binding
of Bacillus thuringiensis var. kurstaki lepidopteran-
specific toxin is an initial step in insecticidal action.
FEBS letters 168:197-202.
Vaeck M, Reynaerts A, Hofte H, Jansens S, Dean C, De Beuckeleer
M, Zabeau M, Van Montagu M and Leemans J (1987) Transgenic
plants protected from insect attack. Nature 328:33-37.
Wolfersberger M., Luthy P., Maurer A., Parenti P., Sacchi V.,
Giordana B. and Hanozet G. (1987) Preparation and partial
purification of amino acid transporting brush border
membrane vesicles from the larval midgut of the cabbage
butterfly (Pieris brassicae). Comp.Biochem.Physiol. 86:301-
308.

Acknowledgements: K. Hendrickx is supported by a grant from the
I.W.O.N.L. of Belgium and was supported by a FEBS travel fund.
*: Further adress: Plant Genetic Systems; Plateaustraat 22;
B 9000 Gent

FUNCTIONAL RECONSTITUTION OF PHOTOSYNTHETIC REACTION CENTRE COMPLEXES FROM RHODOPSEUDOMONAS PALUSTRIS

Douwe Molenaar, Wim Crielaard, Wil N. Konings and Klaas J. Hellingwerf*
Department of Microbiology
University of Groningen
Kerklaan 30
9751 NN Haren
The Netherlands

INTRODUCTION

Reconstitution is a powerful technique in the elucidation of the mechanisms involved in membrane associated biological energy transduction (see for example Racker & Stoeckenius, 1974). For proton motive force (transmembrane electrochemical proton gradient) generation in reconstituted systems several pumps are available (Driessen et al., 1987). Light-driven pumps could have important experimental advantages over chemically driven pumps, since pumping rates, and thereby electrochemical forces, can be simply controlled with light intensity. However until recently the only light-driven pump available, bacteriorhodopsin, reconstitutes in a scrambled orientation and generally generates an inside positive proton motive force.

Here we describe proton motive force generation in the 'in vivo' orientation (negative and alkaline inside) in liposomes in which purified photosynthetic reaction centres (RC's) from Rhodopseudomonas palustris were incorporated. Cyclic electron

*Department of Microbiology, University of Amsterdam, Nieuwe achtergracht 127, 1018 WS Amsterdam, The Netherlands

NATO ASI Series, Vol. H29
Receptors, Membrane Transport and Signal Transduction
Edited by A. E. Evangelopoulos et al.
© Springer-Verlag Berlin Heidelberg 1989

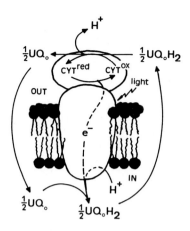

Figure 1. Schematical representation of the reconstituted reaction centre proton pump showing the spatial and electrochemical flow of compounds. Symbols: UQ_0, ubiquinone-0; Cyt c, cytochrome c.

transfer is restored in this system by the addition of the redox mediators cytochrome c and ubiquinone-0 (Figure 1). Ubiquinone-0 is reduced internally by the reaction centre, and oxidized externally by cytochrome c, which in turn reduces the reaction centre again. Ubiquinone-0 thereby provides a vehicle for outward directed proton transport, since its reduction and oxidation are accompanied by protonation and deprotonation respectively, and because it readily diffuses across the membrane due to its moderate hydrophobicity. The driving force for this cyclic redox chain, and consequently for proton transport, is the photon driven electron transfer between cofactors in the reaction centre.

This artificial light-driven pump can be regarded as a short-circuiting of the natural pumping system in which transmembrane proton transport is mediated by the cytochrome b/c_1 complex. Although the liposomes used in this study do not contain the cytochrome b/c_1 complex, the artificial pump is capable of generating a high proton motive force.

MATERIALS AND METHODS

Isolation and reconstitution of reaction centres. Reaction centres with either one or both types of light harvesting (LH) complexes attached (RC/LH$_I$ and RC/LH$_I$/LH$_{II}$) were isolated from Rps. palustris by solubilizing chromatophores with octylgluco-side or a mixture of octylglucoside and deoxycholate and subsequent sucrose density gradient centrifugation in gradients supplemented with these detergents. Reaction centre complexes were reconstituted in liposomes from Escherichia coli phospholipids by detergent dialysis plus a freeze/thaw/-sonication step.

Measurement of the transmembrane electrical potential difference. Reduced horse heart cytochrome c and ubiquinone-0 were added to reaction centre liposomes. The transmembrane electrical potential difference (short: membrane potential or $\Delta\psi$) generated upon illumination was determined from the equilibrium distribution of the lipophilic cation tetraphenyl-phosphonium (TPP$^+$), which was deduced from measurements with an ion-specific (TPP$^+$-specific) electrode. The calculations included a correction for concentration dependent binding of TPP$^+$ to the membrane.

Measurement of cytochrome c redox state. The redox state of cytochrome c in a suspension of reaction centre liposomes was monitored in time in a double beam spectrophotometer by the difference of absorption at 540 and 550 nm. An extinction coefficient of 19.5×10^3 $M^{-1}.cm^{-1}$ was used. The experiments were performed at pH 8 and room temperature. Light intensity was varied with neutral density filters, up to a maximum of approximately 40 W/m^2 of light of uniformly distributed intensity with wavelengths between 600 and 1100 nm.

These and other methods have been described in detail elsewhere (Molenaar et al., 1988).

RESULTS AND DISCUSSION

Characterization of the reaction centre liposomes. An important condition for the validity of calculating membrane potentials ($\Delta\psi$) from the distribution of the lipophilic cation TPP$^+$ is that RC liposome preparations should be homogeneous in RC distribution. This was investigated by sucrose density gradient centrifugation of an RC liposome preparation, and the measurement of RC and lipid markers upon fractionation. As a

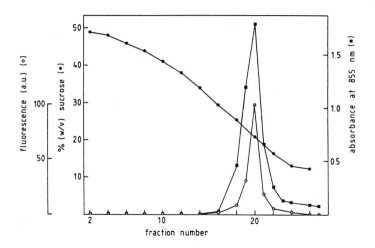

Figure 2. Sucrose density gradient centrifugation of RC liposomes. Circles and squares respectively indicate the presence of fluorescent lipid marker or of reaction centres in the fractions.

lipid marker, a trace amount of octadecyl-rhodamine B chloride, a fluorescent lipid, was added during liposome preparation. As an RC marker we used the intrinsic light absorbing cofactors of the RC, measured at 850 nm. Figure 2 shows that both markers comigrate, indicating homogeneous mixing of lipid and RC's because protein free liposomes would remain on top of the gradient, and solubilized RC's band at 30% sucrose.

The homogeneity of RC distribution in lipid was confirmed by electron microscopy of freeze fracture preparations (not shown). One other feature of these RC liposome preparations was that more than 95% of the RC's were incorporated in the in vivo orientation, i.e. having the cytochrome c binding site towards the external aqueous phase (Molenaar et al., 1988).

Light induced generation of a proton motive force in reaction centre liposomes. The generation of a proton motive force depends on the pumping capacity of the redox cycle, but also on the rates of ion leakage through the liposomal membrane. We found that this leakage varied considerably with the lipid composition of reaction centre liposomes. In liposomes made of E. coli phospholipids a high proton motive force could

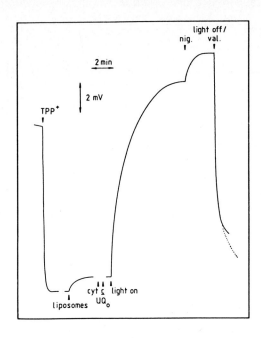

Figure 3. Protonmotive force generation in RC liposomes. The figure shows a recording of the TPP$^+$-selective electrode. Arrows indicate the additions of: An amount of TPP$^+$, doubling its concentration from 1 to 2 μM; RC liposomes; reduced cytochrome $\underline{c}$ (cyt $\underline{c}$, 10 μM); ubiquinone-0 (UQ$_0$, 400 μM); nigericin (nig, 20 nM) and valinomycin (val, 200 nM). The pH was 8.

be generated. However, in soya bean asolectin or egg-yolk phospholipids the proton motive force was very low, although the turn-over rates were comparable in these preparations.

The electrical component ($\Delta\psi$) of the proton motive force generated in RC liposomes was quantified by using the distribution method for TPP$^+$. Figure 3 shows that TPP$^+$ is taken up by RC liposomes upon illumination, due to the generation of an inside negative membrane potential. Addition of the proton-/potassium electroneutral exchanger nigericin leads to the abolishment of the pH gradient between in- and outside, and to an increment of the $\Delta\psi$. After turning off the light, or adding the potassium ionophore valinomycin, the $\Delta\psi$ is dissipated and accumulated TPP$^+$ is released.

We found that the highest $\Delta\psi$ generated was 180 mV which is exceptionally high compared to other reconstituted pumps. The optimal conditions were pH 8, 10 µM reduced cytochrome c, 400 µM ubiquinone-0, an RC content of 1.4 nmol/mg of phospholipid and the presence of nigericin. In subsequent measurements this ionophore was used to dissipate the pH gradient, so that the terms proton motive force and $\Delta\psi$ can be used interchangebly.

Determination of the electromotive force. As said before, the maximal $\Delta\psi$ generated in RC liposomes depends on the pumping rate of the proton pump, as well as on the passive diffusion of ions through the membrane. Additionally it is expected that both pumping rate and passive diffusion depend on the $\Delta\psi$. Leakage will be stimulated by a higher $\Delta\psi$, and pumping rates will be decreased by an opposing $\Delta\psi$ (backpressure effect). Our question was: what is the intrinsic power of the reconstituted RC pump, or more precisely, at which magnitude of the proton motive force or $\Delta\psi$ does it stop pumping protons? This value, called the electromotive force, is the maximal $\Delta\psi$ attainable by the RC pump and should in principle be indepen-

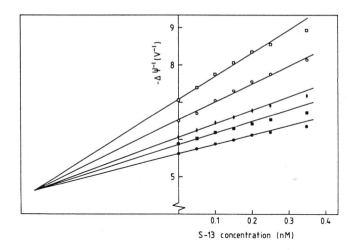

Figure 4. Electromotive force determination of reaction centre complex. The steady state membrane potential was determined under different light intensities and protonophore (S-13) concentrations. (●) 1350, (■) 700, (♦) 350, (O) 160, and (□) 80 W/m². When the inverse membrane potential is plotted against protonophore concentration theory predicts straight lines for each light intensity, all intersecting in one point. The membrane potential at that point (1/4.76 V = 210 mV) equals the electromotive force (see text).

dent of the membrane system used in reconstitution. An approach to such a problem, using nonequilibrium thermodynamics, was presented by Westerhoff et al. (1981, see also refs. 35, 36 of that article). Addressing this question experimentally comes to titration with a protonophore, thereby increasing the leakiness artificially, at different light intensities (so different pumping rates), and measuring the steady state $\Delta\psi$ under all these conditions. If the resulting data are plotted properly one is able to extrapolate to a hypothetical situation in which the membrane is infinitely resistant to passive ion diffusion. In this hypothetical situation the $\Delta\psi$ would always reach its maximal value irrespective of the pumping rate. Figure 4 shows the outcome of this experiment: the extrapolated value for $\Delta\psi$, i.e. the electromotive force equals 210 mV.

Comparison of RC/LH$_I$ and RC/LH$_I$/LH$_{II}$ complexes. It is possible to isolate complexes from <u>Rps. palustris</u> with either the RC surrounded by a ring of LH$_I$ units or surrounded by both LH$_I$ and LH$_{II}$ units. One of the purposes of this study was to compare light saturation characteristics of both types of

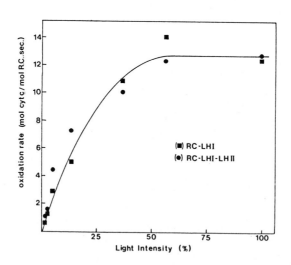

Figure 5. Comparison of light saturation of light-dependent cytochrome <u>c</u> oxidation by solubilized RC/LH$_I$ and RC/LH$_I$/LH$_{II}$ complexes. Cytochrome <u>c</u> and ubiquinone-0 concentrations were 10 and 350 µM respectively. Maximal light intensity was about 40 W/m^2 (see Methods).

reconstituted complexes. Since the presence of LH_{II} antennae in a complex would increase the photon trapping efficiency for that complex, we expected that for $RC/LH_I/LH_{II}$ complexes the pumping rate would saturate at lower light intensities than for RC/LH_I complexes. Figure 5, however, shows that the saturation characteristics, are essentially the same for both types. We conclude that the transfer of excitational energy from LH_{II} to the reaction centre is, for unknown reasons, negligible in these reconstituted complexes.

Kinetics of the reaction centre/redox mediator system. To gain some insight into the dependence of the pumping rate of the RC/redox mediator system on the concentration of redox mediators, we performed initial rate studies of cytochrome $\underline{c}$ oxidation by solubilized reaction centre complexes at different cytochrome $\underline{c}$ concentrations. The inset of figure 6 shows that the system displays Michaelis-Menten kinetics with respect to cytochrome $\underline{c}$ concentration. The experiments shown in the inset were carried out at 2.45 % of maximal light

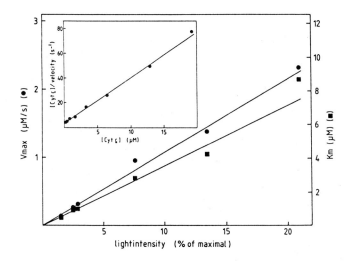

Figure 6. Dependence of the K_M and V_{max} for cytochrome $\underline{c}$ oxidation by solubilized reaction centre complexes on the light intensity. Ubiquinone-0 concentration was 400 μM. The inset is one example of the Hanes plots used for this figure. It shows the rate/concentration data for 2.45% of the maximal light intensity.

intensity (which is about 40 W/m^2, see Materials and Methods). The Michaelis constant (K_M) for cytochrome c found here was 1 μM. The dissociation constant for cytochrome c, that is under non-energized conditions, was reported to be 5 μM. We also performed measurements at other light intensities, and found that the K_M and the V_{max} increased proportionately with light intensity (Figure 6). This behaviour could be explained by a model (described by Molenaar et al., 1988) which integrates knowledge about the detail reactions taking place in the reaction centre to obtain an overall kinetic steady state description. Besides a qualitative similarity there was also a fair quantitative agreement between model and results.

CONCLUDING REMARKS

Since a high proton motive force can be generated in reaction centre liposomes, and the magnitude of the proton motive force can be manipulated with light intensity, this system is very suitable for the study of proton motive force dependent solute transport proteins. As an example we studied the leucine carrier of _Lactococcus_ _lactis_ (Crielaard et al., 1988a). Recent experiments furthermore show that it is possible to measure the electrical field dependent shift of the absorption bands of the field sensitive carotenoids, present in LH$_{II}$ (Crielaard et al., 1988b). These liposomes therefore contain an intrinsic probe of the membrane potential. This simple system will allow a comparison of the carotenoid band shift method for $\Delta\psi$ measurement and other methods like the distribution method for lipophilic ions.

REFERENCES

Crielaard W, Driessen AJM, Molenaar D, Hellingwerf KJ, Konings WN (1988a) Light-driven amino acid uptake in <u>Streptococcus cremoris</u> or <u>Clostridium</u> <u>acetobutylicum</u> membrane vesicles fused with liposomes containing bacterial reaction centers. J. Bacteriol. 170:1820-1824.

Crielaard W, Hellingwerf KJ, Konings WN (1988b) Reconstitution of electrochemically active pigment-protein complexes from <u>Rhodobacter</u> <u>sphaeroides</u> in liposomes. Biochim. Biophys. Acta, accepted for publication.

Driessen AJM, Hellingwerf KJ, Konings WN (1987) Membrane systems in which foreign pumps are incorporated. Microbiol. Sci. 4:173-180.

Molenaar D, Crielaard W, Hellingwerf KJ (1988) Characterization of proton motive force generation in liposomes reconstituted from phosphatidylethanolamine, reaction centers with light-harvesting complexes isolated from <u>Rhodopseudomonas</u> <u>palustris</u>. Biochemistry 27:2014-2023.

Racker E, Stoeckenius W (1974) Reconstitution of purple membrane vesicles catalyzing light-driven proton uptake and adenosine triphosphate formation. J. Biol. Chem. 249:662-663.

Westerhoff HV, Hellingwerf KJ, Arents JC, Scholte BJ, van Dam K (1981) Mosaic nonequilibrium thermodynamics describes biological energy transduction. Proc. Natl. Acad. Sci. USA 78:3554-3558.

Na^+/H^+ EXCHANGE IN CARDIAC CELLS: IMPLICATIONS FOR ELECTRICAL AND MECHANICAL EVENTS DURING INTRACELLULAR pH CHANGES.

F.V. Bielen, S. Bosteels, F. Verdonck
University of Leuven, Campus Kortrijk
8500 Kortrijk
Belgium

ABSTRACT.

Intracellular acidification induced by changing from Hepes to a CO_2/HCO^-_3 buffer results in an increase in the intracellular Na^+ activity (a^i_{Na}). The exchanger is nearly inactive at intracellular pH (pH_i) of 7.2 - 7.0 and is strongly activated at pH_i below 7.0. The rate of rise in a^i_{Na} in conditions of a blocked Na^+/K^+ pump increases by a factor of 3.8 ± 0.9 (n=6) when pH_i drops from 7.2 ± 0.1 in Hepes to 6.8 ± 0.1 in 15 % CO_2. Acid-induced increase in a^i_{Na} and accelerated rate of rise of a^i_{Na} can be blocked by amiloride (2.10^{-3} M) or by decreasing pH_0 to 6.7. The rise in a^i_{Na} is associated with the generation of a Na^+/K^+ pump-dependent outward current. At low pH_0 the increase in outward current is much smaller which demonstrates the absence of secondary pump stimulation when Na^+/H^+ exchange is inhibited. The Na^+/K^+ pump dependent hyperpolarization modifies spontaneous activity. Recovery of contractile force in an acid-loaded cell is related to the gain in a^i_{Na}.

INTRODUCTION.

Recovery from acid loads in cardiac cells occurs via trans-membranar Na^+/H^+ exchange which couples Na^+ influx to H^+ extrusion (Deitmer and Ellis, 1980; Ellis and MacLeod, 1985; Piwnica-Worms et al.,1985; Frelin et al., 1985). Consequently intracellular acidosis is expected to induce a gain in a^i_{Na}. Because of the crucial role of internal Na in driving other transmembranar transport processes,i.e. active Na^+/K^+ transport and Na^+/Ca^{++} exchange, pH_i changes will be

NATO ASI Series, Vol. H29
Receptors, Membrane Transport and Signal Transduction
Edited by A.E. Evangelopoulos et al.
© Springer-Verlag Berlin Heidelberg 1989

associated with secondary changes of the electrical and mechanical activity. When a^i_{Na} increases the Na^+/K^+ pump is stimulated (Glitsch, 1982). In acid-loaded chick embryonic cells a threefold increase in Na^+/K^+-ATPase activity is measured (Frelin et al., 1985). This explains the acid-induced hyperpolarization observed by Piwnica-Worms et al. (1986) in the same preparation.

Besides the effects of a^i_{Na} on electrical activity, internal Na^+ has important consequences for contractility. Twitch tension depends steeply on a^i_{Na} in cardiac Purkinje fibres (Eisner et al., 1984). An increase in a^i_{Na} rises intra-cellular free Ca^{++} via sarcolemmal Na^+/Ca^{++} exchange in cardiac cells (Sheu and Fozzard, 1982). In addition to the indirect effects of pH_i on contractility via a^i_{Na}, pH_i has a direct effect on the sensitivity of contractile proteins to Ca^{++}: acidification reduces and alkalinization increases the Ca^{++} sensitivity (Fabiato and Fabiato, 1978).

The presented experiments integrate results from intracellular ion measurements, voltage clamp experiments and measurement of mechanical activity in rabbit cardiac Purkinje fibres and papillary muscle. They were designed to answer two questions. (a) What is the contribution of Na^+/H^+ exchange to the Na^+ influx during intracellular acidification? (b) What is the role of intracellular Na^+ for the electrical and mechanical responses during intracellular pH changes?

METHODS.

The experiments were carried out on Purkinje fibres and papillary muscle isolated from rabbit heart. The preparations were perfused at 36 °C with a Tyrode solution, containing (in mM): NaCl, 126; $NaHCO_3$, 24; KCl, 5.4; $CaCl_2$, 1.8; $MgCl_2$, 0.5; glucose, 10. In Hepes buffered solutions $NaHCO_3$ was substituted by 10 mM Hepes. The external pH was 7.4. When different concentrations of CO_2 (3, 5, 7 and 15 % mixed with O_2) were used the HCO^-_3 concentration was adjusted to keep the extracellular pH constant. To decrease extracellular pH in a solution containing 15 % CO_2, the HCO^-_3 concentration was

lowered from 72 to 14 mM. Voltage clamp experiments were
performed in Purkinje fibres with the two micro-electrode
technique. a^i_{Na} and pH_i were measured with silanized ion-
sensitive liquid membrane (Fluka) micro-electrodes in rabbit
Purkinje fibres and papillary muscles.
Contractions were measured in papillary muscles with an
isometric force transducer (HP FTA-10-1).

RESULTS.

Effect of intracellular acidosis on a^i_{Na}.

An increase in CO_2 in the external solution produces an intracellular acidification (Boron and De Weer,1976) because CO_2
crosses the membrane faster than bicarbonate. Fig. 1 shows
that a change from Hepes to a 15 % CO_2 solution produces a
rapid fall in pH_i from 7.2 to 6.8. The decrease in pH_i is

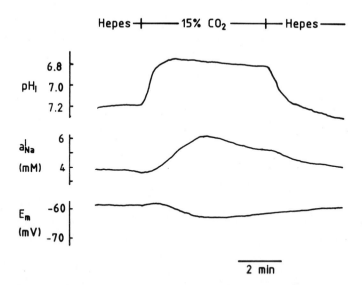

Fig. 1. Effect of alternating a CO_2-buffered Tyrode and CO_2-
free Hepes solution on pH_i, a^i_{Na} and membrane
potential in a rabbit Purkinje fibre at a constant pH_o
(7.4).

associated with an increase in a^i_{Na}. The membrane potential hyperpolarizes by 5 mV after a small initial transient depolarization. pH_i, a^i_{Na} and the resting potential partially recover in the presence of 15 % CO_2. The maximal gain in a^i_{Na} is about 2 mM and is reached within 3 min after the fall in pH_i. When pH_i is varied by changing the CO_2 concentration, maximal a^i_{Na} can be plotted as a function of pH_i (Fig. 2A). Internal Na^+ sharply increases when pH_i is lower than 7.0. The increase in a^i_{Na} can be due to an increased Na^+ influx or a decreased efflux. To investigate the effect of pH_i on the Na^+ influx, the rate of rise in a^i_{Na} (Δa^i_{Na}) was measured in a K^+_o-free solution. In the absence of K^+_o the main efflux pathway for Na^+, i.e. the Na^+/K^+ pump, is blocked. The maximal Δa^i_{Na} is plotted as a function of pH_i in Fig. 2 B. The Δa^i_{Na} starts to increase steeply at a pH_i below 7.0. The curves showing a^i_{Na} and Δa^i_{Na} in 0 K^+_o versus pH_i are very similar. The increase in the rate of gain of a^i_{Na} is blocked by amiloride (2.10^{-3} M), a blocker of Na^+/H^+ exchange. This result illustrates that the increase in a^i_{Na} is largely due to activation of Na^+/H^+ exchange. The amiloride-sensitive rate of rise of a^i_{Na} in 0 K^+_o is increased by a factor of 3.8 ± 0.9 (n=6) when pH_i is decreased from 7.17 ± 0.10 in Hepes to 6.77

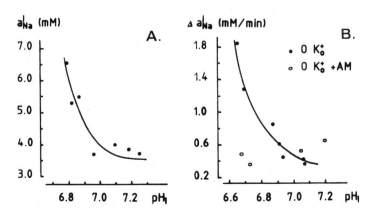

Fig. 2. a^i_{Na} (A) and rate of rise of a^i_{Na} in K^+_o-free solution with (o) and without (●) amiloride (2.10^{-3} M) (B) as a function of pH_i.

± 0.13 in 15 % CO_2. In papillary muscle almost identical
results were obtained.

pH$_i$ and electrical responses.

The initial effect of intracellular acidosis is a transient
depolarization which is followed by a slowly decaying hyperpolarization. The depolarization preceeds the increase in
a^i$_{Na}$, the hyperpolarization coincides with the increase in
a^i$_{Na}$ (Fig.1). Fig. 3 shows the holding current during
intracellular acidification and alkalinization at a holding
potential of -55 mV. After an initial inward shift, the

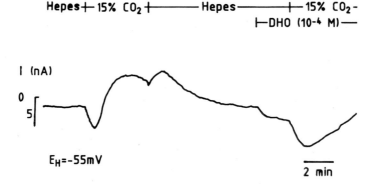

Fig. 3. Holding current during a change of pH$_i$ in the presence
        and absence of dihydro-ouabain (DHO, 10^{-4} M).

holding current changes in the outward direction. In the
presence of dihydro-ouabain (DHO, 10^{-4} M) the inward shift is
still present, but the increase in outward current is much
less pronounced. This demonstrates that the major part of the
outward current upon intracellular acidification is due to
activation of the Na^+/K^+ pump. The increase in outward
current is largely blocked when during intracellular acidosis
the external pH is decreased to 6.7 (not shown). External
acidification depresses Na^+/H^+ exchange (Mahnensmith and
Aronson, 1985). The initial inward current is still present.

Preliminary results indicate that this primary effect of
intracellular acidosis is due to a block of the K^+ background
current.

Since Na^+/K^+ pump stimulation secondary to pH_i regulation
affects membrane potential spontaneous activity will be
influenced. Intracellular acidosis stops the burstlike
rhythmic activity present in Hepes (Fig. 4). In some cases
the rhythmic activity increases during the short phase of
depolarization preceding the Na^+/K^+ pump dependent hyperpolarization.

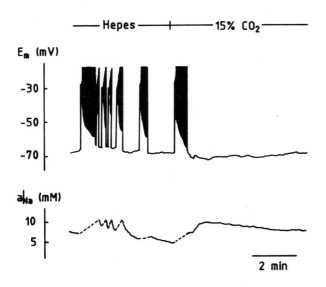

Fig. 4. Effect of intracellular acidification on the membrane
potential and a^i_{Na} of a spontaneously active Purkinje
fibre.

pH$_i$ and tension development.

Peak tension falls drastically after the transition from a
Hepes to a 15 % CO_2 solution (Fig. 5). Contractile force
partially recovers. The initial fall in tension and the
degree of recovery depend on the CO_2 concentration which
determines pH_i (not shown). The recovery of peak tension is
higher in 15 % CO_2 compared to 5 %. When during intracellular

368

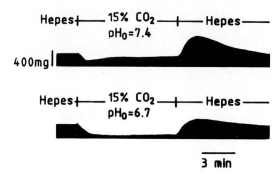

Fig. 5. Mechanical activity of a rabbit papillary muscle
during a change of pH_i at pH_o 7.4 and during simultaneous intra- and extracellular acidification (pH_o=
6.7).

acidification Na^+/H^+ exchange is depressed by an acid external
pH (6.7 in Fig. 5), the recovery is much slower or even
absent. Restoration of pH_i causes a marked overshoot in peak
tension. Reapplication of Hepes after pH_i and pH_o were
simultaneously decreased, results in a smaller overshoot.

DISCUSSION.

Our results show a steep increase in Na^+ influx during
intracellular acidification. A pH_i drop of 0.4 units, induced
by changing from Hepes to a 15% CO_2/HCO^-_3 buffered solution,
increases Δa^i_{Na} almost 4 times. Our findings agree with the
results obtained by other authors in different cardiac
preparations and experimental conditions. Piwnica-Worms et
al. (1985) calculated a 3.5 fold increase in Na^+ influx in
chick embryonic cells when pH_i was decreased by prolonging the
period of NH^+_4 exposure. In the same preparation Frelin et
al. (1985) found a steep dependency of Na^+ uptake on pH_i. In
contrast to our results the Na^+/H^+ exchanger is activated at
more alkaline pH_i and is saturated at 7.0. In sheep Purkinje
fibres H^+ extrusion and amiloride-sensitive Na^+ influx
increased eightfold when pH_i decreases from 7.12 to 6.77
(Kaila and Vaughan-Jones, 1987). The fourfold increase in Na^+
influx in our findings results in relatively small increase in
a^i_{Na}. This suggests that the active Na^+ efflux via the Na^+/K^+

pump is increased. The electrical manifestation of the increased Na^+/K^+ pump rate is the generation of an outward current. By using DHO secondary Na^+/K^+ pump stimulation became evident: the outward shift of the holding current upon intracellular acidification was largely reduced in the presence of the drug. The outward current shift still present can be explained by electrogenic Na^+/Ca^{++} exchange caused by the increase in a^i_{Na} when the Na^+/K^+ pump is blocked. The pump-dependent increase in outward current underlies the observed hyperpolarization. The inward shift of the holding current which can be observed before the rise in a^i_{Na} might be due to an inhibition of the inwardly rectifying K^+ background channels. Intracellular acidosis is found to inhibit K^+ conductance in other preparations (Moody, 1984). The effect of intracellular acidification on the outward current can be largely suppressed when pH_o is decreased simultaneously. In this condition the Na^+/K^+ pump is less activated due to the smaller gain in a^i_{Na}. However, a direct inhibitory effect on the Na^+/K^+ pump cannot be excluded. The stimulated Na^+/H^+ exchange and the subsequent increase in a^i_{Na} has important consequences for the spontaneous activity of cardiac Purkinje fibres. We showed that the burstlike activity which is typical for rabbit Purkinje fibres (Verdonck et al., 1982) slows or stops during intracellular acidosis. If both, intra- and extracellular acidosis occur, spontaneous activity will be increased. Therefore, during ischemia-induced acidosis the stabilizing role of an increase in internal Na^+ is counter- acted by the simultaneously occuring extracellular acidosis. A second consequence of the Na^+/H^+-dependent increase in a^i_{Na} is the recovery of contractile force during intracellular acidosis. A steep relationship does exist between a^i_{Na} and peak force in cardiac tissue (Eisner et al., 1984). The powerful role of a^i_{Na} in determining contractile force is indirectly dependent on the activation of Na^+/Ca^{++} exchange (Sonn and Lee, 1988). The recovery of peak tension which can be observed after the initial acid-induced fall, coincides with Na^+/H^+-dependent increase in a^i_{Na}. The marked overshoot of the contractile response in Hepes can be explained by the gain in internal Na^+ during the preceding period of acidosis and by the restoration of an alkaline pH_i in Hepes. When pH_o is decreased simultaneously with pH_i, a^i_{Na} rises slower and pH_i recovery is depressed. This accounts for the inhibited recovery of contractile force and the diminished contractile response upon reapplication of Hepes. These different effects of intra- and extracellular pH on peak tension may have important consequences for contractile behaviour of cardiac muscle in ischemia and during reperfusion.

In conclusion, the increase in a^i_{Na} during intracellular acidosis via electroneutral Na^+/H^+ exchange stimulates two Na^+-dependent electrogenic transports, which both elicit an outward current: 1) the Na^+/K^+ pump which is responsible for the largest part of the acidosis-induced outward current. This current modifies spontaneous activity in cardiac cells. 2) Na^+/Ca^{++} exchange which may contribute to the a^i_{Na}- dependent increase in outward current. The secondary rise in intracellular free Ca^{++} increases peak tension, counteracting the direct inhibitory effect of low pH_i on the Ca^{++} sensitivity of contractile proteins.

REFERENCES.

Boron WF, De Weer P (1976) Intracellular pH transients in squid giant axons caused by CO_2, NH_3, and metabolic inhibitors. J Gen Physiol 67:91-112
Deitmer JW, Ellis D (1980) Interactions between the regulation of the intracellular pH and sodium activity of sheep cardiac Purkinje fibres. J Physiol 304:471-488
Eisner DA, Lederer WJ, Vaughan-Jones RD (1984) The quantitative relationship between twitch tension and intracellular sodium activity in sheep cardiac Purkinje fibres. J Physiol 355:251-266
Ellis D, MacLeod KT (1985) Sodium-dependent control of intracellular pH in Purkinje fibres of sheep heart. J Physiol 359:81-105
Fabiato A, Fabiato F (1978) Effects of pH on the myofilaments and the sarcoplasmic reticulum of skinned cells from cardiac and skeletal muscles. J Physiol 276:233-255
Frelin C, Vigne P, Lazdunski M (1985) The role of the Na^+/H^+ exchange system in the regulation of the internal pH in cultured cardiac cells. Eur J Biochem 149:1-4
Glitsch HG (1982) Electrogenic Na pumping in the heart. Ann Rev Physiol 44:389-400

Kaila K, Vaughan-Jones RD (1987) Influence of sodium-hydrogen exchange on intracellular pH, sodium and tension in sheep cardiac Purkinje fibres. J Physiol 390: 93-118

Mahnensmith RL, Aronson PS (1985) The plasma membrane sodium-hydrogen exchanger and its role in physiological and pathophysiological processes. Circ Res 57:773-788)

Moody WJr (1984) Effects of intracellular H^+ on the electrical properties of excitable cells. Ann Rev Neurosci 7:257-278

Piwnica-Worms D, Jacob R, Horres CR, Lieberman M (1985) Na/H exchange in cultured chick heart cells. J Gen Physiol 85:43-64

Piwnica-Worms D, Jacob R, Shigeto N, Horres CR, Lieberman M (1986) Na/H exchange in cultured chick heart cells: secondary stimulation of electrogenic transport during recovery from intracellular acidosis. J Mol Cell Cardiol 18:1109-1116

Sheu S, Fozzard HA (1982) Transmembrane Na^+ and Ca^{2+} electrochemical gradients in cardiac muscle and their relationship to force development. J Gen Physiol 80:325-351

Sonn JK, Lee CO (1988) Na^+-Ca^{2+} exchange in regulation of contractility in canine cardiac Purkinje fibers. Am J Physiol 255:C278-C290

Verdonck F, Glitsch HG, Pusch H (1982) Importance of electrogenic sodium extrusion for suppression of spontaneous activity in rabbit Purkinje fibres. Arch int Physiol Biochim 90:36-37

RECEPTOR-MEDIATED INHIBITION OF REPRODUCTIVE ACTIVITY IN A SCHISTOSOME-INFECTED FRESHWATER SNAIL.

P.L. Hordijk, R.H.M. Ebberink, M. de Jong-Brink, and J. Joosse.
Department of Biology
Vrije Universiteit
de Boelelaan 1087
1081 HV Amsterdam
The Netherlands.

In host-parasite relationships, intervention in host reproduction and development by the parasite is frequently observed. This intervention is beneficial for the parasite (in terms of energy and nutrient availability), but not for the host. Parasites have developed various strategies to 'regulate' their host for their own benefit (for reviews, see: Baudoin, 1975; Beckage, 1985).

Schistosome trematode parasites cause the wide-spread tropical disease schistosomiasis (bilharziasis), which affects the health of large human populations and that of cattle and other vertebrates. This disease is transmitted by freshwater snails, the intermediate hosts of these digenetic trematodes. In these snails, asexual reproduction of the parasites takes place, leading to the release of large numbers of cercariae, a free living stage of the parasite. The cercariae are able to penetrate and to infect a vertebrate host. When snails become infected at a juvenile stage, the parasites strongly affect the development of the reproductive system of their intermediate snail hosts, which results in sexually inactive animals. In infected snails the production of hormones from (neuro)endocrine centres involved in the regulation of reproduction is not inhibited, however, the secreted hormones are no longer effective (Sluiters *et al.*, 1984; Sluiters *et al.*, 1984a). Furthermore, giant growth of the snail host can be observed. This growth is not normal; the wet weight of the infected snails increases more rapidly than of the non-infected animals whereas the dry weight initially increases normally, eventually being lower than that of controls. This leads to large individuals which retain an overall juvenile body structure (gigantism, McClelland & Bourns, 1969; Joosse & van Elk, 1986). Presumably, the energy and nutrients not spent on the production of snail eggs will be used by the parasite to produce its own offspring.

Research on the combination of the avian schistosome *Trichobilharzia ocellata* and the hermaphrodite pond snail *Lymnaea stagnalis* has supplied most of our knowledge about parasite-related inhibition of reproductive activity. *L. stagnalis* is the intermediate host of this trematode parasite, which reproduces sexually in the duck. These parasitological studies were

NATO ASI Series, Vol. H29
Receptors, Membrane Transport and Signal Transduction
Edited by A.E. Evangelopoulos et al.
© Springer-Verlag Berlin Heidelberg 1989

considerably promoted by the fact that the physiology and (neuro)endocrinology of *L. stagnalis* have thoroughly been studied (Joosse; 1984; 1988; and Geraerts *et al.*, 1988). The aim of the present study is to achieve a better understanding of the mechanisms involved in the inhibition of reproduction of freshwater snails by schistosome parasites. Hopefully, this leads to the development of a drug capable of eliminating snail and hence parasite populations.

The effects of schistosome infection on the reproductive activity of *Lymnaea stagnalis*.

In *Lymnaea stagnalis*, normally an increase in the weight and size of the secretion stores of the male and female part of the reproductive tract takes place at puberty (Geraerts, 1976a). In infected snails, this increase does not occur. Oogenesis and spermatogenesis appear to be blocked (Sluiters, 1981). This gametogenesis is controlled by gonadotropic hormones, mainly produced in (neuro)endocrine centres in the central nervous system of the snail. A number of studies, in which the effects of gonadotropic hormones in relation to parasitic infection were studied, have led to the conclusion that a factor is present in haemolymph (blood) of infected snails which is able to counteract the effects of these gonadotropic hormones. Important evidence obtained from these studies is presented in the next paragraphs.

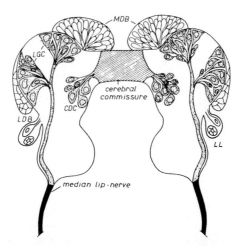

Fig. 1. Diagrammatic transverse section through the cerebral ganglia of Lymnaea *stagnalis* showing the location of the various neuroendocrine and endocrine centres. CDC, caudo dorsal cells; LGC, light green cells; LL, lateral lobes; MDB and LDB, mediodorsal and laterodorsal bodies. The periphery of the cerebral commissure is the neurohaemal area of the CDC's.

One of the female accessory sex glands, the albumen gland, is responsible for adding galactogen and proteins to freshly ovulated and fertilized oocytes. The synthetic activity of this gland is stimulated by the dorsal body hormone (DBH). DBH is a female gonadotropic hormone which is present in extracts of endocrine organs, the dorsal bodies, which are attached onto the cerebral ganglia of the central nervous system of the snail (fig. 1). When incubations of albumen glands were performed in the presence of extracts of the dorsal bodies and haemolymph of infected snails, the stimulatory effect of DBH on galactogen production of the albumen gland was reduced (Joosse & van Elk, 1988). This indicates that haemolymph of infected snails contains a factor that counteracts the effects of the DBH.

During the onset of egg-laying, a set of neuropeptides, derived from a common egg-laying hormone precursor (the CDCH-precursor, fig. 2, Vreugdenhil et al., 1988), is released from two clusters of cells located in the cerebral ganglia of the central nervous system of the snail, the caudo-dorsal cells (CDC, fig. 1). The peptides are released from the interconnecting cerebral commissure and are thought to be responsible for a series of covert (ovulation and egg-mass production) and overt (alterations in locomotion, feeding movements, and bodily postures) behaviours. The release of these peptides coincides with a period of electrical activity of these cells (discharge; Ter Maat, 1987). One of these peptides is the ovulation hormone, caudo dorsal cell hormone (CDCH), responsible for the ovulation of oocytes from the ovotestis (Ebberink et al., 1985).

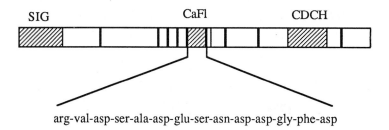

Fig. 2. Structure of the CDCH-precursor of *Lymnaea stagnalis*. SIG: signal peptide; CaFL, calfluxin; CDCH, caudo dorsal cell hormone. I = Potential cleavage site. The primary structure of calfluxin is shown (Dictus & Ebberink, 1988).

One of the other peptides, released simultaneously with CDCH, is named calfluxin (Dictus et al., 1987). Calfluxin stimulates the increase of the calciumconcentration in secretory cells of the albumen gland. This increase is probably related to activation of cellular release or exocytosis of carbohydrates and proteins. Subsequently, the calcium is taken up by the mitochondria. In the mitochondria, calcium deposits can be made visible at the ultrastructural level with the pyroantimonate precipitation technique (Dictus et al., 1987). Incubation of the albumen gland with calfluxin in haemolymph of infected snails reduced the effect of calfluxin to control levels. Since the effect of calfluxin in haemolymph of non-infected snails was normal, a calfluxin-inhibiting factor was apparently present in haemolymph of infected snails (de Jong-Brink, 1988a). Haemolymph of snails, from 3 weeks post exposure to the parasites onwards, appeared to contain this calfluxin-inhibiting factor (de Jong-Brink *et al.*, 1988b). It is not known whether the factor, inhibiting the effect of DBH and the one, inhibiting calfluxin, are identical. We assume that the effects are caused by one and the same substance. The factor is heat stable, pronase sensitive and has been called *schistosomin* (Joosse *et al.*, 1988). Here, we describe the purification and partial characterization of schistosomin.

Development of a receptor-binding assay for schistosomin using membrane-bound calfluxin receptors.

Since the assay described above is too laborious to be effective in a purification procedure, a new assay for calfluxin and schistosomin was developed.

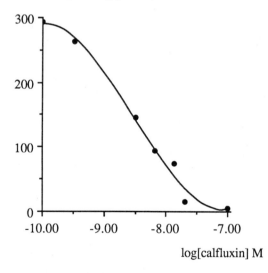

FITC-calfluxin Bound (fmol/mg protein)

log[calfluxin] M

Fig.3. Binding of FITC-calfluxin to membrane-bound receptors. Incubations of albumen gland membranes of *Lymnaea* were performed for 10 min (20°C) in the presence of increasing concentrations of unlabelled synthetic calfluxin. The concentration of unlabelled calfluxin is plotted against the amount of bound FITC-calfluxin/mg protein. Addition of 5 nM unlabelled calfluxin reduced binding to 50%.

Based on the indications that the inhibition of calfluxin by schistosomin occurs at the hormone-receptor level (de Jong-Brink *et al.*, 1988b), a receptor-binding assay was developed, based on the method described by Krodel *et al.* (1979). Synthetic calfluxin was labelled with fluoresceine isothiocyanate (FITC; Strottmann *et al.*, 1983). The labelled calfluxin (FITC-calfluxin) was purified using reverse-phase HPLC and subsequently used as a ligand in receptor-binding studies. Albumen gland membranes, prepared by homogenizing and centrifuging albumen glands of adult snails in Hepes Buffered Saline (HBS ; (mM): NaCl (30); KCl (1,5); $MgCl_2$ (2); $CaCl_2$ (4); HEPES (10); $NaHCO_3$ (8); NaOH (2,5); pH 7.8), were incubated in the presence of 15 nM FITC-calfluxin and the various concentrations of synthetic calfluxin, tissue extracts or haemolymph. After a 10 min incubation (20°C) bound and free ligand were separated by centrifugation, after which the supernatant was assayed for fluorescence (λ_{exc} 492 nm / λ_{em} 518 nm). Using this technique, it was possible to identify specific binding sites for calfluxin on albumen gland membranes (fig. 3). These calfluxin receptors are characterized by a B_{max} of 290 fmol/mg protein and an apparent Kd of 5 nM.

As a test-case for the assay, the localization of native calfluxin in the central nervous system was investigated. Acid extracts of the different ganglia of the CNS of *Lymnaea stagnalis* (prepared as described by Ebberink *et al.*, 1985) were tested for binding activity (fig. 4). It was shown that only extracts from the cerebral ganglia (containing the CDC) and the cerebral commissure (the neurohaemal area of the CDC) contained material replacing FITC-calfluxin from its receptor. These are the production and release site, respectively, of native calfluxin. Subsequently, the hypothesis was tested that schistosomin, present in haemolymph of infected snails, would be able to inhibit binding of FITC-calfluxin to its receptor. Incubations of albumen gland membranes in different concentrations of haemolymph showed that FITC-calfluxin binding is inhibited in the presence of haemolymph of infected snails (fig 5). Haemolymph of non-infected snails had no effect on the FITC-calfluxin binding.

Purification and characterization of schistosomin.

As a rapid and sensitive assay for schistosomin, the method described above was used during the purification. The purification of schistosomin from haemolymph of infected snails consisted of a 5-step procedure, starting with boiling and centrifuging (10 min. each), followed by concentrating and desalting on Seppak C-18 cartridges. Subsequently, the material was subjected to negative ion-exchange chromatography on a PBE 94 column, equilibrated at pH 6. The flow-through fraction was finally applied to two successive RP-HPLC steps using a WP-C18 column. The solvent system used in the first HPLC step consisted of 10 mM Na_2HPO_4/ NaH_2PO_4, pH 6,8 (solvent A) and solvent A / 60% acetonitril (solvent B). Material inhibiting binding of FITC-calfluxin to its receptor eluted at 45% solvent B.

Bound (% of control)

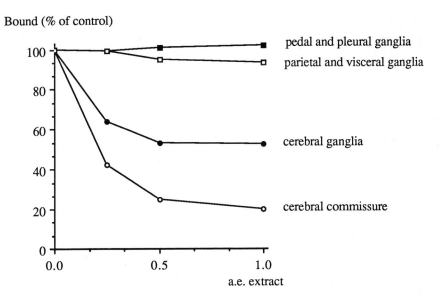

Fig. 4. Specificity of the membrane-bound calfluxin-receptor. Binding of FITC-calfluxin to receptors on albumen gland membranes was measured in the presence of increasing concentrations of extracts from different parts of the central nervous system of *Lymnaea*. Binding is expressed as % of control values.

Bound (% of control)

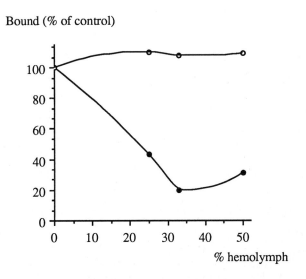

Fig. 5. The presence of schistosomin in the haemolymph of *Lymnaea stagnalis* infected with *Trichobilharzia ocellata*. Albumen gland membranes were incubated in increasing concentrations of haemolymph of non-infected (o) and infected snails (●). FITC-calfluxin binding is expressed as % of control values.

To assure purity, this fraction was rechromatographed, using 7.5 mM trifluoroacetic acid, pH 2 (solvent A) and 7 mM trifluoroacetic acid, 60% acetonitril, 20% 2-propanol (solvent B). Active material eluted at 36 % solvent B (fig. 6).

The question of wether schistosomin is produced by the parasite itself or by the snail upon stimulation by the parasite, was subsequently investigated.

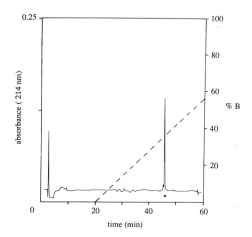

Fig. 6. Elution profile on reverse-phase HPLC of purified schistosomin from haemolymph of *Lymnaea stagnalis* infected with *Trichobilharzia ocellata*. The gradient used is indicated (- - -). Material, inhibiting the binding of FITC-calfluxin to its receptor eluted at 36% solvent B (bar). See text for details.

Evidence obtained in other experiments had indicated that schistosomin is produced by the snail, the most likely centre of production being the CNS of the snail (de Jong-Brink, unpublished data). In order to establish this, CNS of *L. stagnalis* were homogenized, centrifuged and the supernatant was subjected to a similar purification procedure as described above. The elution profile of the final HPLC step is shown in fig. 7.

The pure peptide that inhibits the binding of FITC-calfluxin, coelutes with schistosomin from haemolymph of infected snails. This indicates that schistosomin is indeed produced in the central nervous system of *Lymnaea stagnalis* and thatn it is apparently released into the haemolymph during infection with *Trichobilharzia ocellata*. We do not know whether schistosomin is present (in undetectable amounts) in the haemolymph of non-infected snails.

It could be calculated that, in order to obtain sufficient material to perform sequence analysis and characterization of schistosomin, about 1000 ml haemolymph (from ±1500 individually infected snails) or ±1500 CNS from non-infected snails would have to be purified. We decided to purify schistosomin from central nervous systems from non-infected snails. This resulted in about 15 µg pure schistosomin.

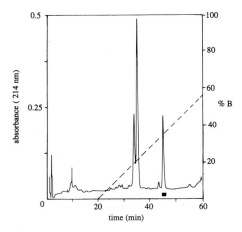

Fig. 7. Elution profile on reverse-phase HPLC of schistosomin, purified from central nervous systems of non-infected snails. The gradient used is indicated (- - -). Material, inhibiting the binding of FITC-calfluxin to its receptor eluted at 36% solvent B (bar). See text for details.

Analysis by SDS-PAGE (Laemmli, 1970) revealed that the molecule has a molecular mass of 12 kD, indicating that the material consists of about 100 amino acids (result not shown). The amino acid ratio (Table 1) of the pure material was determined as described previously (Ebberink et al., 1985).

Table 1. Amino acid ratio of schistosomin

Amino acid	ratio	Amino acid	ratio
Asx	5.1	Val	2.5
Glx	6.1	Phe	2.9
His	–	Ile	1.0
Ser	3.7	Leu	2.2
Arg	2.4	Lys	–
Gly	6.9	Cys	nd
Thr	2.2	Met	nd
Ala	4.0	Pro	nd
Tyr	1.7	Trp	nd

The ratios were calculated using the figure for iso-leucine as the common divider.
nd = not determined.

Additional information about the structure of the schistosome molecule will be obtained, using automated Edman degradation of tryptic fragments of purified material.

Discussion and conclusions.

The results presented in this chapter indicate that the avian schistosome parasite *Trichobilharzia ocellata* uses a very sophisticated mechanism to manipulate its intermediate snail host, *Lymnaea stagnalis*. The parasite appears to be able to induce, in an unknown way, the release of a neurohormone (schistosomin) from the snail host central nervous system, which is responsible for blocking the effects of its gonadotropic (neuro)hormones. One can imagine that employing the host's own regulatory systems by inducing the release of schistosomin, is much more advantageous for the parasite compared to the development of specific anti-gonadotropic antagonists by the parasite itself. The function of schistosomin in normal snails is not yet clear, but it seems likely that the hormone is also released during adverse environmental conditions (starvation, low temperatures etc.). Under these conditions, egg-laying is rapidly inhibited (Dogterom, 1984 a, b).

Although we have elucidated part of the mechanism underlying inhibition of reproductive activity of schistosome-infected freshwater snails, a number of questions remains to be solved. We do not know how the parasite induces the release of schisosomin, and, even more important, the mechanism of action of schistosomin itself is still unclear. Solving these mechanisms will hopefully contibute to combat the spreading of the second important parasitic disease in the tropics, schistosomiasis.

This investigation received financial support from the UNDP/World Bank/WHO Special Programme for Research and Training in Tropical Diseases.

References

Baudoin, M (1975) Host castration as a parasitic strategy. Evolution **29**: 335-352

Beckage, NE (1985) Endocrine Interactions Between Endoparasitic Insects and Their Hosts. Ann. Rev. Entomol. **30**: 371-413

Dictus, WJAG, de Jong-Brink, M, and Boer, HH (1987) A neuropeptide (Calfluxin) is involved in the influx of calcium into the mitochondria of the albumen gland of the freshwater snail *Lymnaea stagnalis*. Gen. Comp. Endocrinol. **65**: 439-450

Dictus, WJAG and Ebberink, RHM (1988) Structure of one of the neuropeptides of the egg-laying hormone precursor of *Lymnaea*. Mol. and Cell. Endocrinol. **60**: 23-29

Dogterom, GE, Loenhout, H van, Koomen, W, Roubos, EW, and Gereaerts, WPM (1984a) Ovulation hormone, nutritive state, and female reproductive activity in *Lymnaea stagnalis*. Gen. Comp. Endocrinol. **55**: 29-35

Dogterom, GE, Hofs, HP, Wagenaar, P, Roubos, EW, and Geraerts, WPM (1984b) Temperature, and spontaneous and ovulation hormone induced female reproduction in *Lymnaea stagnalis*. Gen. Comp. Endocrinol. **56**: 204-209

Ebberink, RHM, van Loenhout, H, Geraerts, WPM, and Joosse, J (1985) Purification and amino acid sequence of the ovulation hormone of *Lymnaea stagnalis*. Proc. Natl. Acad. Sci. USA **82**: 7767-7771

Geraerts, WPM (1976a) Control of growth by the neurosecretory hormone of the light green cells in the freshwater snail *Lymnaea stagnalis*. Gen. Comp. Endocrinol. **29**: 61-71

Geraerts, WPM, Ter Maat, A, and Vreugdenhil, E (1988) The Peptidergic Neuroendocrine Control of Egg-Laying Behavior in *Aplysia* and *Lymnaea*. Endocrinology of Selected Invertebrate Types pp: 141-231

Jong-Brink, M, de, Elsaadany, MM, and Boer, HH (1988a) *Trichobilharzia ocellata*: Interference with the endocrine control of female reproduction of its host *Lymnaea stagnalis*. Exp. Parasitol. **65**: 91-100

Jong-Brink, M, de, Elsaadany, MM, and Boer, HH (1988b) Schistosomin, an Antagonist of Calfluxin. Exp. Parasitol. **65**: 109-118

Joosse, J (1984) Recent progress in endocrinology of molluscs. In "Biosynthesis, Metabolism and Mode of Action of Invertebrate Hormones." Hoffmann, J, and Porchet, M (eds.),pp:19-35 Springer-Verlag, Berlin, Heidelberg

Joosse, J, and Van Elk, R (1986) *Trichobilharzia ocellata*: Physiological Characterization of Giant Growth, Glycogen Depletion and Absence of Reproductive Activity in the Intermediate Snail Host, *Lymnaea stagnalis*. Exp. Parasitol. **62**: 1-13

Joosse, J, van Elk, R, Mosselman, S, Wortelboer, H, and van Diepen, JCE (1988) Schistosomin: a pronase-sensitive agent in the haemolymph of *Trichobilharzia ocellata*-infected *Lymnaea stagnalis* inhibits the activity of albumen glands *in vitro*. Parasitol Res. **74**: 228-234

Krodel, EK, Beckman, RA, and Cohen, JB (1979) Identification of a Local Anesthetic Binding Site in Nicotinic Postsynaptic Membranes Isolated from *Torpedo marmorata* Electric Tissue. Mol. Pharmacol. **15**: 294-312

McClelland, G, and Bourns, TKR (1969) Effects of *Trichobilharzia ocellata* on Growth, Reproduction, and Survival of *Lymnaea stagnalis*. Exp. Parasitol. **24**: 137-146

Sluiters, JF (1981) Development of *Trichobilharzia ocellata* in *Lymnaea stagnalis* and the Effects of Infection on the Reproductive System of the Host. Z. Parasitenkd. **64**: 303-319

Sluiters, JF, Roubos, EW, and Joosse, J (1984) Increased Activity of the Female Gonadotropic Hormone Producing Dorsal Bodies in *Lymnaea stagnalis* Infected with *Trichobilharzia ocellata*. Z. Parasitenkd. **70**: 67-72

Sluiters, JF, and Dogterom, GE (1984a) The Effect of Infection of *Lymnaea stagnalis* with *Trichobilharzia ocellata* on the Presence of, and Reactivity with the Ovulation Hormone of the Host. Z. Parasitenkd. **70**: 477-484

Strottmann, JM, Robinson, JB, and Stellwagen, E (1983) Advantages of Preelectrophoretic Conjugation of Polypeptides with Fluorescent Dyes. Anal. Biochem. **132**: 334-337

Ter Maat, A, van Duivenboden,YA, and Jansen, RF (1987) Neurobiology; Molluscan Models (Boer, HH, Geraerts, WPM, Joosse, J, eds.) North Holland Publishing Company, Amsterdam, Oxford, New York pp: 255-260

Vreugdenhil, E, Jackson, JF, Bouwmeester, T, Smit, AB, Van Minnen, J, Van Heerikhuizen, H, Klootwijk, J, and Joosse, J (1988) Isolation, characterization, and evolutionary aspects of a cDNA clone encoding multiple neuropeptides involved in the stereotyped egg-laying behavior of the freshwater snail *Lymnaea stagnalis*. J. of Neurosci. (in press)

NMR STUDY OF GRAMICIDIN CATION TRANSPORT ACROSS AND INTEGRATION INTO A LIPID MEMBRANE

P. L. Easton, J. F. Hinton, and D. K. Newkirk
Department of Chemistry and Biochemistry
University of Arkansas
Fayetteville, Arkansas 72701
U.S.A

Gramicidin is a 15 amino acid linear polypeptide which forms cation conducting channels in lipid membranes (Meyers and Haydon, 1972; Anderson, 1983). The gramicidin channel in a lipid bilayer membrane is shown in Figure 1. The amino acid sequence of gramicidin is: formyl-L-Val^1-Gly^2-L-Ala^3-D-Leu^4-L-Ala^5-D-Val^6- L-Val^7-D-Val^8-L-Trp^9-D-Leu^{10}-L-Trp^{11}-D-Leu^{12}-L-Trp^{13}-D-Leu^{14}-L-Trp^{15}-ethanolamine. The channel consists of two monomeric beta helices joined by hydrogen bonds at their NH_2 terminal ends. The dimmer's length is 26 Å and forms a 4 Å pore. The three naturally occurring analogs Gramicidin A, B, and C differ by one amino acid, tryptophan, phenylalanine, and tyrosine respectively, at position 11 (Sarge & Witkop, 1965). Here we report the preliminary results of the Na-23 NMR investigation of the thermal integration of gramicidin analogs into large unilamellar vesicles under ionic equilibrium conditions. This technique is potentially useful for exploring the effect of single amino acid substitution in gramicidin on the rate of incorporation in and formation of a viable channel in a lipid membrane. The effect of different lipids on the incorporation process can also be studied using this technique. The only requirement for a peptide to be studied in this way is for the peptide to conduct ions at a rate sufficient to cause a change in the linewidth of the NMR signal of the conducting ion. The incorporation of gramicidin into lysolecithin micelles has been investigated previously by C-13 NMR (Spisni et al., 1979). This technique, however, is less amiable to studying the rate of incorporation. The incorporation of gramicidin into lysophosphatidylcholine micelles was also studied using fluorescence measurements (Cavatorta et al., 1982). However, with our system the peptide does not have to have a fluorescence or absorbance spectrum.

The transport of Na ions by gramicidin across bilayer membranes of large unilamellar vesicles under ionic equilibrium conditions can be studied using Na-23 NMR (Buster et al., 1988). This is possible because the transport of Na ions across the vesicle wall by ac-

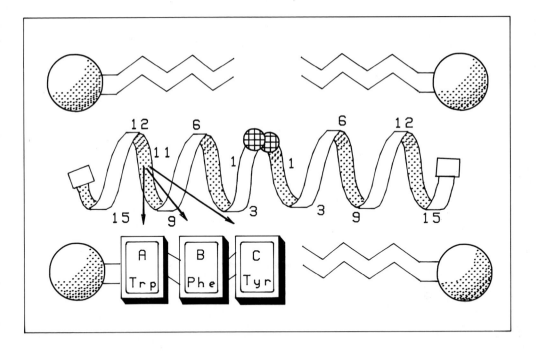

Figure 1. Gramicidin dimmer in a lipid bilayer membrane.

tive gramicidin dimmers results in a change in the linewidth of the NMR signal. The linewidth of the NMR signal is proportional to the square of the gramicidin concentration (Buster et al., 1988). The incorporation of conducting gramicidin channels into the vesicle membrane under ionic equilibrium conditions is not spontaneous but requires an incubation period. In a nonequilibrium system the incorporation of gramicidin is spontaneous, driven by transmembrane potential effects (Buster et al., 1988).

The spectra in Figure 2 show the effect of varying amounts of gramicidin D (a mixture of gramicidins 80%A, 5%B, and 15%C) on the Na-23 NMR signals in the vesicle system when a chemical shift reagent is added to the solution outside the vesicals (Buster et al., 1988). This same effect was seen when the incorporation of gramicidin into the membrane was followed as a function of time. Line broadening occurred as the number of conducting dimmers in the lipid membrane increased with time of incubation. The rate constant can be determined at different incorporation temperatures and then the thermodynamic parameters calculated for the incorporation process.

The experiment was performed in the following manner. Gramicidin, dissolved in

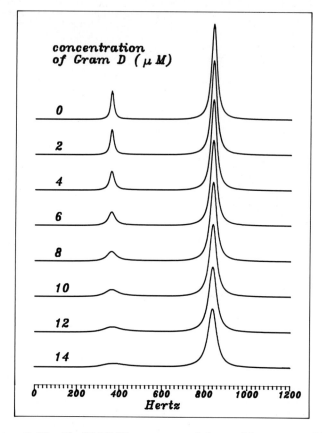

Figure 2. The Na-23 NMR spectrum of the vesicle system with
increasing amounts of gramicidin D.

trifluroethanol, was added to an aqueous solution of phosphatidylcholine(PC)/phos-
phatidylglycerol(PG) vesicles that was thermally equilibrated at the desired temperature
(45-70 °C). Sample aliquots were then taken as a function of time and placed in an ice
bath to halt the incorporation process. A chemical shift reagent was added to each sample.
The Na-23 NMR spectrum was obtained for each sample at a constant temperature of
27°C, where incorporation is negligibly slow. The linewidth at half height was determined
for the Na ions on the inside of the vesicles. A sample without gramicidin was also treated
similarly and used as a blank. The change in linewidth was calculated by subtracting the
linewidth of the blank from the linewidth of the sample. Figure 3 shows the time course
of the increase in the change in linewidth of the inside signal upon heating at different
temperatures.

The incorporation of gramicidin into the lipid membrane involves two steps. First,

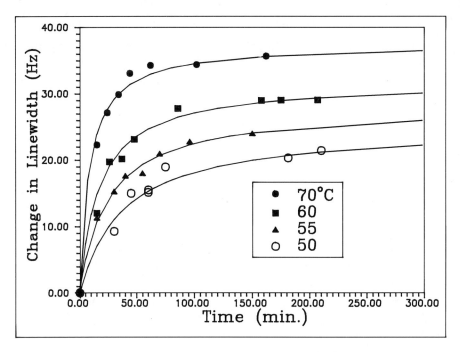

Figure 3. Change in linewidth vs. time at various incubation
temperatures for gramicidin A.

the peptide must penetrate the membrane. Then two monomers must diffuse to a position that will enable them to form a conducting dimmer. The overall reaction sequence for the thermal incorporation of gramicidin into PC/PG vesicles is:

$$2Gr_{as} \xrightarrow{k_1} 2Gr_{in} \underset{k_{-2}}{\overset{k_2}{\rightleftharpoons}} Gr_2$$

It was proposed that prior to heat incubation, gramicidin is associated on the surface of lysophosphatidylcholine micelles (Cavatorta et al., 1982) and phospholipid vesicles (Kemp and Wenner, 1976). Gr_{as} designates this vesical associated state. Gr_{in} represents a final incorporated state. Gr_2 is the active conducting dimmer. The concentration of Gr_{in} is assumed to be in a steady state condition.

Under the conditions of the experiment, the rate for the process may be expressed as:

$$dGr_2/dt = k_1 [\, Gr_{as} \,]^2$$

upon integration the equation becomes:

$$k_1 t = 1/(a-X_t) - 1/a.$$

The forward rate constant for the incorporation process can be determined graphically from a plot of $1/(a - X_t)$ versus time (t). X_t is the change in linewidth at time t. The slope of this line is equal to k_1. a is the intial concentration of gramicidin outside the vesicle. The y- intercept yields $1/a$, from which one may obtain the value of a in linewidth units. An initial a was chosen from the change in linewidth versus time plots. a was then altered to obtain the best fit of the data. Figure 4 shows the second order dependance of the incorporation process.

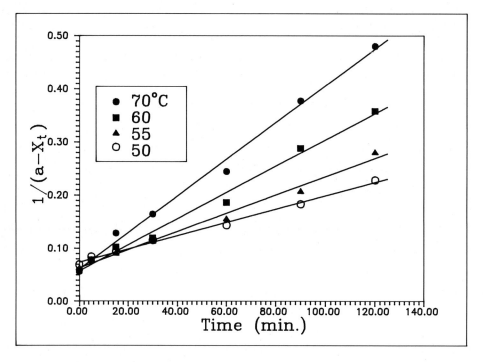

Figure 4. $1/(a - X_t)$ vs time for gramicidin B.

Once the rate constants were determined, the thermodynamic parameters were then calculated. A plot of $\ln(k_1/T)$ versus $1/T$ was made for gramicidin A and B. The ΔH for both analogs was the same within experimental error with gramicidin A equal to 9 $\pm$ 1 kcal/mole and gramicidin B equal to 10 $\pm$ 1 kcal/mole. ΔS for gramicidin A is 46 $\pm$ 6 eu. and for gramicidin B ΔS is 42 $\pm$ 3 eu. It appears that Gramicidin A and B have about the same incorporation properties. Since the two analogs differ by only one amino acid it would appear that this is not enough to greatly influence the rate of incorporation. An analog, however, which is highly polar at either end might show different characteristics. The rate of incorporation is also probably greatly influenced by the type of lipids in the membrane and will be the subject of future investigations.

References

Anderson O (1983) Ion movement through gramicidin channels. Biophys J 41:119-133

Buster DC, Hinton JF, Millett FS, Shungu DS (1988) [23]Na-nuclear magnetic resonance investigation of gramicidin-induced ion transport through membranes under equilibrium conditions. Biophys J 53:145-152

Cavatorta P, Spisni A, Casali E, Lindner L, Masotti L, Urry DW (1982) Intermolecular interactions of gramicidin A transmembrane channels incorporated into lysophosphatidyl-choline lipid systems. Biochim Biophys Acta 689(1):113-120

Kemp G, Wenner C (1976) Solution, interfacial, and membrane properties of gramicidin A. Arch Biochem Biophys 176:547-555

Meyers VB, Haydon DA (1972) Ion transfer across lipid membranes in the presence of gramicidin A. Biochim Biophys Acta 274:313

Sarges R, Witkop B (1965) Gramicidin VIII. The structure of valine- and isoleucine-gramicidin C. Biochem 4:2491-2494

Spisni A, Khaled MA, Urry DW (1979) Temperature-induced incorporation of gramicidin A into lysolecithin micelles demonstrated by ^{13}C NMR. FEBS Let 102:321-324

NATO ASI Series H

Vol. 1: Biology and Molecular Biology of Plant-Pathogen Interactions.
Edited by J. A. Bailey. 415 pages. 1986.

Vol. 2: Glial-Neuronal Communication in Development and Regeneration.
Edited by H. H. Althaus and W. Seifert. 865 pages. 1987.

Vol. 3: Nicotinic Acetylcholine Receptor: Structure and Function.
Edited by A. Maelicke. 489 pages. 1986.

Vol. 4: Recognition in Microbe-Plant Symbiotic and Pathogenic Interactions.
Edited by B. Lugtenberg. 449 pages. 1986.

Vol. 5: Mesenchymal-Epithelial Interactions in Neural Development.
Edited by J. R. Wolff, J. Sievers, and M. Berry. 428 pages. 1987.

Vol. 6: Molecular Mechanisms of Desensitization to Signal Molecules.
Edited by T. M. Konijn, P. J. M. Van Haastert, H. Van der Starre, H. Van der Wel, and
M. D. Houslay. 336 pages. 1987.

Vol. 7: Gangliosides and Modulation of Neuronal Functions.
Edited by H. Rahmann. 647 pages. 1987.

Vol. 8: Molecular and Cellular Aspects of Erythropoietin and Erythropoiesis.
Edited by I. N. Rich. 460 pages. 1987.

Vol. 9: Modification of Cell to Cell Signals During Normal and Pathological Aging.
Edited by S. Govoni and F. Battaini. 297 pages. 1987.

Vol. 10: Plant Hormone Receptors. Edited by D. Klämbt. 319 pages. 1987.

Vol. 11: Host-Parasite Cellular and Molecular Interactions in Protozoal Infections.
Edited by K.-P. Chang and D. Snary. 425 pages. 1987.

Vol. 12: The Cell Surface in Signal Transduction.
Edited by E. Wagner, H. Greppin, and B. Millet. 243 pages. 1987.

Vol. 13: Toxicology of Pesticides: Experimental, Clinical and Regulatory Perspectives.
Edited by L. G. Costa, C. L. Galli, and S. D. Murphy. 320 pages. 1987.

Vol. 14: Genetics of Translation. New Approaches.
Edited by M. F. Tuite, M. Picard, and M. Bolotin-Fukuhara. 524 pages. 1988.

Vol. 15: Photosensitisation. Molecular, Cellular and Medical Aspects.
Edited by G. Moreno, R. H. Pottier, and T. G. Truscott. 521 pages. 1988.

Vol. 16: Membrane Biogenesis. Edited by J. A. F. Op den Kamp. 477 pages. 1988.

Vol. 17: Cell to Cell Signals in Plant, Animal and Microbial Symbiosis.
Edited by S. Scannerini, D. Smith, P. Bonfante-Fasolo, and V. Gianinazzi-Pearson.
414 pages. 1988.

Vol. 18: Plant Cell Biotechnology.
Edited by M. S. S. Pais, F. Mavituna, and J. M. Novais. 500 pages. 1988.

Vol. 19: Modulation of Synaptic Transmission and Plasticity in Nervous Systems.
Edited by G. Hertting and H.-C. Spatz. 457 pages. 1988.

Vol. 20: Amino Acid Availability and Brain Function in Health and Disease.
Edited by G. Huether. 487 pages. 1988.

NATO ASI Series H

Vol. 21: Cellular and Molecular Basis of Synaptic Transmission.
Edited by H. Zimmermann. 547 pages. 1988.

Vol. 22: Neural Development and Regeneration. Cellular and Molecular Aspects.
Edited by A. Gorio, J.R. Perez-Polo, J. de Vellis, and B. Haber. 711 pages. 1988.

Vol. 23: The Semiotics of Cellular Communication in the Immune System.
Edited by E.E. Sercarz, F. Celada, N.A. Mitchison, and T. Tada. 326 pages. 1988.

Vol. 24: Bacteria, Complement and the Phagocytic Cell.
Edited by F.C. Cabello und C. Pruzzo. 372 pages. 1988.

Vol. 25: Nicotinic Acetylcholine Receptors in the Nervous System.
Edited by F. Clementi, C. Gotti, and E. Sher. 424 pages. 1988.

Vol. 26: Cell to Cell Signals in Mammalian Development.
Edited by S.W. de Laat, J.G. Bluemink, and C.L. Mummery. 322 pages. 1989.

Vol. 27: Phytotoxins and Plant Pathogenesis.
Edited by A. Graniti, R.D. Durbin, and A. Ballio. 508 pages. 1989.

Vol. 28: Vascular Wilt Diseases of Plants. Basic Studies and Control.
Edited by E.C. Tjamos and C.H. Beckman. 590 pages. 1989.

Vol. 29: Receptors, Membrane Transport and Signal Transduction.
Edited by A.E. Evangelopoulos, J.P. Changeux, L. Packer, T.G. Sotiroudis,
and K.W.A. Wirtz. 387 pages. 1989.

Vol. 30: Effects of Mineral Dusts on Cells.
Edited by B.T. Mossman and R.O. Bégin. 470 pages. 1989.